Nursery Production and Orchard Management

Volume 2: Temperate and Sub-tropical Fruits

The Editor

Dr Wasakha Singh Dhillon has significantly contributed in the field of education, research, extension and research management in horticultural sciences. He did his graduation in agriculture, masters and doctorate in horticulture from PAU, Ludhiana. He worked on various research management positions as Director, Fruit Research Station Gangian; Director, Punjab Horticultural Post-harvest Technology Centre, PAU campus; Assistant Director General Horticulture in ICAR, New Delhi, and after retirement working as Consultant NHB, GOI. He guided 12 M Sc and 7 Ph D students. He has written 9 books namely : Viticulture in north-west India; Impact of climate change on fruit crops; Fruit tree physiology; Fruit production in India; Doubling farmers income through horticulture; Horticulture for food and environment security; Plantation crops in India; Post-harvest management of Fruits and plantation crops (Two Volumes *i.e.* Basic concepts, and Applied concepts). His technological pursuits led to the development of 10 fruit varieties, 1 root-stock and 38 technologies. Dr Dhillon published 152 original research papers in national and international journals of repute, 25 book chapters, 4 review papers, 20 full proceeding papers, 60 technical articles, and 1 technical bulletin. He attended 60 national and international seminars. He has been awarded with prestigious Fellowships by Indian Academy of Horticultural Sciences (IAHS); Confederation of Horticulture Association of India (CHAI); and Society for Horticultural Research and Development. CHAI also awarded Dr Dhillon with Life Time Achievement Award-2024, and Life Time Achievement Award in Sports by PAU. He was also awarded with Dr JC Anand Gold Medal, and Shri GL Chadha Memorial Gold Medal by IAHS; Meritorious Teacher Award, S GS Nihalsingwala Award, Letter of Appreciation for Agriculture Development by PAU; Distinguished Scientist Award by High-tech Horticultural Society. Apart from research, academics and management, Dr Dhillon was an outstanding athlete of PAU during his student days. He was best athlete of the College, University and Ludhiana district, and won numerous medals like Gold, Silver and Bronze at College, University, District, State, and Open National level competitions. For recognition of achievements in sports he was honored with College and University colors, and merit certificates.

Nursery Production and Orchard Management

Volume 2: Temperate and Sub-tropical Fruits

– Editor –

W.S. Dhillon

Former Assistant Director General, Horticulture, ICAR, New Delhi

Former Director, Punjab Horticultural Post-harvest Technology Centre

Former Director PAU-Fruit Research Station, Gangian

Former Professor of Horticulture, PAU, Ludhiana

2026

Daya Publishing House®

A Division of

Astral International Pvt. Ltd.

New Delhi – 110 002

ISBN 978-93-5919-884-2

Published by : **Daya Publishing House®**
A Division of
Astral International Pvt. Ltd.
– ISO 9001:2015 Certified Company –
4736/23, Ground Floor, Ansari Road
Darya Ganj, New Delhi-110 002
Ph. 011-43549197, 8130496929
e-mail: info@astralint.com
Website: www.astralint.com

Printed at : **Replika Press Pvt Ltd**

This volume of the book is dedicated to

Dr Jagir Singh Samra, FNAS

Former CEO, NRAA, GOI; DDG NRM, ICAR

Currently working as Senior Advisor Renewable Energy of CRIDD, Chandigarh

He played very important mentor role to groom me professionally

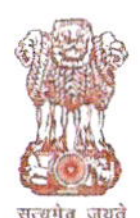

सत्यमेव जयते
डॉ. एम. एल. जाट
सचिव (डेयर) एवं महानिदेशक (भाकृअनुप)

Dr M. L. Jat
SECRETARY (DARE) & DIRECTOR GENERAL (ICAR)

भारत सरकार
कृषि अनुसंधान और शिक्षा विभाग एवं
भारतीय कृषि अनुसंधान परिषद
कृषि एवं किसान कल्याण मंत्रालय, कृषि भवन, नई दिल्ली 110 001

GOVERNMENT OF INDIA
DEPARTMENT OF AGRICULTURAL RESEARCH & EDUCATION (DARE)
AND
INDIAN COUNCIL OF AGRICULTURAL RESEARCH (ICAR)
MINISTRY OF AGRICULTURE AND FARMERS WELFARE
KRISHI BHAVAN, NEW DELHI 110 001
Tel.: 23382629; 23386711 Fax: 91-11-23384773
E-mail: dg.icar@nic.in

Foreword

Horticulture is a rapidly growing sector in agriculture. Apart from providing livelihood to millions of people in India, these crops support many industrial and commercial entrepreneurships. In general, the fruit crops generate much more employment per unit area than other agricultural crops. It has become a sustainable and rewarding venture for progressive as well as marginal farmers. This sector has emerged as potential enterprise and sustaining agro-industries in different parts of India, with its varied climatic and soil conditions provides opportunities for growing a large number of temperate, tropical, arid, humid tropical and sub-tropical fruit crops. In recent years, India has emerged as one of the leading producer and exporter of fresh, processed and dry fruits.

In the last century, several aspects of fruit production have been innovated which were hitherto unknown. There has been a substantial improvement in different aspects of fruit cultivation, particularly production of high-quality disease free, clean planting material. The canopy architecture and orchard management practices have also changed the face of fruit industry in India. With these efforts, India has become second in the world in fruit production, producing 114.5 million metric tonnes of fruits from 7.04 million hectares area. Despite producing large quantities of fruits, India is still lacking in per capita consumption that is required for nutritious diet recommended by the World Health Organization (WHO). To fulfil the minimum dietary requirement per head per day, the country requires more than 150 million tonnes of fruits every year. To achieve this mission, concerted efforts by the scientists, government agencies, policy support and fruit growers are being made to enhance the fruit productivity per unit area.

The cultivation of fruit crops is a cornerstone of global agriculture, supporting nutrition, economic development, and ecological stability. At the heart of this endeavour lie two critical processes; nursery production focusing on the propagation and early development of high quality, disease free clean planting material, and orchard management that encompasses the long-term care and optimization of fruit trees in the field. Together, these practices determine the success of fruit production systems.

The second volume of the book on "*Nursery Production and Orchard Management: Temperate and Sub-tropical Fruit Crops*" deals with the protocols to be followed for higher productivity, especially the nursery production and canopy management.

I compliment Dr. Dhillon, former Assistant Director General Hort., ICAR, New Delhi for this important publication and hope that it will be useful to the academicians, policy makers, extension workers and other stakeholders.

(M. L. Jat)

Dated the 4th August, 2025
New Delhi

Preface

Incorporation of fruit cultivation in horticulture farming systems is a cornerstone of global agriculture, providing nutrition, economic development, and ecological sustainability. At the centre of this endeavour lie two critical components: nursery production of fruit plants and orchard management. Nursery production mainly deals with propagation, production and supply of high-quality, disease free planting material. Nursery production is a science and art of propagating fruit plants under controlled conditions and best agro-horticultural practices until they are ready for transplantation. It serves as the foundation of orchard establishment by ensuring uniformity, vigour, and genetic fidelity in planting material. Poor planting materials usually lead to low quality yield. Whereas, orchard management encompasses the long-term care and to maintain proper health of fruit trees in the field. In unison, these high precision practices decides the success of fruit production systems.

India is the second largest producer of fruits in the world. The average productivity of fruits in India is, however, low as compared to many developed countries. The main reasons for low productivity are small orchards, low yielding traditional varieties, poor orchard efficiency and inadequate technological upgradation and adoption by the growers. The high-density planting technique represents a contemporary approach to fruit cultivation that involves planting trees densely, resulting in smaller or dwarf trees with modified canopies. Effective orchard management systems impact fruit quality, output, and development by using best practices in orchard culture, which also preserve moisture, reduce weed competition, and increase nutrient availability.

India's diverse climate and topography contribute to its rich fruit diversity and have potential for feeding the growing global population. Despite the tremendous potential, challenges like the abiotic and biotic stresses due to climate change during the last two decades have resulted in productivity decline of fruit crops. Optimizing canopy management is essential for improving both the yield and quality of fruit.

The advances made in nursery production techniques, fast multiplication and production of disease free planting material and canopy architecture and management has totally changed fruit production systems. The second volume of the book solely deals with the Temperate and Sub-tropical Fruits.

I humbly acknowledge the contribution of contributors, who are experts on their own specialized fields that helped me getting this book (Volume: 2) in the present shape. No professional and academic journey is complete without constructive feedback and comments from the readers of this book. I urge upon the scholars and readers to share their critical views on wasakhasingh@yahoo.com regarding the coverage of chapters in this book so that it would value for the improvement in future editions.

W.S. Dhillon

Contents

Chapter 1

Orchard Planning and Layout

G.S. Sidhu and M.K. Batth

School of Agricultural Biotechnology, PAU, Ludhiana–141004, Punjab

Orcharding involves the organized cultivation of fruit plants to ensure sustained economic yield over many years. Establishing a fruit orchard is a significant long-term investment, typically requiring around three years to produce a commercial crop, followed by continued economic viability for 15 to 30 years. Strategic planning is crucial to maximize production, profitability, and the longevity of the orchard, especially given the high costs associated with land acquisition, preparation, and the extended lifespan of the orchard. Therefore, thoughtful site selection is imperative to ensure optimal outcomes.

The success of an orchard heavily relies on the initial establishment of its plants. A well-established orchard not only prevent many growth and performance issues but also mitigates potential risks through effective establishment practices. Key steps include proper land preparation. Fields selected for orchards should undergo deep ploughing followed by harrowing to eliminate perennial weeds, roots of bushes, and large clods. On bushy or minor lands, clearing of bushes and forest debris is crucial before deep ploughing begins.

In hilly or steep sloped areas (30-40 per cent gradient), contour trenching is preferable over deep ploughing as it provides suitable soil tilth for healthy root development. After harrowing, the land is levelled and gently sloped in one direction to aid irrigation and prevent waterlogging during rains. Areas prone to poor drainage should be trench-treated from the outset to prevent potential harm to young plants from water stagnation.

On slopes up to 10 per cent, planting trees on mounds or in trenches is advisable to ease management tasks and improve drainage. Steeper slopes require terracing, despite the challenge it poses to ploughing. Additionally, it's beneficial to have grass or stone-lined irrigation ditches to curb soil erosion.

Deep ploughing in dry conditions before planting ensures uniform growth, particularly in soils previously under cultivation where compact layers tend to form. While not mandatory, using cover crops before and during the initial 3-4 years after planting helps suppress weeds, shield young trees from wind, and provide on-site mulching. These measures collectively optimize orchard establishment and foster long-term productivity. Following are the steps which should be followed during planning, establishment and layout of an orchard.

Soil Reclamation

The first step before planning of an orchard is the reclamation of problematic soils. In both normal and alkaline soils, plants typically do not require liming. However, if soil analysis reveals a pH lower than the optimal range (5.6-6.0), lime should be applied. In cases of magnesium deficiency, dolomite can be used to improve soil pH. For very acidic soils, substantial lime applications may be necessary.

To prepare the planting area, two-thirds of the recommended lime quantity should be evenly spread over the soil surface, mixed with the topsoil, and ploughed deeply at least 9 to 12 months prior to planting. This allows calcium (lime) to slowly penetrate into the root zone. If a lighter lime application (2-4 t/ha) is needed, it can be applied at least three months before planting.

In alkaline soils with a pH ranging from 7 to 8, repeated applications of sulphur at a rate of 500 kg/ha can be employed until the pH drops to 5.6. Trees can thrive in alkaline soils if proper soil reclamation practices are followed and sufficient micro-nutrients are incorporated into the soil. These soil management strategies ensure optimal soil conditions for successful fruit tree cultivation, taking into account specific pH requirements and nutrient deficiencies.

Proper Layout of an Orchard is Necessary

Proper orchard layout is essential for efficient management and optimal plant growth. On level ground, establishing a straight baseline adjacent to a fence or roadway is straightforward. Perpendicular lines are then marked at both ends and possibly in the middle using a simple method involving three ropes of 3:4:5 proportions (Figure 1.1). This ensures precise right angles for accurate layout. Stakes are positioned along these lines to extend them for further marking.

Once the boundaries are defined, the orchard area can be divided into squares or rectangles with sides equal to the desired planting distance. For hexagonal planting patterns, two chains of equal length to the planting distance are joined with rings.

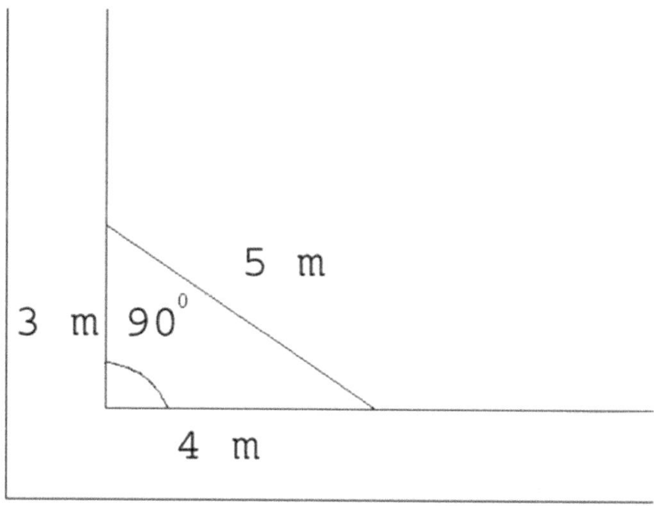

Figure 1.1: Checking Right Angle during Layout.

Stakes are placed at intervals along the baseline, and the chains are anchored at each end with rings, ensuring tautness. The midpoint ring is then marked with a stake (Figure 1.2), establishing the first row. Subsequent rows can be marked using this as a baseline. Triangular planting follows a similar method to square or rectangular systems, except rows in even-numbered positions are staggered between those in odd-numbered rows.

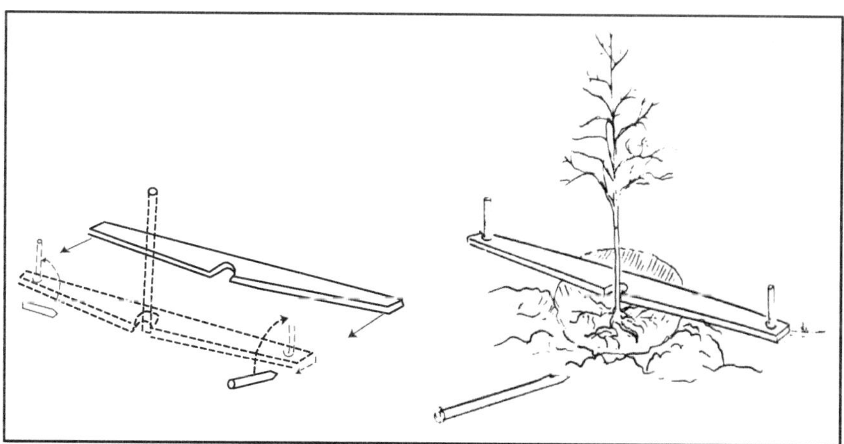

Figure 1.2: Marking Centre of Pit with Planting Board.

On sloping land, contour layout requires a surveyor's level and rod. The highest row is staked out on a level line, maintaining uniform elevation along its length. From this point, the next row is positioned down slope at a distance determined by the minimum row spacing. As the slope gradient lessens, rows are spaced wider apart. Short contour rows may be added where the distance between adjacent rows doubles, maintaining efficient land use and erosion control.

In orchard production, one notable change has been the adoption of high-density planting (HDP) and the rationalization of tree shapes. This approach aims to create fruit-bearing canopies that are more compatible with modern management and harvesting techniques. HDP in orchards focuses on achieving early yields and improving overall efficiencies. These efficiencies target optimizing the relationship between capital investment and specific inputs relative to the outputs of the production system, whether in terms of yield levels, fruit quality, or financial returns.

For instance, some orchards have transitioned to cultivating shorter stature trees that require less pruning, training, and labor during harvesting. This reduction in labor needs enhances the competitiveness of high-density production systems by streamlining operations and improving cost-effectiveness.

When planting on contours, a staggered planting method maximizes natural resources and enhances production efficiency. These systematic approaches ensure that orchards are laid out effectively for optimal growth, harvest, and management practices.

Planting Methods

There are different methods of planting orchards, each suited to different needs and conditions:

1. **Square planting**: In this method, the orchard is divided into squares, with trees planted at the corners. Rows run at right angles, facilitating easy cultural and harvesting operations (Figure 1.3).

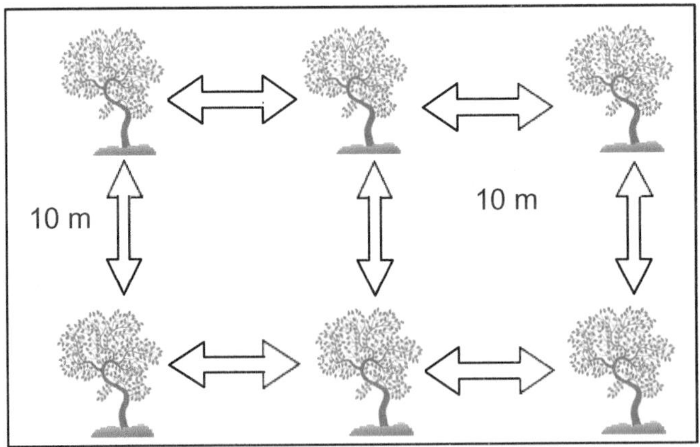

Figure 1.3: Square System of Planting.

2. **Rectangular planting**: Trees are planted in straight rows that intersect at right angles, with closer spacing within rows than between them. This layout is adaptable for high-density planting, allowing flexibility by removing alternate trees for additional drive row width (Figure 1.4).

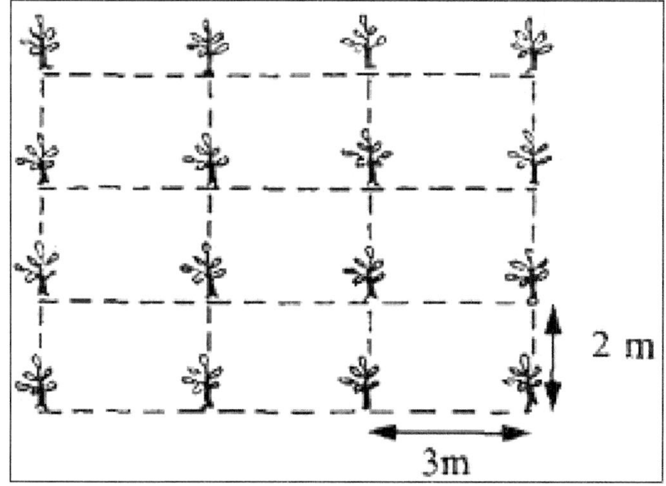

Figure 1.4: Rectangular System of Planting.

3. **Quincunx (Five Spot)**: Similar to square or rectangular patterns, but with an additional central tree. This central tree can serve as temporary filler or for growing short-statured plants until main trees mature. It optimizes space utilization beyond traditional layouts (Figure 1.5).

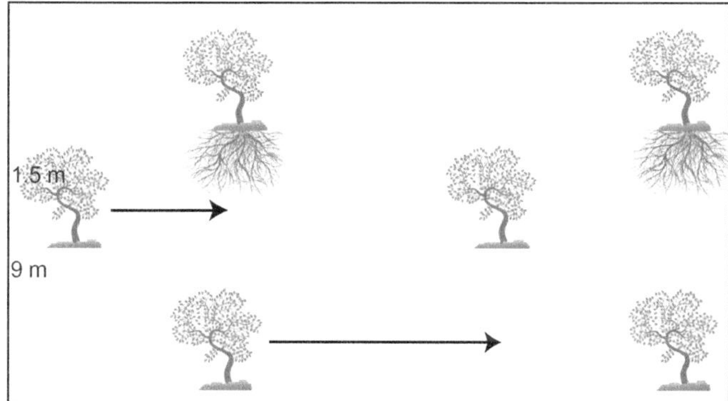

Figure 1.5: Quincunx System of Planting.

4. **Hexagonal**: Trees are planted in alternating rows in an equilateral triangle pattern, maximizing land use efficiency by accommodating approximately 15 per cent more plants per unit area compared to square planting of the same size (Figure 1.6).

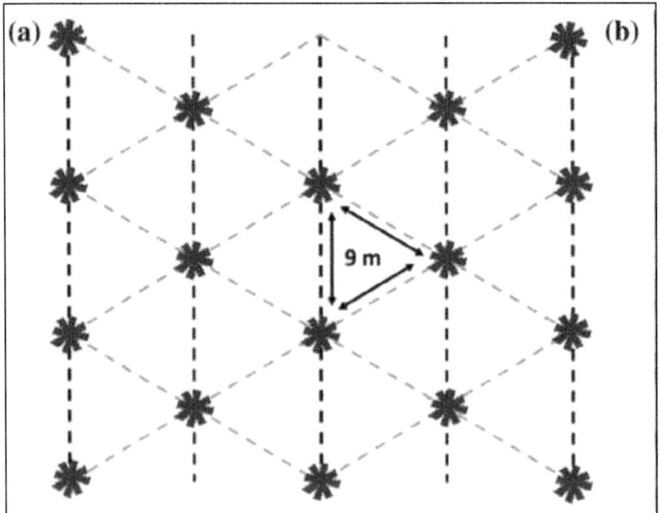

Figure 1.6: Hexagonal System of Planting.

5. **Triangular planting**: Similar to hexagonal planting but with alternate rows offset by half the distance between plants in a row. While easier to layout than hexagonal patterns, it results in 9 per cent fewer plants than the equivalent square or rectangle (Figure 1.7).

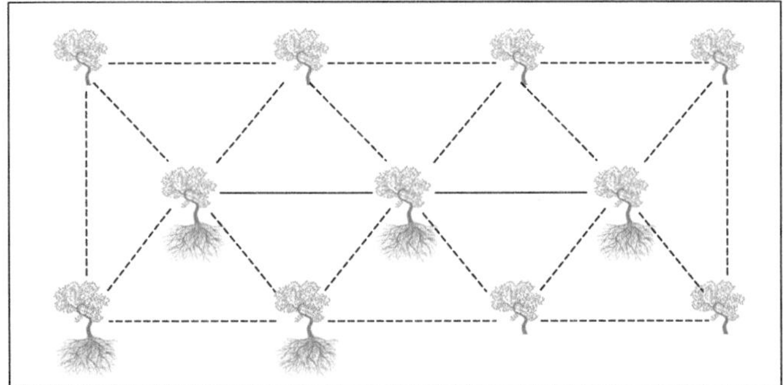

Figure 1.7: Triangular System of Planting.

6. **Single Hedge Row planting**: Trees are closely spaced in rows, with the distance between trees in a row typically half to one-third the distance between rows. This method is used for high-density planting, particularly in experimental stages for orchards, showing potential for early fruiting (Figure 1.8).
7. **Double Hedge Row System**: This involves planting trees at half the usual spacing or adding extra rows to a square or rectangular system initially. As trees grow and begin to crowd, additional plants are removed to maintain

spacing, eventually reverting to the original layout. It optimizes land use during initial growth stages (Figure 1.9).

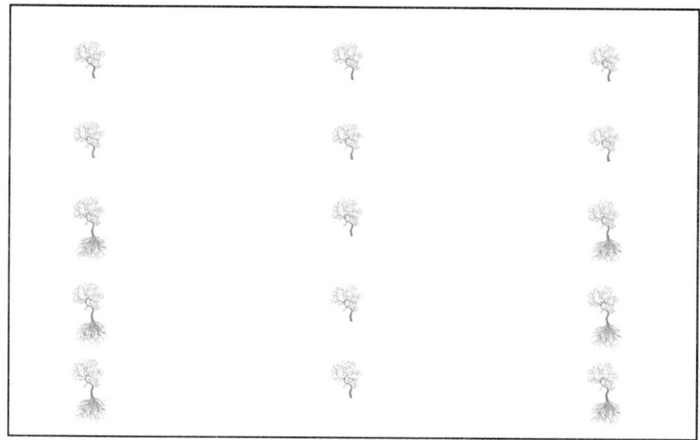

Figure 1.8: Single Hedge Row System of Planting.

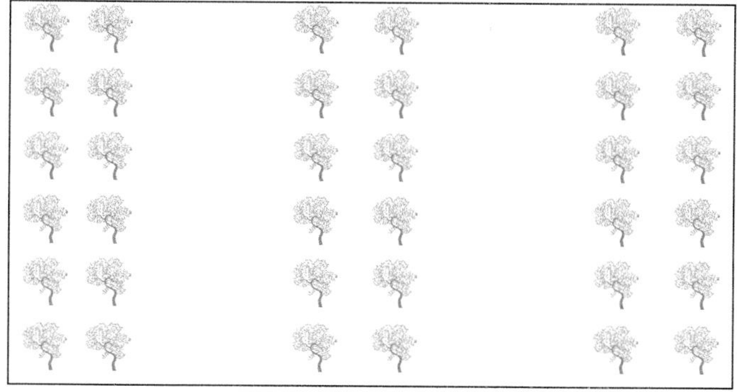

Figure 1.9: Double Hedge Row Planting.

8. **Contour Planting**: Suitable for sloping or hilly terrain, contour planting utilizes natural terracing to prevent soil erosion. Rows or groups of rows follow the natural contour of the land, ensuring irrigation water remains within terraces. This method allows productive use of otherwise challenging terrain (Figure 1.10).

Each planting method offers distinct advantages depending on factors like land topography, desired planting density, and long-term management goals, ensuring efficient orchard establishment and management

Digging of Pits

Once the planting layout is determined and the land prepared, the pits for transplanting saplings should be dug one to two months prior to planting. The primary goal of pit preparation is to create optimal conditions for plant growth and development, particularly during the crucial establishment phase.

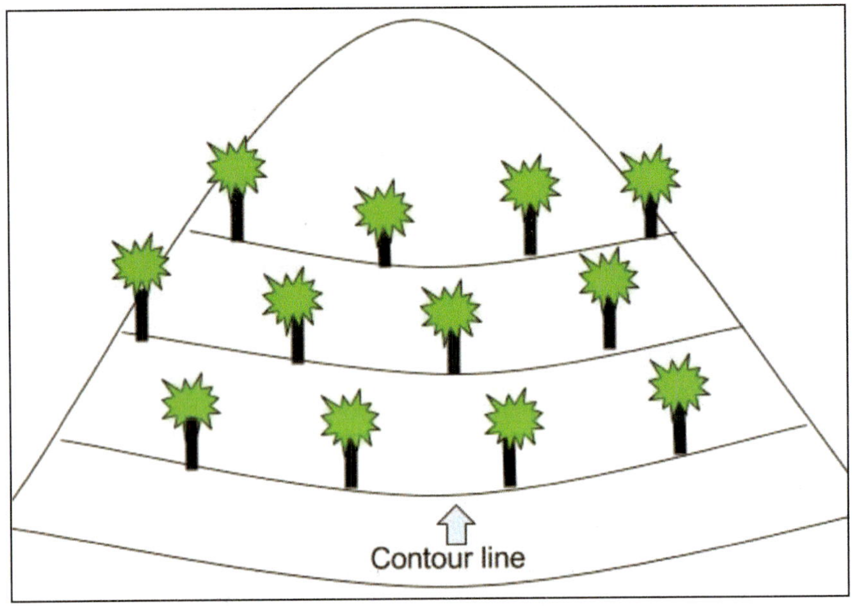

Figure 1.10: Contour System of Planting.

The process begins by digging pits of appropriate size depending on soil type and location. For light soils, 60 x 60 x 60 cm pits are sufficient, while heavy soils require larger pits, typically 90 x 90 x 90 cm. In gritty (murrum) soil, pit size may be increased to 150 x 150 x 100 cm, with 60-75 per cent of the dug soil replaced with a mixture of good soil and compost. In loamy and deep soils, smaller pits of 50 x 50 x 50 cm may suffice, spaced at desired intervals.

On hill slopes, pits should be dug in a 1 x 1 x 1 m size along the contour. In sloping areas, pits are shaped like trenches by cutting the upper side of the slope and levelling the lower side to create a levelled platform near the pit. Pits should be aligned correctly to ensure saplings are planted in the center.

During pit preparation, the upper half of excavated soil is kept separate from the lower half under normal soil conditions. In murrum soil, the entire excavated soil is kept aside and only topsoil is used for filling. The dug-out soil and open pit are left exposed for 2-4 weeks during summer months to allow sunlight and heat to naturally disinfect and condition the soil.

In stony soils, it is advisable to remove all stones from the excavated soil and mix the remaining soil with manure and field scrapings. Using a JCB machine for pit digging, especially in hard and murrum soils, is recommended as the bucket pressure can break through hard layers, facilitating better root penetration for trees.

By following these steps, the pit preparation ensures an optimal start for saplings, promoting healthy growth and establishment in diverse soil conditions and terrains.

Filling of Pits

The initial growth and vigor of plants depend significantly on the quality of the substrate used to fill the planting pits. Therefore, it is crucial to fill the pits with a nutrient-rich mixture. Common components of the substrate include farmyard manure, cakes (such as neem or karanj), vermicompost, Trichoderma-enriched manures, bio-fertilizers, soil from the root zone, and VAM (vesicular-arbuscular mycorrhiza) culture.

In rocky and murrum soils, which typically lack inherent nutrient capacity, the composition of the initial substrate in the pit plays a crucial role in determining the performance of plants. As a standard application, one pit in normal soil requires approximately 30-40 kg of Trichoderma-enriched farmyard manure, 2 kg of neem or karanj cake, and 250g of single super phosphate (SSP) or bone meal. In problematic soils, the pit soil may need to be replaced entirely with good quality soil.

In highly acidic conditions, adding 1-2 kg of slaked lime, along with 250 g each of zinc and boron (agricultural grade), helps restore the vigour of the plants. If magnesium deficiency is anticipated, incorporating 500 g of dolomite into the pit filling mixture is recommended. In sandy or infertile soils, SSP can be substituted with a balanced fertilizer mixture like 12-12-17, ideally applied two weeks before planting or up to five months earlier if possible.

To manage white ant infestations, fenvalerate dust is mixed with both upper and lower layers of the pit soil. The upper layer, along with the substrate, is filled first followed by the lower soil mixture. Using high-quality soil mixture, particularly from beneath well-established trees with potential mycorrhiza cultures, enhances the initial establishment and growth of saplings.

Filling the nutrient-rich mixture at the bottom of the pit encourages root proliferation in the lower two-thirds of the pit. The upper level of the pit should be kept approximately 15 cm above ground level. After filling, the pits should be irrigated if rain is not expected to settle the soil.

By following these guidelines for pit preparation and substrate filling, plants are set up for robust initial growth and development, ensuring favourable conditions for their long-term health and productivity.

Selection of Nursery Stocks

Establishing an orchard is a long-term endeavour that requires meticulous planning and care from the outset. The success of the orchard hinges significantly on the quality of the planting material used. It should be true-to-type, healthy, and vigorous. Optimal nursery practices are crucial as they directly influence the future performance and profitability of the orchard.

It is advisable to procure planting materials from registered nurseries to ensure authenticity of the variety and health of the plants. Alternatively, growers can produce their own planting materials using standardized procedures well in advance, allowing them to reach the appropriate stage for field planting.

When selecting planting materials, careful inspection is essential. Avoid purchasing overaged saplings exceeding two years old. Ideal saplings should be between 40 to 60 cm in height with an adequate number of leaves for photosynthesis. Look for glossy, dark green leaves as indicators of good plant health. The presence of new leaves with a coppery red color is normal and should not be mistaken for nutrient deficiency. Conversely, deformed or discoloured leaves suggest underlying issues such as incorrect nursery practices or pest and disease damage. Avoid saplings with compressed internodes or a rosette of leaves at the shoot apex, as these may indicate malnutrition or stunted growth.

For nurseries, use polythene bags or pots measuring at least 20 x 15 x 45 cm. Only select vigorous, hardened plants for field planting. Hardening can be achieved by gradually exposing nursery stocks to direct sunlight and adjusting water application accordingly, which enhances their resilience and success rate during field establishment.

Regarding planting time, it can be done throughout the year, except during extreme hot, cold, or windy conditions. Optimal planting times are on cloudy, humid days, typically occurring in spring, late summer, early monsoon, or early autumn. Since requires ample water during its establishment phase, planting during the onset of monsoon is advantageous, especially where irrigation facilities are limited. In frost-prone areas, spring planting is preferable.

Across India, the ideal planting time is during the monsoon (July to August), when atmospheric moisture is abundant. In North India, monsoon planting ensures optimal conditions for establishment. Alternatively, in areas with assured irrigation, planting can be done during the spring season (February to March), although precautions are needed due to the subsequent hot, dry period which can lead to higher plant mortality.

Planting of the winter deciduous crops is done in winter when buds are not open. Such type of plants are transplanted bare rooted.

For regions with heavy rainfall, planting at the end of the rainy season is recommended, while in drier areas, early monsoon planting promotes better establishment. Avoid planting during bright sunlight hours to prevent excessive water loss and ensure higher humidity conditions prevail, ideally planting during cloudy or hazy sunlight periods. Evening planting is also beneficial when humidity levels are naturally higher.

By adhering to these guidelines for selecting planting materials and choosing the appropriate planting time, growers can optimize the establishment and long-term success of their orchards.

Planting Distance

The optimal planting distance in orchards plays a crucial role in maximizing land use efficiency and ensuring profitability. It is designed to effectively utilize light, water, and nutrients, thereby maximizing the total yield potential within a compact area.

The spacing between trees is determined by several factors such as soil type and fertility, vigor of the cultivar, local growth conditions, and the orchard's intended purpose. In less fertile soils where plants grow slowly, they require less space, whereas in fertile soils with vigorous growth, more spacing is necessary. A mature tree, typically after 25 years, can develop a canopy diameter of up to 12 meters, requiring adequate aeration and light around its entire circumference for optimal yield.

Due to relatively slow growth rate, there is a growing preference for higher density plantations. For instance, planting at 5 meters by 5 meters can accommodate approximately 400 plants per hectare. Such dense planting requires intensive management practices including regular and severe pruning to maintain the trees' canopy within the designated space.

When determining the planting distance for orchards, considerations should also include local climatic conditions, specific cultivar requirements, and the intensity of crop management practices. By carefully assessing these factors, growers can optimize their orchard layout to achieve maximum productivity and economic returns over the long term.

Care during Planting of Saplings

Under no circumstances should plants be carried by holding the stem. When transplanting, dig a hole slightly larger than the size of the root ball at the designated planting spot. Before removing the polythene bag from the root ball, moisten it with water for about 5 minutes to help compact the earth ball. Then, carefully cut around the bottom edge of the polythene bag to remove it. Place the plant, still with its intact earth ball, into the hole ensuring it remains straight and that the collar portion of the plant is level with or slightly above the ground level.

Next, make two longitudinal cuts on the polyethylene bag from bottom up and remove it. Fill the planting hole with soil, gently pressing the moist soil around the root ball to firmly secure the plant in place. This method minimizes root damage that can occur from breaking or splitting of the potting media block. For precise placement of saplings in the pit, use a planting board (Figure 1.2).

While planting, take care not to press the soil too firmly, which could damage the earth ball. Immediately after planting, create a small basin around the plant for irrigation. The frequency of irrigation during the first month should be adjusted based on weather conditions.

Ensure the plant is not planted too deeply or too high, as this can harm the roots. Use a stake placed close to the plant and secure it carefully. Regularly inspect the plants to detect any issues such as soil settling, tilting, or cracking in the irrigation basin. Maintain sufficient moisture until the plants are established and begin new growth in the field.

Irrigation Management

Water is essential for establishing a successful orchard. Ensuring a reliable water supply and choosing the right irrigation system are critical for commercial viability. Various irrigation methods such as flood, basin, check basin, furrow, drip, sprinkler, and microjet systems can be selected based on water availability and budget. Among these, drip or micro sprinkler systems are preferred under trees due to their efficient water use, although they require significant power, technical knowledge, and have high initial costs.

In upland areas with limited irrigation infrastructure, plants can survive with life-saving irrigation from natural water stored in plastic-lined pots, Jal kundas, or open ponds. In such scenario, *in situ* water harvesting and moisture conservation techniques near the plant roots are crucial. Mulching is also beneficial as it moderates soil temperature, controls weeds, prevents soil drying, and reduces erosion. Simple mulching materials like dry grasses can effectively be used around the plants. Providing appropriate shading helps protect plants from intense sunlight, supporting their healthy growth and development.

Role of Mulching

One of the crucial step in the planting process is applying mulch, which is highly recommended for young fruit trees. Any type of mulch material suffices, as its primary role is to suppress grass and weeds that compete vigorously for water and nutrients with the young tree, hindering proper establishment.

When mulching, extend it outwards from the trunk to about 1 meter. Unless your area is prone to water-logging, consider building up the mulch around the edges to naturally channel water towards the tree.

After applying the mulch, water the tree with approximately 10-15 litres of water. During summer or from July to September, water again after 4-5 days if there has been no rainfall. This regimen helps ensure the tree receives adequate moisture crucial for its initial growth and development.

Impact of Drainage

Drainage of excess water poses challenges in orchards, particularly in heavy soils or low-lying flood-prone areas. In these regions, inadequate drainage leads to poor growth, unhealthy appearance, and even death of young plants. Persistent high water tables can severely impair root growth and function, detrimentally affecting the establishment and productivity of trees.

Effective drainage systems are essential to address these issues. They aim to eliminate excess water from the active root zone, especially during rainy seasons, and maintain the water table in the orchard below the active root zone (approximately 1.80-2.40 meters below ground level).

However, it's important to note that excessively sloped soils as they are susceptible to erosion, and overly efficient drainage in such soils can lead to rapid moisture depletion, necessitating more frequent irrigation.

Proper management of drainage is critical to ensure optimal growing conditions for trees, supporting healthy growth, robust flowering, and sustained orchard productivity.

Tree Guard and Staking

When newly planting trees, it is advisable to install a tree guard made of bamboo or wood, shaped in round, rectangular, or square forms, around the tree. This guard should be placed approximately 50 cm away from the tree trunk and extend to a height of about 1.5 m. Its primary purpose is to shield the main stem and branches from strong winds, intense sunlight, and frost damage, while also offering partial protection against small animals.

However, it is important to remove this tree guard within one year of planting. During the initial year after planting, it may be necessary to stake the trees to prevent damage from occasional strong winds. Stakes and ties contribute to the initial cost of establishing the plant and require maintenance.

When staking, it is crucial to use a short stake with a loose tie. This allows the plant to sway in the wind, which stimulates natural root growth and encourages the trunk to develop a stronger diameter at its base. If the stake and tie are too tight, the plant may become dependent on them, increasing the risk of snapping or toppling once they are removed.

Position the stake on the windward side of the tree, near the root ball without causing damage, and ensure it reaches no more than one third of the tree's height. Use a rope to loosely attach the tree to the stake, forming a figure-eight pattern or leaving a gap to prevent rubbing against the plant.

Regularly inspect the plant and stake, and remove the rope at the start of the second growing season. By this time, the plant should have established sufficient root anchorage. If not, additional measures may be required to promote healthy growth. Alternatively, double stakes can be employed for added reliability and to protect against accidental damage from mowing. These can be set up by planting four stakes in a square formation, each one meter high and spaced at one-meter intervals, with synthetic cloth interwoven around them up to the height of the plant

Live Fences

Protecting orchards from encroachment, theft, and animal grazing is a common concern addressed through fencing, which can be achieved in two main ways: using cement poles and wire fencing or erecting a boundary wall. An alternative method involves live fencing with specific plant species, an age-old practice known for its multiple benefits. Live fences not only safeguard plants from stray animals and extreme weather but also provide fuelwood, fodder, and enrich the soil.

Live fencing is particularly valuable in areas where cold and hot winds pose challenges, acting as effective windbreaks. Traditional farmers possess extensive knowledge about live fencing practices and suitable species. Typically, tall trees are planted in the first row with small, bushy thorny plants staggered in the second row along the orchard boundary. On sloppy lands, windbreaks are strategically planted along contours to protect plants at higher elevations.

Establishing a live fence requires patience as it takes two to three years to develop fully. Ideal species for live fencing are thorny, inedible, and non-grazable by cattle and goats, resilient to local conditions, fast-growing, and capable of generating additional revenue. Plants like Gliricidiasepium, Moringa, and Flame tree are popular choices due to their durability against termites and decay.

For orchards where grazing animals are present, initially combining traditional wire fences with wooden poles is advisable, gradually replacing them with live fence plants as they mature. Gliricidia live posts, for instance, can be established by planting large stakes within existing wire fences, which root quickly and require periodic pruning to maintain shape and density.

Thick, dense hedge plants can also serve as effective barriers without additional support like barbed wire, reducing initial establishment costs for orchards. Some farmers opt for a combination of poisonous and unpalatable species to further deter trespassers.

In essence, establishing a robust fencing system, whether through traditional methods or live fences, is essential for safeguarding orchards and ensuring their long-term productivity and sustainability.

Windbreak

To mitigate the impact of wind on trees' fruiting, tall trees are planted along the orchard border to create a windbreak. These windbreaks typically consist of trees with dense foliage that help maintain humid conditions around the orchard. They play a crucial role in reducing wind velocity and minimizing the adverse effects of high temperatures, which can lead to excessive fruit drop in young plants.

Commonly used trees for windbreaks include seedling mango, jamun, mulberry, moringa, jackfruit, carambola, shisham, teak, gamhar, and bamboo. These trees are often planted close together to form an effective shield. Tall species like shisham and jamun are spaced approximately 6 meters apart, whereas lower-growing trees such as mulberry and carambola are planted at intervals of 7 meters.

To prevent nutrient competition from windbreak trees affecting plants, it is advisable to dig a trench about one meter deep between the rows of fruit plant and the windbreak. This trench helps to separate the root systems and ensures that the trees receive adequate nutrients without competition from the windbreak species.

When establishing a live fence, it is best to plant just before the monsoon season and ensure regular watering after the rainy season ends to promote optimal growth. Promising species for live hedges along orchard boundaries include Agave, Sisal, Boradi, Karonda, and Mehandi, among others. These plants contribute not only to fencing but also provide additional benefits such as soil enrichment and habitat for beneficial insects.

Replanting/Gap Filling

Despite careful efforts, some plants may perish within the first two years after transplanting. Typically, such losses do not exceed 5 per cent to a maximum of 10 per cent of the total planted population when adequate care is provided. These gaps are typically filled by replacing with new saplings of the same cultivar. Replanting can occur during the planting season or off-season, excluding severe winter or summer months.

The primary reasons for establishment failures in newly planted trees are improper irrigation and inadequate weed control, which contribute more significantly to plant loss than any other factors during the initial years. It is crucial to meticulously manage irrigation and implement effective mulching practices to ensure plant health and survival.

When an old orchard is retired, it is generally not replanted. This decision stems from the fact that over the orchard's long lifespan, the soil tends to become depleted, and pests and diseases specific to them can establish firmly. Therefore, replanting on the same site would pose uncertain success for the subsequent establishment of trees.

Integrated Fruit Production

The trend towards reducing the use of pesticides and fungicides, driven by concerns over ecosystem impact and production costs, has led to new production systems. Buyers, consumers, and governmental agencies all have preferences regarding which products they find acceptable. Governments, in particular, bear responsibility for national biosecurity and public health. Fruit shipments must adhere to specific standards concerning the presence of certain organisms and the safety of the product, with particular scrutiny on the presence and levels of chemicals deemed hazardous.

Producers face decisions regarding their production methods and inputs, which must align with the expectations of buyers, consumers, and regulatory bodies. This shift towards minimizing chemical inputs not only benefits ecosystems by reducing disturbance but also lowers production costs.

Juvenility Mechanism in Fruit Plants

The juvenile period is the period elapsing between seed germination and first flowering of the seedling. During the juvenile phase of plant development meristems acquire reproductive competence, becoming able to sense and respond to signals that induce flowering. Within a species, the onset of flowering can vary tremendously, either because of differences in the environment or because of genetic differences. Fruit tree seedlings in their juvenile period show a number of anatomical and morphological characters which disappear or change with time. Leaves differ from those in the adult phase as to size (smaller), width (narrower), serration (sharper), cell size (larger). Beside this, thorns are present during the juvenile period of the tree and the angle between side shoots and main stem is wide. However, the attainment of the flowering stage and the disappearance of juvenile symptoms do not necessarily occur simultaneously in all seedlings of a given progeny.

Methods to Break Juvenility in Pre-Bearing Stage

There exists a transition period between juvenile and adult period. The end of the juvenile period is indicated by the attainment of the ability to flower and the actual production of flowers is the first evidence that plant is in the adult phase. This period of transition is also defined as the adult vegetative phase. The adult vegetative phase is the most important phase for growers as during this phase most floral-inducing techniques are applied. Observations on several fruit tree species have shown that seedling vigour and juvenile period are inversely related. This inverse correlation exists also for trees on rootstocks, *i.e.* between stem diameter and unproductive period. It was shown that adult seedlings, mature shoots with flowering signals after having budded on a rootstock flowered sooner when the juvenile period has been shorter.

To reduce juvenility in fruit crops use of plant growth regulators, training, pruning including branch bending is being manipulated. PGRs like cytokinins and Ehtylene (ethephon) at optimum doses help to break juvenility in fruit crops. Triggered change may occur by regulating the biosynthesis, metabolism or translocation of plant hormone. Sometimes the applied PGR may increase or decrease the endogenous levels of plant hormone when their endogenous levels are lower or higher than their need for changing plant growth and development. Physiological responses regulated by the precise use of PGRs are promotion of branching, increase or inhibit flower bud formation, thinning of fruits or flowers, reduce biotic and abiotic stress, retard pre-harvest drop, improve overall fruit quality, increase fruit set and fruit colour, advance or delaying fruit ripening, enhance rooting, suppress growth of water sprouts, improve stress tolerance and many more.

Training refers to the judicious removal of plant part/parts to develop proper shape of a plant capable of bearing a heavy crop load. It facilitate orchard cultural operations, admit more light and air to the centre of the tree and to expose maximum leaf surface to the sun for increasing production and for complete colour development, protect the tree trunk from sunburn injury and secure a balanced distribution of fruit bearing parts on the main limbs of the plant. Different specialized training systems like Espalier, Palmette, Slender spindle, Spindle bush and Tatura V trellis have been reported to induce precocity and increase in cumulative yields being used in high density planting which typically aim at increasing light interception by the canopy. Although these has high cost of installation but more responsive in temperate fruit crops like apple, grapes, peach and plums.

Pruning refers to the removal of any vegetative plant part to regulate the production and maintain the balance between vegetative and reproductive growth. It reduces tree size, control tree shape, make trees structurally strong, improve fruit quality, improve light penetration thereby improves fruit colour, flower bud initiation and flowering in the consequent season. Knowledge of fruit bearing habit is a necessary to prune fruit plants to harvest the good of pruning like apple and pear bear fruit on spurs; Peach, nectarines and some of the sour cherry varieties bear crop on the shoots produced during the preceding summer; Grape, bramble, persimmon and quince do not have separate fruit buds during preceding winter. These crops bloom and bear fruit on current season growth which arises from lateral buds formed the preceding summer. Ber and guava also bear on current year growth. Induction of flowering in guava by pruning under Meadow Orcharding has been recommended. Tip pruning for synchronous flowering in mango under warm tropics has been observed.

REFERENCES

Bal JS, 2019. *Fruit Growing*, Kalyani Publisher, New Delhi, 454p

Chadha KL, 2019. *Handbook of Horticulture:* (*Facts and Figures for Farmers, Students and All Interested in Horticulture*), Directorate of knowledge Management in Agriculture, Indian Council of Agricultural Research, New Delhi, 1218p

Jackson D, Looney NE and Morley-Bunker M, 2011. *Temperate and sub-tropical fruit production*, CABI, 316p

Ravishankar H, Singh VK, Misra AK and Mishra M, 2014. *Souvenir, National Seminar-cum-Workshop on Physiology of Flowering in Perennial Fruit Crops,* Lucknow, Uttar Pradesh, 295p

Singh G, Nath V, Panday SD, Ray PK and Singh HS, 2012. *The Litchi*, Food and Agriculture Organization of the United Nations, New Delhi, India, 319p

Singh J, 2018. *Basic Horticulture*, Kalyani Publisher, New Delhi, 570p

Singh R and Saxena SK, 1998. *Fruits*, National Book Trust, New Delhi, 221p

Chapter 2

Propagation Techniques in Temperate Fruits

Wasim H. Raja, M.K. Verma, O.C. Sharma, J.I. Mir, S.U. Nabi, Iqra Qureshi, R. Sumaira and R. Kavitha

ICAR–Central Institute of Temperate Horticulture, Old Airfield Rangreth Srinagar–191132, J&K

Temperate fruit crops play a vital role in global agriculture and horticulture, providing essential nutrients, economic opportunities, and environmental benefits. These crops are fruit-bearing plants that thrive in regions with distinct seasons, including cold winters and warm summers. The significance of temperate fruit crops extends beyond their nutritional value, as they contribute substantially to the agricultural economy of many countries and sustain rural livelihoods. Temperate fruit crops encompass a diverse range of species, each with unique characteristics and cultivation requirements. Some of the most economically important temperate fruit crops include apples (*Malus domestica*), pears (*Pyrus communis*), peaches (*Prunus persica*), plums (*Prunus domestica*), cherries (*Prunus avium*), and grapes (*Vitis vinifera*) (Jackson, 2011). These crops are cultivated extensively in North America, Europe, parts of Asia, and temperate zones of the Southern Hemisphere. The economic importance of temperate fruit crops cannot be overstated. According to the Food and Agriculture Organization (FAO), global production of temperate fruits reached over 87 million tonnes in 2019, with apples alone accounting for nearly 87 million tonnes (FAO, 2021). This substantial production volume translates into significant economic value, sustaining countless jobs in agriculture, processing, and related industries.

Temperate fruit crops offer numerous health benefits, contributing to a balanced diet rich in vitamins, minerals, and antioxidants. For instance, apples are a great source of dietary fiber and vitamin C, while cherries are known for their anti-inflammatory properties (Hyson, 2011). The consumption of these fruits has been associated with reduced risks of various chronic diseases, including cardiovascular disorders and certain types of cancer (Boeing *et al.,* 2012). In addition to their nutritional and economic value, temperate fruit crops play a crucial role in environmental sustainability. Furthermore, many temperate fruit crops support pollinator populations, essential for maintaining biodiversity and ecosystem health (Garibaldi *et al.,* 2013). As global demand for fruit continues to rise, driven by population growth and increasing health consciousness, the significance of efficient and sustainable propagation techniques for temperate fruit crops becomes ever more critical. Consequently, when a tree raised from seed showed desirable characteristics the only way of increasing it in number was by asexual (vegetative or clonal) propagation. Traditionally, cultivars of most tree fruit species propagate poorly on their own roots, irrespective of whether layering or cutting techniques are used. These techniques not only ensure the continued availability of high-quality planting material but also contribute to the genetic improvement of crops, enhancing their resistance to pests, diseases, and environmental stresses (Hartmann *et al.,* 2014). Propagation techniques play a crucial role in the cultivation and improvement of temperate fruit crops.

Plant propagation refers to the controlled process of increasing the number of plants while preserving their essential characteristics. This process aims to (i) enhance plant numbers and (ii) maintain the key traits of the plants. Propagation is fundamentally divided into two main types:

Sexual Propagation

Sexual propagation refers to the multiplication of plants through seeds formed by the union of male and female gametes. During this process, meiosis occurs, reducing the chromosome number by half, and restored to normal after fertilization. This method offers several advantages, but it also has its disadvantages. Effective sexual propagation requires careful management of germination conditions, and facilities, and an understanding of the specific requirements of different seeds. Plants propagated by seeds include most annuals, biennials, and many perennial fruit plants, as well as vegetable crops, plantation crops, aromatic and medicinal plants, ornamental flowering plants, and shade-providing shrubs and trees. Generally, self-pollinated plants, considered homozygous, are multiplied by seeds. In sexual propagation, the sex organs of the flower are involved in processes like pollination and fertilization, leading to seed formation. Seeds produced through sexual reproduction may exhibit different characteristics from their parent plants.

Examples of plants propagated by this method rising of seedling rootstocks of temperate fruits, Apricot, Plum, Pear, Apple *etc.*

Asexual Propagation

Asexual propagation is a process independent of sexual reproduction, occurring through mitosis. Mitosis continues in the shoot tip, root tip, and cambium. When a portion of the plant is wounded, mitosis enables the formation of new cells. During mitosis, chromosomes divide longitudinally to form two daughter cells, creating the foundation for asexual propagation. Plants raised through this method are genetically identical to the parent plant. Common techniques of asexual propagation include cutting, budding, grafting, division, and layering. Additionally, cloning and apomixis are part of asexual propagation (Hartman *et al.*, 2009). With changing living standards, dietary needs have evolved, and fruits now play a significant role in people's diets. Consequently, the area dedicated to fruit cultivation has increased rapidly over the past decades. However, the production of high-quality planting materials for various temperate fruits remains limited to a few registered nurseries that produce quality plants. High-quality fruit plants are essential for successful and profitable fruit production. Therefore, modernizing fruit nurseries in temperate growing regions has become urgent to meet the rising demand for quality plants. Plant propagation has evolved into a highly specialized industry, requiring significant investment in certified skills and advanced technology.

Methods of Propagation of Temperate Fruits

Propagation by Seeds

The majority of rootstocks for many temperate fruit and nut species, including peaches, nectarines (*Prunus persica* L.), apricots (*P. armeniaca* L.), and Asian pears (*Pyrus pyrifolia* (Nakai)), are raised from seed. For nurserymen, propagation by seed offers significant advantages, being simpler and cheaper than vegetative methods. However, the benefits of seedling rootstocks to fruit growers are less clear. In many cases, trees on seedling rootstocks are significantly inferior to those on clonal rootstocks. In species such as apples, where viruses are typically not transmitted through seeds, seedling rootstocks can be advantageous, particularly in regions where maintaining virus-free clonal rootstocks is challenging. Seedling propagation also helps avoid the transmission of root-borne diseases like crown gall (*Agrobacterium tumefaciens*), which can be transferred to new scion trees and sites from infected stool or layer beds. Seedling stocks raised in soils free of crown gall do not face this problem. Clonal rootstocks offer benefits such as reducing the vigor of growth, inducing early and plenty cropping, and conferring resistance to soil-borne pests and diseases. Initially, all rootstocks were raised from seed. Fruits were collected from indigenous populations of native fruit species; the seeds were extracted, germinated, and grown for use as rootstocks. Suitable wild populations of apple (*Malus domestica* Borkh.), pear (Pyrus communis L.), plum (Prunus domestica L.), sweet cherry (Prunus avium L.), and sour cherry (Prunus cerasus L.) were abundant in many parts of Europe. During the early growth stages, the possible benefits of some of these rootstock and effects on scion growth would have been identified and the first clonal rootstocks selected and propagated. These advantages make them superior to seedling rootstocks in many instances. However, seedlings of many heterozygous, out-crossing fruit species show extreme variability in performance as rootstocks. One clear benefit of switching to clonal rootstocks is the improved uniformity of growth and cropping in the scion. Despite these benefits, nurserymen often resist switching from seedling to clonal rootstocks due to economic and logistical reasons. The uniformity of performance in seedling rootstocks can be improved to some extent by:

1. Using seed from a single clonal variety (*e.g.*, 'Red Delicious' apple or 'Bartlett' pear) or seed from a self-fertile cultivar grown in a monoculture.
2. Using seed collected from virus-free mother orchards planted in isolation (*e.g.*, Pontavium and Pontaris *Prunus avium* lines of Mazzard rootstocks available in France).
3. Using seed from apomictic rootstock selections.

Overall, while propagation by seed is advantageous for nurserymen, but clonal rootstocks offer significant benefits in terms of uniformity and performance for fruit growers.

Propagation by Vegetative Structure

Vegetative propagation involves the use of vegetative plant parts (such as roots, stems, or leaves) to produce new plants. This method ensures that the new plants are genetically identical to the parent plant, maintaining uniformity and desirable traits. Vegetative propagation is widely used in horticulture, agriculture and forestry to maintain and reproduce plants with desirable characteristics efficiently and reliably. Example: Citrus, mango, apple *etc.*

Propagation by Cuttings

A cutting is a piece of vegetative tissue (stem, root or leaf) that, when placed under suitable environmental conditions, regenerates the missing parts and develops into a self-sustaining plant. This method involves taking a part of the plant, usually a stem with a few buds, detaching it from the parent plant, and placing it under favorable conditions to develop into a complete plant that resembles the parent in all characteristics. This technique is commonly used for plants that root easily and readily, making multiplication quick and inexpensive. Fruit plants like and grapes, pomegranate and figs are commercially propagated by cuttings.

Cuttings are prepared from vegetative portions of plants such as stems, roots, or leaves and are classified according to the plant part used. The classifications and descriptions are as follows:

Stem Cutting

Stem cuttings are taken from the main shoot or any side shoot of a plant. This is the most commonly used method to propagate many woody plants. There are several types of stem cuttings:

Hardwood Cuttings

- **Preparation**: Prepared during the dormant season from one-year-old shoots of the previous season's growth. For deciduous plants, cuttings are taken after pruning.
- **Characteristics**: Typically 15-20 cm long with 3-5 buds.
- **Cutting Technique**: A straight cut below the node at the base and a slanting cut 1-2 cm above a bud at the top.
- **Examples**: Grape, fig, pomegranate, mulberry, kiwifruit, olive, quince, hazelnut, chestnut, plum, gooseberry, and apple, rootstock clonal.

Leaf Bud Cutting

Relatively few plants can be reproduced using a leaf or portions of a leaf to produce a new plant, as leaf cuttings must regenerate both new root and bud tissues (a capability not many plants possess). Most plants propagated by leaf cuttings are houseplants with thick, fleshy leaves. Depending on the species, leaf cuttings may involve the entire leaf blade, a leaf with its petiole, or just portions of a leaf. Leaf-bud cuttings differ from regular leaf cuttings as they include a portion of stem tissue and, most importantly, a bud. This bud, situated at the junction of the leaf petiole and the stem, is a preformed growing point. Leaf cuttings are best prepared during the growing season, as buds that enter dormancy can be challenging to activate. The plant material selected for leaf cuttings should be healthy, actively growing, and free from insect or disease problems. Examples of plants that can be propagated this way include blackberry and raspberry.

Root Cutting

Root cuttings can be used to propagate a variety of herbaceous perennials in late autumn or early winter when the plants are dormant. This method is ideal for plants that naturally produce suckers (new shoots) from their roots. Root cuttings do not require the special aftercare needed for aerial cuttings and can generate large numbers of new plants from each parent plant. Additionally, plants derived from root cuttings are relatively large and vigorous. Another advantage is that these plants are free from foliar pests and pathogens that might affect their parents, such as stem and leaf nematodes. Examples of plants propagated by root cuttings include blackberry, fig, cherry, and raspberry.

However propagation through cutting has certain advantages and disadvantages.

Advantages

- Quick and cost-effective method of multiplication of plants.
- Produces plants identical to the parent.

Disadvantages

- Plants difficult to root does not root easily.
- The benefits of specific rootstocks on desired varieties cannot be exploited.

Propagation by Grafting

Grafting is a vegetative propagation method where two plant parts, the stock and the scion, are joined together so they unite and continue to grow as a single plant. Various grafting methods include tongue grafting, cleft grafting, approach grafting, side grafting, and veneer grafting.

Tongue Grafting (Splice/Whip Grafting)

Tongue grafting is a simple and popular method used in apples and pears, especially when the stock and scion have equal diameters. The scion sticks should have at least two to three sets of buds. Identical diagonal cuts are made at the top of the rootstock and bottom of the scion, fitting them together. The scion, with 2 to 3 buds, is tightly fitted with the rootstock, ensuring the cambium layers of at least one side unite. The graft is then wrapped with a polyethylene strip. Tongue grafting is done in March–April in high hills and dry temperate zones, and in February-March at lower elevations.

Cleft Grafting

Also known as wedge grafting, cleft grafting is useful when the rootstock is thicker than the scion. The rootstock, up to 8 cm in girth, is decapitated 45 cm above the ground and split about 5 cm deep. The scion, 15-20 cm in size, is taken from a terminal shoot and wedge securely. The scion is slipped into the split of the stock, ensuring the cambium layers match. It is used for avocado, apple, pear, plum, walnut *etc.*

In contrast, various grafting techniques are utilized for the vegetative propagation of commercial apple cultivars, with tongue and cleft grafting being the most prevalent. Apple plantations often utilize seedling rootstocks derived from seeds of cultivars such as Crab apple, White Dotted Red, Yellow Newton, Golden Delicious, and Northern Spy, noted for their good seed viability, germinability, and seedling growth. Srivastava (1966) observed that grafting generally outperformed budding, with tongue grafting yielding the best results. However, cleft grafting is simpler and can be performed in late winter when bark slipping has not yet begun. Despite their effectiveness, these methods require significant quantities of budwood and can lead to variable canopy growth due to differences in budwood maturity. Moreover, there is a higher risk of transmitting pathogens & viruses through grafted scion wood.

Propagation by Budding

Budding is a propagation method where a single bud from the desired scion is inserted into the rootstock. The ideal time for budding is when the bark on both the stock and scion starts slipping, indicating active cambium growth. Key budding methods include shield (T-) budding, patch budding, chip budding, and ring budding.

Shield or T- budding

Shield or T-budding involves reducing the scion to a single bud. A branch with several suitable buds, known as a bud stick, is used. For successful T-budding, the scion material should have fully-formed, mature, dormant buds, and the rootstock should be actively growing, with the bark slipping easily. The buds are cut from the bud stick with a sharp budding knife, leaving a small sliver of wood underneath. The cut should start about 1/2 to 3/4 inch below the bud and finish 1/2 to 3/4 inch above it. A perpendicular cut across the top separates it from the bud stick. The bark on the rootstock is carefully peeled back to create a pocket for the bud shield. The bud is placed into this pocket, and the bark flaps are wrapped tightly with budding rubber, grafting tape, or another suitable material. The wrapping must be removed in 2 to 3 weeks after the union has healed to prevent girdling. This method is used for apple, pear, apricot, plum, peach, cherry, and almond.

Patch Budding

Patch budding involves removing a rectangular patch of bark from the stock and replacing it with a similar patch containing a bud from the desired variety. This method is typically performed during the growing season when the bark separates readily from the wood along the cambial layer. It is particularly effective for species with thick bark, such as walnut and pecans.

Chip Budding

Chip budding is useful when the bark of the stock does not slip easily. A chip of bark and wood is removed from a smooth surface between the nodes of the stock. A chip of similar shape and size is then removed from the bud wood of the desired cultivar. A 2-3 cm long downward cut is made through the bark and slightly into the wood of the stock. A second cut, about 2.5 cm long, intersects the first cut at a 30-45° angle, removing a chip of wood from the stock. The bud chip is then inserted into the place on the rootstock from which the chip was removed. Different budding techniques are employed for the vegetative propagation of commercial apple cultivars, with chip budding emerging as a popular alternative widely adopted in various countries. This method optimizes the economical use of budwood by ensuring precise alignment of the cambium layers of the scion and rootstock, leading to rapid union within weeks of budding. Plants propagated through chip budding exhibit vigorous growth, characterized by the development of extensive lateral branches, which is beneficial both for nursery sales and fruit production (Howard *et al.,* 1974). Chip budding also reduces the incidence of bud loss due to union cankers from Nectria galligena, as it avoids including budwood below the rind. Compared to other budding methods, chip budding is noted for its efficiency and speed of execution (Howard, 1974). Research has shown that chip budding results in superior growth (Dikson, 1991), rapid formation of graft union (Sahu, 1993), higher success rates in bud-take (Kviklis, 1996), and a high ratio of successful budding (Asma and Guloglu, 1999). Howard (1971) reported that chip budding produces taller trees with more lateral branches compared to other methods.

Czarnecki (1987) observed that chip budding promoted more vigorous growth of maiden apple trees of cv. Macspur on M 26 rootstock compared to T-budding, while yielding an equal number of maiden trees. Ananda and Negi (1998) recommended chip budding over shield budding or tongue grafting for its ease of application, high success rate, rapid formation of bud union, uniformity in plant development, production of high-quality trees, and reduced labor requirements. In evaluating chip budding for apple propagation, Howard *et al.* (1983) reported high success rates, strong graft unions, rapid bud union formation, and efficient production of marketable plants. The success of budding techniques is significantly influenced by the timing of propagation, with Porebski *et al.* (2002) recommending mid-August to early September as optimal for chip budding in apples (Lee and Yoon, 2001).

Overall, chip budding stands out as a preferred method for its efficiency, robust results, and practical advantages in apple propagation.

Ring Budding

Ring budding involves removing a complete ring of bark from the stock, completely girdling it. A similar ring of bark containing a bud is removed from the bud stick and inserted onto the rootstock. Both the scion and stock should be of the same size. Ring budding is utilized for peach, plum, mulberry, and other similar plants.

These budding techniques allow for efficient and effective propagation, ensuring the continuation of desirable plant characteristics.

Propagation Techniques of Clonal Rootstocks

Clonal rootstocks are widely used to avoid and overcome adaptability problems. Clonal rootstocks are rootstocks that are grafted onto rootstocks which are genetically similar clones of a parent rootstock. These rootstocks are unique because they have particular desirable traits including tolerance to winter cold, poorly drained soil, heavy clay or salty soil, or very dry soil. The ultimate tree size can be manipulated by the use of rootstock. By selecting the right rootstock for the requirements of the area, the grower greatly enhances the productivity of the fruit trees. For many years only those clones that could be propagated easily from layering/stooling techniques were selected as clonal rootstocks. It is only relatively recently that attributes other than ease of propagation have become important considerations in clonal rootstock selection. Nowadays, with much greater emphasis being put on rootstock attributes, such as their ability to dwarf scions and their resistance to soil-borne pests and diseases, some rootstock clones have been selected which are quite difficult to propagate. This has necessitated changes and improvements in traditional rootstock propagation techniques.

Propagation by Layering

Trench Layering

Trench layering involves establishing a permanent row of plants to be propagated. A branch is laid horizontally in a shallow trench to promote the growth of multiple new shoots. As these shoots grow, soil is filled around them, and roots eventually develop. The mother plants are planted at the base of a trench at a 45-degree angle in rows. The long and flexible stems are pegged down to form a continuous line of layered plants. The young shoots that emerge from these plants are gradually covered with soil to a depth of 15-20 cm. This method is used for apple rootstocks (M16 and M25), cherry, plum, *etc.*

Tip Layering

In tip layering, the tip of a plant stem is bent to the ground and covered with soil so that roots and new shoots develop. Rooting takes place near the tip of the current season's shoot.

Serpentine/Compound Layering

Serpentine or compound layering is similar to simple layering but involves burying multiple sections of the stem, resulting in multiple plants from one stem. This technique is ideal for long, vine-like stems. A one-year-old branch is alternately covered and exposed along its length. This method is used for plants such as the American grape.

Mound Layering/Stooling

Mound layering is a propagation technique commonly employed for clonal rootstock production in temperate fruit crops such as apple, pear, and occasionally guava. Here's a detailed outline of the process for multiplying rootstocks through mound layering in apple:

1. **Establishing the Stool Bed**: Prepare a stool bed on loose, fertile, well-drained soil at least one year before propagation begins. Space mother plants 30-35 cm apart in rows spaced 1-2 m apart, leaving adequate room between rows for cultivation.

2. **Dormant Season Preparation**: In early spring, before new growth begins, prune all plants back to 2.5 cm above ground level. This encourages the emergence of new sprouts within approximately two months.
3. **Sprout Development**: Once new sprouts emerge, girdle each sprout near the base. Apply rooting hormone (IBA) mixed in lanolin paste to the upper portion of the girdled area. Typically, a concentration of 3,000 to 5,000 ppm of IBA is used, depending on the species. Allow two days for proper absorption of the rooting hormone before covering the girdled area with moist soil.
4. **Hilling Operations**: As the sprouts grow, perform hilling operations:
   - ☆ When the sprouts reach 8-12 cm in height, mound soil around each sprout up to half its height.
   - ☆ Repeat the hilling when sprouts grow to 20-25 cm tall.
   - ☆ Perform a final hilling in midsummer when sprouts reach a total length of 45 cm. Cover the bases of the sprouts with soil to a depth of 15-20 cm.
5. **Root development**: Roots typically emerge within 30 to 40 days after the final hilling operation. Allow the layered shoots to root sufficiently throughout the season.
6. **Separation and transplanting**: After 60 to 70 days, when the rooted shoots are well-established, sever them from the mother plants and transplant them into the nursery for further growth or the rooted plants are uprooted during winter months to be grafted in spring season.

Studies have shown promising results in stooling, particularly with apple clonal rootstocks treated with IBA at a concentration of 2,500 ppm, ensuring robust rooting and successful propagation (Pathak, 1991). Similarly, research in cherry clonal rootstocks demonstrated an 80 per cent rooting percentage when treated with 5 mM IBA over a six-week period (Gulen *et al.*, 2004). These methods highlight effective techniques for enhancing rootstock production and maintaining desired genetic traits in temperate fruit crops.

Innovative Techniques Developed by ICAR–Central Institute of Temperate Horticulture, Srinagar (J&K)

Technology for the Vertical Expansion of Nursery under Protected Conditions using Soilless Rooting Medium

The experiment on the vertical expansion of the nursery was started during the year 2019, to utilize the available space in the greenhouse and to exploit the vertical growth of plants. Success was achieved in apple clonal rootstock MM-106, (M9, Pajam, T337, T339, MB-9, P-22, M-27) buy using the same technique. Those plants having a diameter of (5 mm and above) at 30 cm (1.0 ft) above ground level were selected and wounding/incision has been given by sharp knife/blade and rooting hormones IBA (2500 ppm) was applied to the wounded portion. The wounding is done to the targeted region to expose the inner stem for applying the rooting compound. The operation was started from the second week of June till the last week of August. Small polybags filled with rooting medium have been fastened at the points where rooting needs to be initiated. A lightweight substrate (Cocopeat) having high water holding capacity was used. Staking was done with the help of bamboo sticks to hold the bags in a proper position. Watering at regular intervals was done to keep the rooting media moist. Sufficient rooting has been recorded in all the rootstocks. This technology will be very useful in promoting the vertical expansion of the nursery in greenhouse conditions and no. of plants per unit area can be increased 3-4 times without utilizing any extra inputs. One more additional benefit of this technology is that under greenhouse conditions a plant is attained a sufficient girth, above (5mm) and all the daughter plants are suitable for budding operation. The budding has been done to about 45 per cent of daughter plants with an almost 95 per cent success rate. This technology not only produced the additional 2-4 plants but also the budded plants which added the further advantage to this technology that budded plants with well-developed root system are produced in one year of the nursery cycle.

Technology for Propagation of Clonal Rootstocks of Apple through Cutting under Greenhouse in Soilless Beds

To reduce the cost of production ease in operations and to make the technology more feasible for Nurserymen an experiment was designed to plant the cutting in soilless beds. The cuttings one-year-old and greenhouse are necessary for successful propagation by stem cuttings for maintaining high humidity around the cutting is critical particularly in early stage. Sunken beds of dimensions 2.5 0 ft (76cm wide) and 1.0 ft (30cm) depth beds were prepared. The beds were prepared by digging out the upper 1.0 ft soil to prepare a sunken bed. After preparing the sunken beds the beds were filled with sand and cocopeat. The lower 6 inches (15cm) of the bed were filled with sand and the upper 6 inches (15cm) with cocopeat. The purpose of digging out soil from the filled it with sand and cocopeat to make soilless beds for conducive conditions for cuttings and to avoid weed problem and reduce the water requirement considerably. Low lower 6 inches of the bed were filled with well coarse sand to facilitate proper drainage and the upper portion of the bed 6 inches with cocopeat for planting the cuttings in it. The width of the beds was kept 2.5 ft and the length of beds as per the available length of the inside greenhouse. The width of the beds was kept less to facilitate important operations like pinching, weeding, watering, and spray of plant protection chemicals. The cuttings of 30 cm size and pencil thickness (5-6 mm) with 8-9 buds were selected and planted in the first week of March. The lower portion of the cuttings was given a horizontal cut and light cut on either side of the cut stem at the base to expose more of the cambium for callusing. The cuttings were planted in the first week of March after fungicidal treatment and rooting hormone Indole Butyric Acid IBA (2500 ppm) for 2-3 minutes. The cuttings were planted in the soilless beds at a spacing of 3 inches

Figure 2.1: Different Stages of Air Layering Technique in Apple Clonal Rootstocks for Vertical Expansion of Nursery.

(7.62 cm) both in rows and cutting to cutting accommodation 52 cuttings per sq. meter area. The cuttings will start sprouting within two weeks. After planting the growth start and all active buds will sprout from base to top. When these shoots on cuttings will attain 3-5cm length, only vigorously growing uppermost shoot was kept rest were removed. The beds are regularly irrigated twice or once a week based on the temperature inside the greenhouse. Water was given by hose pipe at low intensity to avoid disruption of cuttings and growing medium. Initially up to June watering twice a week was sufficient afterward daily irrigation was given to avoid extreme temperatures in the greenhouse. Shade nets also have been installed to avoid drying of the cuttings and succulent leaflets inside the greenhouse. Although very little weeds growth was observed, weeds emerged if any were removed along with roots without disturbing the cuttings. The insects (aphids and Larval Fungal Gnats) and diseases (powdery mildew) were managed by plant protection chemicals. Important parameters like Percentage of rooted cuttings, plant height, root length, root diameter no of adventitious roots, Root fresh weight and root dry weight was recorded at the end of the growth after leaf fall. From the data, it can be revealed that 84.3 per cent of rooting in cuttings was recorded. Among the cuttings average plant height (115.93 cm) (3.77 ft), plant diameter (8.41mm), and average root length (32.13cm), average number of adventitious roots (12.31). fresh weight (17.40 g) and dry weight (8.80 g) were recorded. More than 90 per cent of rootstocks produced were suitable for grafting having a well-developed root system and clipper size above (6.0 mm). Among the total cuttings, more than 50 per cent have got the calliper size of above (5 mm) during the month of August-September, so budding has also been done for about 25 per cent of plants with a success per cent of above 90 per cent. The purpose of this was to reduce the cost of scion wood and prepare the plants for the coming season. The operation of budding inside the greenhouse can be extended up to two to three weeks due to extended growing conditions available under protected conditions. By this technology, 40-45 plants with well-developed root system were harvested from per square meter area of the nursery. This technology will revolutionize the production of rootstocks that are produced from the thrown-away part of the plant. By adopting this technology, the cost of rootstock/planting material will reduce and dependence on import for quality planting material will also reduce drastically.

Propagation by Apomictic Seedlings

Apomixis is a form of asexual reproduction that allows plants to produce seeds without fertilization. In this process, seeds are formed from the mother plant cells, bypassing the typical sexual reproduction mechanisms involving the fusion of male and female gametes. Thus, the offspring are genetically identical to the parent plant. Apomictic propagation is useful for producing uniform and true-to-type plants, particularly in fruit crops where maintaining specific genetic traits is crucial for quality and consistency. Apomictic seedlings, known for their uniformity, are widely used in citrus tree propagation. Some success has also been achieved in selecting apomictic types of Malus rootstock for apples (Sax, 1949; Luckwill and Campbell,

Figure 2.2: Different Stages of Nursery Development using by Cuttings in Soilless Beds under Protected Condition.

1954; Schmidt, 1982), but these lines proved unpopular with nurserymen and fruit growers. One issue was that apple apomicts were of the facultative type, producing both zygotic and apomictic seeds within their fruits. Culling the zygotic seeds at the nursery stage was difficult. Further, these rootstocks exhibited strong vigor, incompatibility with some scions, and sensitivity to virus infections, which further hindered their acceptance. Compared to vegetative propagation, relatively little research is now conducted into seedling rootstock propagation techniques for temperate fruits. However, the methods of after-ripening and dormancy-breaking, essential for seeds of many Rosaceae, are well understood. Techniques such as scarification and stratification are widely adopted by commercial nurserymen to aid dormancy breaking. Although hormone treatments are less frequently used, some treatments have shown promising results. For instance, treating stratified peach seeds with gibberellic acid (GA3) and benzyl adenine (BA) at 20 mg/liter enhanced germination, whereas thiourea treatment was ineffective (Siyapananont, 1990). Shatat and Sawwan (1985) also found that germination of Prunus mahaleb seed significantly improved with Promalin (a mixture of GA4+7 and BA) at 3000 mg/lt. Research in Poland demonstrated that soaking plum rootstock seeds in a 500 ppm ethephon solution before sowing improved germination. Additionally, autumn sowing of partially stratified seeds produced better results than spring sowing of fully stratified seeds (Grzyb and Czynczyk, 1990). While recent research on the biochemistry of dormancy in temperate fruit tree seeds is limited, studies by Kar and Bhartiya (1987) revealed that the quantity of ortho-diphenols in apple seeds is associated with their depth of dormancy. Current innovations in seedling rootstock propagation focus on developing improved machinery for land preparation, seed sowing, seedling management, and lifting operations.

Key Features of Apomictic Propagation

1. **Genetic Uniformity**: Since apomictic seeds are clones of the parent plant, they maintain the genetic characteristics of the parent, ensuring uniformity in the offspring.
2. **Consistency**: This method is advantageous for preserving desirable traits in horticultural crops.
3. **No Need for Pollination**: Apomictic propagation does not require pollination, which can be beneficial in environments where pollinators are scarce or conditions are not conducive to pollination.
4. **Disease Resistance**: Plants can be selected for resistance to specific diseases and these traits will be consistently passed on to all progeny.

Modern Propagation Techniques

Modern propagation techniques have revolutionized plant breeding and production in horticulture, agriculture, and forestry. Among these techniques, micropropagation and tissue culture stand out as powerful tools for rapid and efficient plant multiplication. These methods have gained significant importance in recent years due to their ability to produce large numbers of genetically similar plants in a controlled environment, free from pests and diseases. Micropropagation, also known as in vitro propagation, is a technique that involves the production of plants from very small plant parts, tissues, or cells grown aseptically in a nutrient medium (George *et al.*, 2008). This method is based on the principle of totipotency, which states that a single plant cell has the potential to give rise to a whole plant. Tissue culture, on the other hand, is a broader term that encompasses various techniques for maintaining and growing plant cells, tissues, or organs under sterile conditions on a nutrient culture medium of known composition (Hussain *et al.*, 2012). The process of micropropagation typically involves several stages. It begins with the selection of a mother plant with desirable traits, followed by the sterilization of explants (small pieces of plant tissue) taken from the mother plant. These explants are then placed on a nutrient medium containing essential minerals, vitamins, and plant growth regulators. Under controlled environmental conditions, the explants develop into small plantlets through various phases of growth, including shoot multiplication, root induction, and acclimatization (Debergh and Zimmerman, 1991).

One of the primary applications of micropropagation and tissue culture is the quick multiplication of elite plant varieties. This approach is particularly useful for species that are hard to propagate through conventional methods or have long generation times. In horticulture, these techniques have found extensive applications in the propagation of ornamental plants, fruit trees, and vegetables. They have been particularly beneficial for the production of virus-free plants, especially in crops like potatoes, strawberries, and bananas, where viral infections can cause significant economic losses (Faccioli and Marani, 1998).

Compared to traditional propagation methods, micropropagation and tissue culture offer several advantages. First and foremost is the high multiplication rate. While conventional methods might produce a few dozen plants from a single parent in a year, micropropagation can potentially yield thousands or even millions of plants in the same timeframe (Kozai *et al.*, 1997). This rapid multiplication is particularly valuable for newly developed plant varieties, allowing for quick market introduction and return on investment. Another significant advantage is the production of disease-free plants. By starting with sterile explants and maintaining aseptic conditions throughout the process, it is possible to produce plants free from viruses, bacteria, and fungi. This is especially important for perennial crops, where infections can persist for years and cause long-term economic losses (Cassells, 2012). Micropropagation also allows for year-round production, independent of seasonal constraints.

Furthermore, these techniques enable the propagation of plants that are difficult or impossible to propagate by conventional means. This includes plants that do not produce viable seeds, have extremely low germination rates, or do not respond well to cutting or grafting techniques (Nehra *et al.*, 2005). Despite these advantages, it is important to note that micropropagation and tissue culture techniques also have some limitations. They require specialized facilities, and skilled labor, and can be more expensive than traditional methods for some plant species. Additionally, there is a risk of somaclonal variation - genetic or epigenetic changes that can occur during the in vitro culture process - which may lead to undesired traits in the propagated plants (Larkin and Scowcroft, 1981). Modern propagation techniques have revolutionized the field of horticulture, particularly in the cultivation of temperate fruit crops. These advanced methods have significantly enhanced our ability to produce high-quality plants with desired traits, improve crop yields, and overcome various limitations associated with traditional propagation methods.

Somatic embryogenesis is a remarkable propagation technique that has gained considerable attention in recent years. This process involves the development of embryos from somatic cells, rather than from the fusion of gametes. The ability to generate whole plants from single cells or tissues has opened up new possibilities in plant propagation and genetic improvement. Somatic embryogenesis offers several benefits, including the potential for large-scale propagation, genetic uniformity, and the production of disease-free plants (Germanà and Lambardi, 2016). In temperate fruit crops, somatic embryogenesis has shown promising results. For instance, a study conducted by Petri and Scorza (2010) on peach (Prunus persica) demonstrated successful induction of somatic embryos from immature cotyledons. The researchers achieved a high frequency of embryo formation and subsequent plant regeneration, highlighting the potential of this technique for mass propagation of elite peach cultivars. Another noteworthy case study involves the application of somatic embryogenesis in apple (Malus domestica). Researchers at the Julius Kühn-Institut in Germany developed an efficient protocol for somatic embryogenesis in apples, using leaf explants as the starting material. This process allowed for the rapid multiplication of desirable apple genotypes, including those with resistance to important diseases such as fire blight (Flachowsky *et al.*, 2021).

Molecular breeding and genetic engineering have also played crucial roles in advancing plant propagation techniques. These approaches allow for the precise manipulation of plant genomes to introduce or enhance desirable traits. In the context of propagation, molecular tools can be used to identify and select superior genotypes, improve rooting ability, and enhance stress tolerance in propagated plants. A significant breakthrough in this area was reported by Wisniewski *et al.* (2015), who used genetic engineering to develop apple trees with increased cold hardiness. By over-expressing a peach dehydrin gene in apple, the researchers were able to enhance the trees' ability to withstand freezing temperatures, which is particularly important for temperate fruit production in colder regions. As research in modern propagation techniques persists to advance, it is crucial to maintain a balance between innovation and responsible implementation. Ongoing dialogue between scientists, policymakers, and the public is essential to ensure that these powerful tools are used in ways that maximize benefits while minimizing potential risks.

Technological Advancements in Propagation

Plant propagation has experienced significant changes in recent years, driven by technological advancements that have revolutionized traditional practices. These innovations have not only enhanced efficiency but also improved the quality and sustainability of plant production. This essay explores the cutting-edge technologies that are reshaping the field of plant propagation, focusing on automation, precision agriculture, and controlled environments. Automation and precision agriculture have emerged as key drivers of progress in plant propagation. The integration of drones and robots has dramatically altered the landscape of modern horticulture and agriculture. Drones equipped with high-resolution cameras and sensors are now routinely employed to monitor plant health, assess growth patterns, and identify potential issues in large-scale propagation operations (Zhang *et al.,* 2019). These aerial devices provide real-time data on factors such as crop stress, nutrient deficiencies, and pest infestations, allowing growers to respond swiftly and precisely to emerging challenges. Robots, on the other hand, are increasingly being utilized for tasks that were once exclusively performed by human workers. Advanced robotic systems can now handle delicate operations such as transplanting seedlings, pruning, and even harvesting with remarkable precision. The role of data analytics and artificial intelligence (AI) in propagation cannot be overstated. These technologies have enabled growers to harness vast amounts of data to make informed decisions and optimize their propagation processes. Machine learning algorithms can analyze historical data on plant growth, environmental conditions, and yield outcomes to predict optimal growing parameters for different plant species and varieties (Liakos *et al.,* 2018). This predictive capability allows propagators to fine-tune their approaches, resulting in higher success rates and more consistent quality in plant production. AI-powered systems are also being employed to automate complex decision-making processes in propagation. For example, researchers at Wageningen University have developed an AI model that can determine the optimal time for harvesting plant cuttings based on a combination of visual cues and environmental data (Van Henten *et al.,* 2021). Such advancements not only improve the efficiency of propagation operations but also contribute to the standardization of practices across the industry.

The concept of smart greenhouses and controlled environments represents another significant leap forward in plant propagation technology. These advanced facilities offer unprecedented control over environmental parameters such as temperature, humidity, light intensity, and CO_2 levels, creating ideal conditions for plant growth and development. The benefits of controlled environments are numerous and far-reaching. They allow for year-round production, regardless of external weather conditions, and can significantly reduce water usage and pesticide application compared to traditional outdoor cultivation methods (Shamshiri *et al.,*

2018). Moreover, controlled environments enable propagators to optimize resource utilization and maximize yield potential. By carefully managing inputs such as nutrients and water, these systems can produce plants with consistent quality and characteristics, which is particularly valuable in commercial propagation settings. Research by Kozai (2018) has shown that plants grown in controlled environments can achieve growth rates up to 2.5 times faster than those in conventional greenhouses, with significantly higher survival rates during the critical early stages of propagation. Successful implementations of smart greenhouse technologies abound in the horticultural industry. AeroFarms in the United States has developed a sophisticated aeroponic system for leafy green propagation, which uses 95 per cent less water than traditional field farming while achieving yields up to 390 times higher per square foot (Benke and Tomkins, 2017).

Challenges and Solutions in Propagation

Plant propagation is a fundamental aspect of horticulture and agriculture, but it faces numerous challenges that require innovative solutions. Biotic and abiotic stress management is crucial for successful plant propagation. Biotic stresses, such as diseases and pests, can significantly impact plant health and yield. To address these challenges, researchers have focused on developing disease and pest-resistant cultivars through various breeding techniques. For instance, marker-assisted selection (MAS) has been employed to identify and incorporate resistance genes into commercially important crops (Collard and Mackill, 2008). This approach has led to the development of disease-resistant varieties in crops like rice, wheat, and tomatoes, improving their resilience against pathogens and reducing the need for chemical interventions. Abiotic stresses, including drought, salinity, and extreme temperatures, pose significant challenges to plant propagation. To combat these issues, breeders have been working on developing stress-tolerant varieties through both conventional and molecular breeding approaches. For example, the use of genomic selection has accelerated the development of drought-tolerant maize varieties, enhancing crop productivity in water-limited environments (Cerrudo *et al.,* 2018). Additionally, techniques such as induced polyploidy have been explored to enhance stress tolerance in horticultural crops, as polyploid plants often exhibit improved adaptability to environmental stresses (Sattler *et al.,* 2016). While these advancements in stress management are promising, economic and market challenges present additional hurdles in plant propagation. The cost-effectiveness of advanced propagation techniques is a significant concern for both researchers and producers. Advanced methods like somatic embryogenesis and micropropagation offer numerous advantages, including rapid multiplication and disease-free plantlets. However, their implementation often requires substantial initial investments in laboratory equipment and skilled labour, which can be prohibitive for small-scale producers (Yadav *et al.,* 2013). To address this issue, researchers are exploring ways to optimize these techniques and reduce production costs, such as developing automated systems for in vitro propagation and improving culture media formulations.

Market acceptance and consumer preferences also play a crucial role in the success of propagated plants. Consumers may have specific preferences regarding plant characteristics, such as appearance, flavor, or production methods. For instance, there is a growing demand for organically propagated plants, which presents both challenges and opportunities for producers (Yue *et al.,* 2011). Meeting these preferences while maintaining economic viability requires a delicate balance between innovation and market-driven production strategies. Furthermore, the introduction of new varieties or propagation methods may face resistance from consumers who are accustomed to traditional products. Educating consumers about the benefits of new varieties or propagation techniques is essential for market acceptance. Overcoming these barriers requires a combination of demonstration trials, extension services, and targeted marketing strategies.

Future Trends and Innovations in Plant Propagation

The field of plant propagation is at the point of a technological revolution, with emerging approaches and innovations poised to change traditional practices. This evolution is driven by advancements in genetic engineering, biotechnology, and sustainable agriculture, promising to reshape the way we propagate and cultivate plants in the coming years. Next-generation propagation techniques are at the forefront of this transformation, with CRISPR gene editing technology leading the charge. CRISPR, which stands for Clustered Regularly Interspaced Short Palindromic Repeats, has opened up new possibilities for precise genetic modifications in plants. This tool allows scientists to make targeted changes to plant DNA, potentially enhancing desirable traits or removing undesirable ones with unprecedented accuracy (Shan *et al.,* 2020). The applications of CRISPR in plant propagation are vast, ranging from improving crop yields and nutritional content to enhancing disease resistance and stress tolerance. Synthetic biology is another cutting-edge field that is making significant inroads in plant propagation. This interdisciplinary area combines principles from biology, engineering, and computer science to design and construct new biological parts, devices, and systems. The integration of these advanced propagation techniques with other agricultural practices is crucial for realizing their full potential. Moreover, these new propagation methods are not developing in isolation but are finding synergies with other farming practices. Precision agriculture, which uses data-driven approaches to optimize crop management, can be enhanced by plants specifically propagated for compatibility with these systems.

Similarly, vertical farming and controlled environment agriculture can benefit from plants propagated to thrive in these artificial conditions, maximizing space efficiency and yield (Benke and Tomkins, 2017). The integration of next-generation propagation techniques with traditional breeding methods is also showing promise. While genetic engineering offers rapid and precise modifications, traditional breeding still plays a crucial role in developing new plant varieties. Combining these approaches can lead to more robust and versatile plant varieties that meet both agronomic and consumer needs (Hickey *et al.,* 2019). However, as with any emerging technology, these advancements in plant propagation come with challenges and concerns. Ethical concerns surrounding genetic modification, potential ecological impacts, and regulatory hurdles are all factors that need to be carefully addressed as these technologies move forward (Whelan and Lema, 2015). Ensuring public understanding and acceptance of these new technologies will be crucial for their successful implementation and widespread adoption, the future of plant propagation is marked by exciting possibilities and potential paradigm shifts. As next-generation techniques like CRISPR and synthetic biology continue to evolve, and as they become more integrated with sustainable and precision agriculture practices, we can expect to see significant changes in how plants are propagated and cultivated. These advancements hold the promise of addressing global challenges such as food security, climate change adaptation, and environmental sustainability. However, realizing this potential will require ongoing research, responsible implementation, and thoughtful consideration of the broader implications of these technologies.

Conclusion

Propagation techniques for temperate fruit crops have evolved significantly, blending traditional practices with modern innovations. Key techniques like grafting, budding, and tissue culture have been refined to improve success rates and efficiency. These advancements are crucial for meeting the growing demand for high-quality fruit trees. Looking ahead, research should focus on developing more resilient rootstocks and improving propagation speed without sacrificing plant health. There's also potential in exploring gene editing to enhance desirable traits. Policymakers could help these efforts by increasing funding for horticultural research and creating incentives for nurseries to adopt cutting-edge propagation methods. By continuing to innovate in fruit tree propagation, we can ensure a sustainable and productive future for temperate fruit crops.

REFERENCES

Ananda SA and Negi KS, 1998. Chip budding in deciduous fruit trees. *Agricultural Reviews* Karnal. **19** (1): 1-5.

Asma BM and Guloglu U, 1999. The determination of suitable budding methods and budding times for walnut in ecological conditions of Malaya research report, Ankara University (Turkey). **1058**: 639-641.

Benke K and Tomkins B, 2017. Future food-production systems: vertical farming and controlled-environment agriculture. *Sustainability: Science, Practice and Policy,* **13**(1): 13-26.

Boeing H, Bechthold A, Bub A, Ellinger S, Haller D, Kroke A, Leschik-Bonnet E, Müller MJ, Oberritter H, Schulze M, Stehle P and Watzl B, 2012. Critical review: vegetables and fruit in the prevention of chronic diseases. *European Journal of Nutrition,* **51**(6): 637-663.

Cassells AC, 2012. Pathogen and biological contamination management in plant tissue culture: phytopathogens, vitro pathogens, and vitro pests. *Methods in Molecular Biology,* **877**: 57-80.

Cerrudo D, Cao S, Yuan Y, Martinez C, Suarez EA, Babu R, Zhang X and Trachsel S, 2018. Genomic selection outperforms marker assisted selection for grain yield and physiological traits in a maize doubled haploid population across water treatments. *Frontiers in Plant Science,* **9**: 366.

Collard BC and Mackill DJ, 2008. Marker-assisted selection: an approach for precision plant breeding in the twenty-first century. *Philosophical Transactions of the Royal Society B: Biological Sciences,* **363**(1491): 557-572.

Czarnecki B, 1987. Comparative study of two methods of apple budding. *Annals of Warrow. Agrilcultural University* SGGW. AR-Horticulture. **14**: 3-7.

Davison J and Ammann K, 2017. New GMO regulations for old: Determining a new future for EU crop biotechnology. *GM Crops and Food,* **8**(1): 13-34.

Dickson GH, 1991. A study of the extent to which apple orchard cultivation may be reduced. *Sci. Agric.,* **17**: 670-677

Faccioli G and Marani F, 1998. Virus elimination by meristem tip culture and tip micrografting. In: Hadidi, A., Khetarpal, R.K., Koganezawa, H. (eds) Plant Virus Disease Control. APS Press, St. Paul, MN, pp. 346-380.

Flachowsky H, Hanke MV and Peil A, 2021. Somatic embryogenesis in apple (Malus × domestica). In Somatic Embryogenesis in Ornamentals and Its Applications (pp. 437-450). Springer, Cham.

Food and Agriculture Organization (FAO), 2021. FAOSTAT Statistical Database. Rome: FAO.

Garibaldi LA, Steffan-Dewenter I, Winfree R, Aizen MA, Bommarco R, Cunningham SA, Kremen C, Carvalheiro *et al.,* 2013. Wild pollinators enhance fruit set of crops regardless of honey bee abundance. Science, **339**(6127): 1608-1611.

George EF, Hall MA and De Klerk GJ, 2008. Plant Propagation by Tissue Culture (3rd ed.). Springer.

Germanà MA and Lambardi M, 2016. In vitro embryogenesis in higher plants. New York: Humana Press.

Grzyb ZS and Czynczyk A, 1990. Attempts to improve seed germination and seedling growth of Wangenheim Prune generative rootstock. *Fruit science reports.* **17**(3): 101-110.

Gulen H, Erbil Y and Eris A, 2004. Improved rooting of Gisla 5 softwood cuttings following banding and IBA application. Hort science **39**: 1403-1405

Hartmann HT, Kester DE, Davies FT and Geneve RL, 2014. Hartmann and Kester's Plant Propagation: Principles and Practices (8th ed.). Pearson.

Hartmann HT, Kester DE, Davies FT and Geneve RL, 2009. Plant propagation: principles and practices. 7th edition Prentice Hall, Saddle River.

Hickey LT, Hafeez AN, Robinson H, Jackson SA, Leal-Bertioli SC, Tester M, and Wulff BB, 2019. Breeding crops to feed 10 billion. *Nature biotechnology*, **37**(7): 744-754.

Howard BH, 1967. Rootstock propagation by hard wood cuttings. A progess report for nurserymen. *Report of the East Mailing Research station for 1966,* p.202.

Howard BH and Kene S, 1974. The effect of different grafting methods up on the development of one-year-old nursery apple trees. *Journal of Horiculture. Science,* **49**: 287-295.

Hussain A, Qarshi IA, Nazir H and Ullah I, 2012. Plant Tissue Culture: Current Status and Opportunities. In: Leva, A. and Rinaldi, L.M.R. (eds) Recent Advances in Plant in vitro Culture. IntechOpen Debergh, P.C., and Zimmerman, R.H. (1991). Micropropagation: Technology and Application. Springer Netherlands.

Hyson DA, 2011. A comprehensive review of apples and apple components and their relationship to human health. Advances in Nutrition, 2(5): 408-420 in ecological condition of Malaya Research Report, Ankara university (Turkey). **1058**: 639-641.

Jackson JE, 2011. Biology of Apples and Pears. Cambridge University Press.

Kar PL, Bhartiya SP, 1987. Effects of phenolic compounds on the dormancy and\ germination of crab apple seed (*Malus baccata* Borkh). *Progressive horticulture,* **19**: 105-108

Kozai T, 2013. Plant factory in Japan—current situation and perspectives. *Chronica Horticulturae,* **53**(2): 8-11.

Kozai T, 2018. Smart plant factory: the next generation indoor vertical farms. Springer.

Kozai T, Kubota C and Jeong BR, 1997. Environmental control for the large-scale production of plants through in vitro techniques. *Plant Cell, Tissue and Organ Culture,* **51**(1): 49-56.

Kuzma J, 2016. Reboot the debate on genetic engineering. *Nature,* **531**(7593): 165-167.

Kviklis WW, 1996. Overcoming the problem of citrus production in Malaysia. *Extn. Bull.,* **36**: 9.

Larkin PJ and Scowcroft WR, 1981. Somaclonal variation — a novel source of variability from cell cultures for plant improvement. *Theoretical and Applied Genetics,* **60**(4): 197-214.

Lee JS and Yoon TM, 2001. Some Chip Budding Techniques for Improving the Nursery Performances in Apple Trees. *Horticultural Science and Technology,* **19**(3): 352-357

Liakos KG, Busato P, Moshou D, Pearson S and Bochtis D, 2018. Machine learning in agriculture: A review. *Sensors,* **18**(8): 2674.

Luckwill LC and Campbell AI, 1954. The use of apomictic seedling rootstocks for apples. pp. 47- 52 *in:* Progress report. *Annual report of Long Ashton Research Station for 1953.* mahaleb cherry seeds. *Dirasat* **12**: 7-12.

Nehra NS, Becwar MR, Rottmann WH, Pearson L, Chowdhury K, Chang S, Wilde HD, Kodrzycki RJ, Zhang C, Gause KC, Parks DW and Hinchee MA, 2005. Forest biotechnology: Innovative methods, emerging opportunities. *In Vitro Cellular and Developmental Biology - Plant,* **41**(6): 701-717

Pathak RK, 1991. Propagation of fruit plants, ICAR, New Delhi. pp. 45-67.

Petri C and Scorza R, 2010. Factors affecting adventitious regeneration from in vitro leaf explants of 'Improved French' plum, the most important dried plum cultivar in the USA. *Annals of Applied Biology,* **156**(1): 79-89.

Porebski S, Rzennicka B and Poniedzialek W, 2002. Comparison of two methods of walnut grafting. *J. Fruit and Ornamental Pl. Res.* **10**: 55-62.

Rao SR and Ravishankar GA, 2002. Plant cell cultures: Chemical factories of secondary metabolites. *Biotechnology Advances,* **20**(2): 101-153.

Ricroch A, Clairand P and Harwood W, 2016. Use of CRISPR systems in plant genome editing: toward new opportunities in agriculture. *Emerging Topics in Life Sciences,* **1**(2): 169-182.

Sahu NC, 1993. Vermicomposting – A Simple but Unique Technique. *Indian Farmer's Digest,* September, 30-33.

Sarasan V, Cripps R, Ramsay MM, Atherton C, McMichen M, Prendergast G and Rowntree JK, 2006. Conservation in vitro of threatened plants—progress in the past decade. *In Vitro Cellular and Developmental Biology - Plant,* **42**(3): 206-214.

Sattler MC, Carvalho CR and Clarindo WR, 2016. The polyploidy and its key role in plant breeding. *Planta,* **243**(2): 281-296.

Sax K, 1949. The use of Malus species for apple rootstocks. *Proceedings of the American Society for Horticultural Science,* **53**: 219-220.

Schmidt H, 1982. Evaluation for rootstock characters in apomictic apple selections. XXIst International Horticultural Congress, Vol. I. 1199. The Hague, Netherlands.

Shamshiri RR, Kalantari F, Ting KC, Thorp KR, Hameed IA, Weltzien C, and Shad ZM, 2018. Advances in greenhouse automation and controlled environment agriculture: A transition to plant factories and urban agriculture. *International Journal of Agricultural and Biological Engineering,* **11**(1): 1-22

Shan Q, Wang Y, Li J and Gao C, 2020. Genome editing in crops: from bench to field. *Nature Plants,* **6**(9): 1009-1011.

Shatat F and Sawwan J, 1985. Effect of promalin and gibberellic acid 9GA3) on germination on Siyapananont, V. 1990: Breaking dormancy of native peach seeds. *Acta horticulturae,* **279**: 481-485.

Smanski MJ, Zhou H, Claesen JB, Fischbach MA and Voigt CA, 2016. Synthetic biology to access and expand nature's chemical diversity. Nature Reviews Microbiology, **14**(3): 135-149.

Srivastava RP, 1966. Research on horticultural crops at Chaubattia. *Indian Horticulture,* **10**: 9-11.

Van Henten EJ, Bac CW, Hemming J and Edan Y, 2021. Robotics in protected cultivation. Annual Review of Control, Robotics, and Autonomous Systems, **4**: 313-337.

Whelan AI and Lema MA, 2015. Regulatory framework for gene editing and other new breeding techniques (NBTs) in Argentina. *GM crops and food,* **6**(4): 253-265.

Wisniewski M, Nassuth A and Arora R, 2015. Cold hardiness in trees: a mini-review. *Frontiers in Plant Science,* **6**: 1070.

Yadav SK, Katikala S, Yellisetty V, Kannepalle A, Narayana JL, Maddi V, Mandapaka M, Shanker AK, Bandi V and Bharadwaja KP, 2013. Optimization of Agrobacterium mediated genetic transformation of cotyledonary node explants of *Vigna radiata*. *SpringerPlus,* **2**(1): 1-7.

Yue C, Dennis JH, Behe BK, Hall CR, Campbell BL and Lopez RG, 2011. Investigating consumer preference for organic, local, or sustainable plants. *HortScience,* **46**(4): 610-615.

Zhang C and Kovacs JM, 2012. The application of small unmanned aerial systems for precision agriculture: a review. *Precision Agriculture,* **13**(6): 693-7

Zhang Y, Malzahn AA, Sretenovic S and Qi Y, 2019. The emerging and uncultivated potential of CRISPR technology in plant science. *Nature Plants,* **5**(8): 778-7

Chapter 3

Orchard and Canopy Management in Temperate Fruit Crops

J.I. Mir, M.K. Verma, O.C. Sharma, W.H. Raja, S.U. Nabi, M.A. Sheikh and M.A. Kuchay

ICAR–Central Institute of Temperate Horticulture, Old Air Field, Rangreth, Srinagar–191132, J&K

Canopy architecture in fruit crops regulates the size, shape, and orientation of the branches, resulting in the production of fruits with uniform dimensions as well as attractive form, colors, and secondary metabolites. A three-dimensional strategy for canopy management is used to maximize the output of fruit trees by taking advantage of the land, light interception, and climate-related factors. A wide range of practices that would help to maintain the ideal tree canopies, including training, pruning (dormant, summer, and root pruning), branch orientation (bending), scoring, girdling, bark inversion, rootstock selection, use of plant growth regulators, appropriate use of fertilizers, deficit irrigation, and the use of genetically modified plants with altered architectural characteristics (Ahmed *et al.*, 2016).

The term canopy refers to the physical structure of plants, which consists mostly of stems, branches, twigs, and leaves, while canopy density is defined by the amount and size of leaves. Its architecture is defined by the number, length and direction of the trunk, branches and twigs, while the volume of the canopy is determined by the spread and height of the canopy (Pathak, 2009). The parameters used to define tree architecture are mostly concerned with primary growth. They are typically categorized into four groups: kind of growth (rhythmic vs. continuous), branching pattern (sympodial vs. monopodial), morphological differentiation of branches (orthotropic vs. plagiotropic), and position of reproductive structures on scaffold branches (lateral vs. terminal). Green leaves of fruit trees convert solar energy into chemical energy through the photochemical process of photosynthesis. It is the primary source of energy that provides food, fodder, fiber and fuel. The fruiting capacity of a fruit tree is determined primarily by its architecture, canopy density, and photosynthetic efficiency (Kallow *et al.,* 2005).

Essential Features of Ideal Canopy

- Strong frame of primary branches
- Wider crotches in scaffold branches
- Healthy and well distributed secondary and tertiary branches
- Sufficient fruiting terminals in most productive areas
- Healthy foliage with high photosynthetic efficiency to maximize the solar radiation use efficiency
- Enough space for air circulation inside canopy
- Should support adequate to nurture and protect the fruits from sunburn

The goal of canopy management is to provide more uniform light distribution and maintain proper ventilation within the canopy to maximize fruit quality, productivity, simplify cultural practices, and reduce pest and disease burdens. In new plantations, early training and pruning are used to establish a strong tree structure, while in older plantations canopy management aims to reduce tree height and remove abnormal biomass

Goals of Canopy Management

- To increase yield efficiency
- To increase quality of produce
- To make high density planting successful
- To rejuvenate old orchard
- To facilitate cultural practices
- To increase input use efficiency
- To reduce pest and disease incidence
- Efficient use of pesticide and fungicide
- To facilitate mechanization
- To Reduce cost of production

Principle of Canopy Management

The fundamental concept behind fruit tree canopy management is to make the greatest possible use of available land and sunlight to increase productivity.

Light is an essential component of fruit production. The arrangement of the canopy optimizes the plant model in such a way that it intercepts the maximum amount of light. Fruit production involves capturing and converting light energy into fruit biomass (dry matter). Plant biomass production and distribution are related to source strength and fruit sink strength, respectively (Ke *et al.,* 2023). The main determining factors are as follows:

- Intensity of incoming radiation
- Proportion of radiation intercepted by the tree
- Degree to which the tree converts this energy into fruit

The amount of radiation received is determined by both climate and the length of the growing season. Apart from the initial choice of orchard location, the grower has no control over any of these. The length of the growing season is also important in deciduous plants, which shed their leaves in autumn and thus have a limited growing season. Growers have the most control over the percentage of radiation a tree receives, so decisions about block size and planting density are important. A substantial proportion of the incoming radiation in low-density plantings is absorbed by the grass between the trees rather than by the fruit trees. Therefore, the light-blocking capacity of the garden canopy as a whole is very important, and this capacity is relevant to the production of all types of biomass. The fundamental ideas of canopy management are:

- Maximal light consumption
- Avoidance of the formation of a microclimate favourable to disease and pest
- Ease of carrying out cultural operations
- Maximizing productivity and quality
- Cost-effectiveness in obtaining the requisite canopy architecture

Light Interception

Light interception is the amount of available light intercepted by the tree canopy and not striking the orchard floor. Interception then is a function of plant density, canopy shape and size, and leaf area index (LAI, m^2 leaf area/m^2 ground area) within the canopy. The amount of light that is accessible but is blocked from reaching the ground by the tree canopy is referred to as light interception. Therefore, within-canopy plant density, canopy size and shape, and leaf area index are all factors that affect interception. Biological yields in many crops are increased by enhancement of seasonal total light interception. These improvements are usually due to improved cultural practices that affect leaf development, leaf duration, and/or canopy display The length of the growing season is the primary factors that determine the overall amount of light intercepted by a fruit orchard system (Palmer, 1989; Wagenmakers, 1991).

Light Interception by Orchard Canopies

Orchard canopy, unlike other crops, a large portion of the land area uncovered by trees due to the alleyways established between the rows (Jackson, 1970). Light interception is low when the orchard is planted alone and increases as the orchard grows in relation to total LAI. Bush trees that are widely spread and large at maturity intercept very little light while they are young. They lead excessive internal shade and poor LAIs when they are fully grown. Larger tree densities are being planted as a result of the awareness that a large amount of land and light resources are lost during the early stages of an orchard. Low-density hedge row planting of the 'Golden Delicious' variety intercepted approximately half of the incident light at maturity, and yields peaked at 40 t ha^{-1}. The high-density hedgerow plants, on the other hand, yielded more than 70 percent of the available light on M.9 (3300 trees ha^{-1}), and M.2 (2260 trees ha^{-1}). The largest single factor affecting light interception in discontinuous orchard canopies is LAI. Increase in tree density had a greater effect on third-year yield and light inhibition of 'Granny Smith' than training methods or rootstocks. Thus light interception in orchards can be raised by increasing the number of trees ha^{-1}, increasing the height of the trees relative to the clear alley width, increasing the density of foliage in the canopy orienting rows in north-south direction and by summer pruning (Wunsche and Lakso, 2000).

Tree Form and Light Distribution

To increase the light dispersion in tree canopies, two methods have been utilized. One is to employ relatively natural tree forms, such as multiple leader, central leader, vertical axis, or slender spindle forms (Lespinasse and Delort, 1986). Another strategy is to create fewer, larger, permanent apertures for light to enter canopies that are restricted to geometric shapes. Thin, constrained foliage planes like hedgerows, tree walls, and A, V, or T forms are few examples.

Round crown tree: As foliage depth increased in large round-crowned trees, light intensity decreased rapidly, and the lowest light intensity reached central parts of the tree ranging from 6 to 30 per cent of full sunlight (Verheij and Verwer, 1973). The effective depth of light penetration in unrestricted apple canopies is 1 m in a round crown tree. The minimum amount of light that apple canopies receive is 30 per cent of full sun. The region of the tree that received the most shade as the tree size decreased.

The central leader tree: Large "Delicious apple" trees experienced a 58 per cent reduction in light intensity when trained on a central guide system about 2 m from the tree top (Barritt *et al.,* 1991) Inner canopy light of central leader trees of the "Delicious apple" species changed with the seasons. The basal parts of the canopy experienced a dramatic decrease of 20 per cent in total sunlight by early June. Summer pruning, enhanced light distribution in the centre section of the canopy trained on a central leader system.). Use interstock (Empire, M.9, MM.111), and greater relative distance of the 1 m between scaffold branches improve light distribution within canopy (Lakso *et al.,* 1994)

The slender spindle tree: Empire variety of Apple trained to slender spindle trees onrootstock M.9 exhibited very good light exposure in most canopies at 14 DAFB. However, in 44 DAFB, light exposure levels had dropped significantly in all but the top of the tree, indicating that the gaps between limbs had closed. At 74 DAFB there was a large, poorly illuminated canopy area. Therefore, summer pruning of slender spindle trees is necessary for best fruit color (Robinson *et al.*, 2014).

Thin Restricted Plane Canopy Trees

Planar canopies have been developed to overcome the problems of light penetration in thick canopies. However, low light transmission through the top layer of trellis resulting in excessive shading in the lower layers (Tustin *et al.*, 1989). These horizontal canopy systems are prone to excessive straight shoot growth from the top of the trellis. This increases the amount of shading, which can result in the space between the tiers being completely closed. Red apple cultivar 'Empire' shows poor fruit color with horizontal trellises due to flat limb orientation. Vertical planar canopies have good light exposure on both sides of the canopy but have a vertical stack of fruiting branches that loses strength to the lower part of the tree. Planar canopies (for example, Y-trellis) overcome the horticultural issues of horizontal terraces by providing light exposure on both sides of the Y-trellis. Furthermore, fruiting branches are not stacked vertically.

Air Circulation and Humidity

The microclimate is influenced by air circulation and humidity within the tree canopy. High humidity below the canopy of an ill-trained tree with a spinning canopy results in limited air circulation, increasing insect and disease attack. The ideal canopy architecture for both high and low density plants must be designed to maximize air circulation and humidity.

Effect of Light Interception and Distribution on Fruit Plants

Fruit yield and quality are greatly affected by tree vigor and climatic parameters such as light interception, temperature, relative humidity and air movement. Adequate solar exposure of plants and fruits, as well as prevailing temperature, play an important role in plant growth, photosynthetic efficiency, flower initiation, fruit set, fruit bud differentiation, ripening, fruit color, fruit quality, *etc.* Fruit trees with a large 'shade to sun' ratio have photosynthetically inactive under deep shade leaves that act as non-productive sinks. Shaded leaves are thinner and have fewer palisade layers. The size and color of the fruit are more than 70 per cent associated with a high dose of radiation. Large fruits are best colored in the canopy's outer periphery where irradiance is high, but smaller poor color fruits are borne in the inner shaded region of the tree (Awasthi and Mehta, 2004). Light exposure to Cox's Orange Pippin tree was reduced to 37 per cent by shading in early June, resulting in a 36 per cent reduction in the number of flowering spurs and flower bud production and symptoms of fruit immaturity, such as a decrease in fruit color, fruit dry matter, fruit soluble solids, but increased fruit firmness. Fruit quality has been reported to suffer as a result of canopy shading in peach and cherry. Shade also reduced the starch content of root tissue and reduced flower by nearly 50 per cent, (Proietti and Tombesi, 1996).

Technique for Canopy Management

Canopy management represents the physiology that governs the relationship between growth and productivity. The orchard production system is a combination of variety, rootstock, tree spacing, training, and pruning, which has a huge impact on orchard design. Training, pruning (dormant, summer, and root pruning), branches orientation (bending), scoring, girdling, bark inversion, selection of proper rootstock, use of plant growth regulators, appropriate fertilizer use, deficit irrigation, and use of genetically engineered plants with altered architectural character would help in maintaining the ideal canopies of trees.

Training

The main goals of the training are to manage plant vigor, increase light exposure to fruits/leaves, promote air movement to reduce disease spread, and simplify harvesting, harvesting and mechanization practices. Whatever system is used, the primary goal of training and pruning is to maximize tree life and yield. Training is an important exercise done in the early years to obtain the benefits of light and good air circulation by keeping ideal forms and strong structures. The use of vertical and horizontal space should be considered when setting up a training system.

Orchard Planting System

Various interrelated factors affect light reception and dispersion within tree canopies. One factor is cultivar genotype or architectural pattern, which ranges from non-branching, columnar to spreading, non-porous. Additionally, the size of the canopy of a particular cultivar rootstock is very important. The second component is the planting system, which includes the arrangement of trees in the and pruning training, as well as limb orientation. Training affects the percentage and distribution of different shoot categories in space, such as long versus short shoots and vegetative versus fruiting shoots. Based on tree canopy geometry, four types of plantation systems are spherical canopy forms, conical canopy forms, flat fan canopy forms, and Y or V canopy forms shape.

Spherical-Shaped Canopy Systems

Traditionally, trees of this canopy type are globe-shaped, with a height of 6-8 m and a trunk length of 2 m. In the late 1800s and early 1900s, it was the most frequent tree form in traditional apple orchards in Europe and North America (Walker, 1980). Introduction of clonal rootstocks in apple in the 1930s, a small fraction of the globe-shaped large tree developed on semi-dwarf rootstocks in Holland and England, called the bush tree (the bush-tree system).

Conic-Shaped Canopy Systems

Large globe-shaped trees result in less light distribution inside the canopy, many researchers recommended conic or pyramidal canopy. In this tree form several planting systems developed like Central-leader system, Mini-central-leader system, Palmette-leader system, Slender-spindle system, North Holland spindle, Slender-spindle multi-row or bed system, Vertical-axis system, SolAxe system, Slender-pyramid system, HYTEC system, Super-spindle system *etc.* and now it is a dominant tree form in commercial orchards in most part of the world. The main benefit of this tree form is that it has good light distribution and light interception in the early stages of orchard establishment. The majority of apple growers worldwide have adopted conic-shaped systems. The central-leader system dominated the world in the 1960s and 1970s, while in Europe, the slender spindle and vertical axis systems predominated in the 1970s and 1980s. The vertical axis and its variants, together with the thin spindle and its variants, are the most often used systems. Most slender-spindle trees have grown taller over the past ten years, approaching 2.75 m, making them closer in size to vertical-axis trees. Slender spindle orchards can now receive more light, produce more fruit, and do so with less pruning to keep trees to a shorter 2m height. Similar to slender spindle orchards, vertical axis orchards have shown an increase in planting density over the last ten years, increasing from 1500 to 2000 trees ha^{-1}.

Flat Planar Canopy Systems

In order to keep tree canopies to a two-dimensional plant, European apple growers have used trellises since very early. Among these geometric shapes, espaliers were most frequently developed around walls and fences. Additionally, to increase labor productivity, Italian fruit growers created the first commercial apple orchard with a two-dimensional plane canopy in the middle of the 1950s. This technique was called the palmette training system. A number of variations of this system, including the "Regular" palmette system, "Free" palmette system, Penn State thin-wall trellis system, Lincoln canopy system, Ebro trellis system, Solen system, *etc.*, have been developed over time using dwarf rootstocks. The main benefits of flat planar systems are that they increase labour efficiency and permit even lighting throughout the canopy.

V-Shaped Canopy Systems

Chalmers and Van Den Ende (1975) popularized V-shaped fruit trees for commercial orchards using the tatura trellis growing technique. This technique was initially developed for peaches to increase yield and enable automated harvesting (Chalmers *et al.*, 1978). It was then swapped for apples. Many variations of Tatura trellis, including Mini-Tatura trellis system, Geneva Y-trellis system, Mikado and perforation system, MIA trellis system (A-shaped trellis), Mini-V-trellis system, Gutingen V thin-spindle system, *etc.*, have been developed in response to dwarf rootstock growth and to improve light interception and penetration in the canopy core.

General Rules of Training of Fruit Plants

Train Tree by Limb Positioning Instead of Pruning

Positioning of limbs can be done by tying young branches using spreaders. Bending nearly vertical limbs 45 to 60° from vertical usually stimulates fruit production later in the life of the tree. In some training systems, tying is limited to the first level of branches (vertical axis, thin pyramid and HYTEC), while, in other systems, upper branches are also tied down (SolAxe).

Spreading of Branches

Tissues in narrow crotches are slower to mature in autumn and may be injured by low temperatures. The attachment will be weak due to inclusion of bark when angles between the scaffold branch and main trunk are less than 35°. Narrow curves are usually further weakened by water, snow, rotting organisms and canker sores. While the tree is young, spread branches that make very narrow angles with the main trunk.

Retention of Branches

Leave more branches in spur-type cultivars than in a tree of standard growth habit.

Removal of Unwanted Shoots in Summer

Remove unwanted shoots when they are small and before the base of shoot has lignified.

Training Programme

Follow the training program consistently, as frequently as necessary, to complete training properly as soon as possible.

Different Training Systems

There is only one planting system to suit all conditions. Factors such as soil, cultivar, rootstock, management system and socio-economic conditions will determine the best combination for each orchard. Here are some of the most common systems in India as well as in the temperate region of the world as shown in Figure 3.1.

Central Leader System

This system is commonly used for training apple, pear, cherry, plum and pecan trees. A central leader tree is characterized by a main, straight trunk, called a leader. Usually the trees have 3 to 4 tiers of branches along the trunk at a height of 4-5 meters. The lower tier of branches is 60 to 90 cm above the ground and has 4 to 5 branches. The second tier of branches is usually 60 cm to 1 m from the first tier and has 4 branches. The third and fourth tiers are about 60-80 cm apart and usually have 3 branches.

Figure 3.1: Different Training Systems for Apple

Espalier Training System

Cordon System

Vertical Axis

Tall Spindle

Slender Spindle

Super Spindle

SolAxe System

Bi-Axis System

Spreaders are placed between the trunk and the branches; to spread the branch angles to the desired extent of about 60 degrees. The lowest scaffold whorl branches will be the longest and the highest scaffold whorl branches will be progressively shorter to allow as much light as possible throughout the tree. It is a medium density system with tree density ranging from 300 to 700 trees ha^{-1}. Semi-vigorous rootstocks such as MM.106, MM.111, M.7 *etc.* are suitable for this training system in apple

Modified Central Leader System

Also known as a multi-leader system, is similar to a central leader system in that numerous vertical branches are taught instead of a single central leader. The technique works best with three to five leaders to ensure that the tree is adequately balanced. Rather than removing competing vertical branches in the first two years after planting, the desired numbers of leaders are left on the tree. This approach is commonly used to train apple, pear, plum, and pecan trees.

Open Centre System

It is also known as open-headed or vase shape because the stem is cut back after a year of planting and allowed to grow to a height of 75 cm. After removing the central leader, three to five evenly spaced scaffold branches emerge in a vase shape from the stem. This system is less satisfactory since it lacks strong curves and a weak framework. However, in fruits such as peaches, nectarines, and plums, it may be desired to tolerate more sunlight for improved fruit coloring.

Mini-Central Leader System

It is similar to the central leader in which no permanent branches are permitted to grow above the second tier of branches, but little fruiting branches on the leader are kept and allowed to cut down and bend horizontally. Semi-dwarf apple rootstock is appropriate for training systems with densities of 500 to 1000 trees ha^{-1}.

Dwarf Pyramid

Dwarf Pyramid is a compact, low-headed central leader tree that works well with most dwarf and semi-dwarf rootstocks. The lowest branch emerges 30-40 cm above ground level, and subsequent branches grow at 15-20 cm intervals along the central leader, which gradually reduces in length from bottom to top, giving a pyramidal structure. This enables for the distribution of 20-25 primary branches in all directions at an appropriate position. Tree height is limited to 2 meters, with a canopy of 1.5 meters.

Head and Spread

It is also known as a two-level training method and performs well with dwarf and semi-dwarf rootstocks. The first tier (4-5 scaffolds) in this training system is allowed to rise 45-60 cm above ground level. After 60 cm of the first layer system, a second tier system (5-6 scaffolds) develops. Fruit development occurs in the inner section of the canopy, resulting in the best color of the fruit.

Palmette-Leader System

It is a variation of the central leader system that is designed to increase light distribution in the tree canopy because as trees mature, the upper limbs develop stronger than the lower limbs. It has a tendency to cause excessive shade in the bottom portion of the trees due to better light exposure. The palmette-leader training system involves removing upper east and west-growing branches from a mature central-leader or mini-central-leader tree, resulting in a flat north-south-oriented palmette (*i.e.* fan-shaped) top with good light penetration into the tree center throughout the season.

Spindle Bush

Spindle bush (or free spindle) is best suited to densities of up to 2,000 trees ha^{-1} at a height of 2-3m and a spacing of 3-4m x 1-2m. For spindle bush systems, well-branched two-year-old nursery trees are generally recommended. At planting, a number of laterals are selected to form part of the permanent scaffolds in the bottom tire of the tree. As the leader grows, more scaffolds are selected and spaced equally. Leader dominance is vital, and if it is lost, the tree canopy will collapse. If it becomes too strong, background growth and development will be reduced. This system can stand alone, but is usually supported by two to three wires or individual supports.

Vertical Axis

In the vertical axis, trees are built on a single vertical trunk, which supports relatively 'weak' or small-diameter fruiting branches. Tree density ranges from 1,000 to 2,500 trees ha-1 at a spacing of 4-5m x 1-2m and can reach up to 3m in height. The most commonly used rootstocks in apple are M.9 and M.26. Maintaining apical dominance in the vertical axis system is important to ensure weak fruiting branches, especially during the early stages of development. As a result, the leader is not heading back at planting or during the development years. Fruiting branches are kept intact along the trunk from 1m above. A healthy balance between vegetative growth and fruiting is usually reached with 12-16 fruiting branches on M.9 apple trees (Lauri and Lespinasse, 2006).

SolAxe System

The SolAxe system of training is developed by modifying the vertical axis system to avoid branch bending and renewal pruning to achieve a more weeping canopy. This system combines the bending of the central axis and fruiting branches. Systems with free-growing fruiting branches and removal of competing plant branches from vertical axis systems. The tree is prepared by bending the wings tightly in an arc along the trunk when planting. Branches less than 1.2m are removed. The leader is left unheaded and allowed to grow to the top of the support system, where it is bent horizontally. Upper fruiting branches bend downwards from the horizontal. The optimal balance between vegetative growth and fruiting on a mature SolAxe tree, like with the vertical-axis method, is obtained with 12-16 fruiting branches placed spirally down the trunk. Annually, upright stems are removed from each fruiting branch while the pendant half is kept.

Slender Spindle

The training is designed to improve early yield and management efficiency by planting with higher tree densities and reducing tree height so that all management is done from the ground. 1500 to 4000 trees ha^{-1} in either single, double, triple or multiple row beds. Planting distance is 3-3.5m x 1-1.5m and tree height is usually limited to 2-3m. The tree is a narrow, fully dwarfed, conical shape with a distinct, supporting, vertical central leader branch. Strong leaders and lateral development at the top of the trees should be avoided if possible. When the leader becomes too powerful, it usually returns to a weaker side. The most frequent rootstock used in apples is M.9, however other dwarfing level stocks such as B.9 and G.16 are also utilized.

Slender-Pyramid System

This training system was developed by Tustin and colleagues. In designing this system, he considered the physical principles of light interference, light distribution and harvesting, as well as the vertical axis, slender spindle system and McKenzie central leader (McKenzie and Mouat, 1963). This training system can be used with both low tree densities (600 trees ha^{-1}) and fairly high tree densities (2000 trees ha^{-1}) using MM.106 rootstock. A prominent central leader, a well-developed, spreading base tier of four branches bearing fruiting laterals, and a narrow pyramidal canopy above the basal tier characterize this training system.

Super Spindle

The super spindle technique is used for extremely high-density orchards on weaker rootstocks, such as Quince C in pear. There is a row spacing of 3m and a tree distance within the row of 0.8m, giving a density of about 4,000 trees ha^{-1}. The average tree height is 2-3m. This system features compact trees with short fruiting wood or spurs uniformly placed along the center leader.

Tall Spindle Planting System

The tall spindle planting system uses highly feathered nursery trees planted at a density of 2,500-3,300 trees ha^{-1} at 3.0-3.5m between rows and 0.9-1.2m within rows. It is a combination of slender spindle, verticalaxis, super spindle, and solAxe system. This system uses a higher tree density concept than the slender spindle system, but uses a lower planting density (2,500-3,300 trees ha^{-1}) than the super spindle. Trees are minimally pruned after planting but the wings are tied down horizontally immediately after planting. The tree quickly grows to 3.3 m tall with no leader crown and short pruning for the first 4-5 years after which a side branch is cut each year to limit the tree's height to 3.3 m. A mature tree is pruned annually by cutting 2-3 large limbs (> 2 cm diameter) in the canopy into 2-3 cm long stubs using limb renewal pruning. The system is intended to produce high light interception (70–75%) by growing the trees to a height of 90–100 per cent of the between-row spacing (3–3.5 m), but it also aims to produce good light distribution throughout the canopy (Robinson *et al.*, 1991).

Cone System

Conical system of training is developed by (Gandev, 2009). This is a new training method and non-spur apple cultivar grafted on MM106 rootstock in intensive orchards. This includes pruning activities, which are applied continuously to form a crown cone shape and different number of scaffold branches loaded with fruits depending on the age of the trees and vigour.

Trellis System

Trees can be supported by trellising to improve their bearing surface which in turn increases their output. The trellis is made up of a network of wires supported by posts. Limbs and branches are placed on this wire fence to encourage fruit production. Apple cultivars on M.27, Bud.9, M.9, or M.26 as well as pears on Quince C rootstock can be trellis trained. This system is specially adopted for high density planting on clonal rootstocks to improve yield by capturing 70 per cent of available sunlight. This allows for more fruiting wood than structural wood and yield at early stages of plant/orchard life.

Palmette System

There are three types of palmette namely, horizontal arm (espalier), oblique arm (palmette) and combination of horizontal and oblique arm. In the palmetto system of training, the tree is trained up four to six wire vertical trellises 2.5-3 m long. A tree has a central trunk and regular tiers of branches that are trained along a two-dimensional vertical plane. Branches at each tier are tied either at an oblique angle (30°), so that they grow across the wires (oblique palmette) or horizontally along the wire at each tier (horizontal palmette or espalier). The first two branches like across the wire and then along the wire (combination of diagonal and horizontal). The Espalier system of training is suitable for apple and pear varieties that produce short fruiting spurs (not tip-bearing). Trees should be planted at a distance of 3-6 meters according to their strength. MM.106 is best suited for small apple trees with three or four tiers but M.26 can also be used, however MM.111 is best for large apple trees and poor soil. Quince A is suitable for pear cultivars.

V or Y Shaped Canopy Systems

Canopies generally have two basic shapes, Y-shaped trees with a vertical trunk and two opposite tree arms trained on either side of the trellis, and V-shaped trees where the entire tree leans to one side of the trellis while the next tree in the row is bent to the other side.

V Hedge

Planting distance in V hedge is 3.5 x 1.25 m which is equal to 2057 trees ha^{-1}. These systems are planted using two-year-old nursery trees with good foliage. Four branches are placed as fruiting branches and four central leaders are considered on one stem. The height of the tree is maintained at 2m with a V opening of 1.4m. Light blocking can be avoided by implementing these systems more precisely.

Open Tatura Trellis

This is a modification of the original Tatura Trellis system. IT has a narrow strip of about 50 cm that separates the diagonally planted trees alternately within each row. This system is planted 4-4.5m x 0.5-1m with 2,000 to 5,000 trees ha^{-1}. Trees can be trained in a variety of ways. The three most common are: single leader, double leader and, more recently, cordon. Single-leader OTT is similar to planting a slender spindle-type system, with root systems spaced approximately 0.5 m apart and leaders spaced 1 m apart. Double leaders involve training two leaders per tree (about 1 m apart), to establish a high density of fruiting units at a low cost per tree. Open Tatura Trellis with cordon allows a medium-dense orchard of about 2,000 trees ha^{-1} with about 8,000 fruiting units growing on the strings. Nursery trees are bent over when planted and trained horizontally. Fruiting units are then encouraged at regular intervals along each cordon.

Cordon System

The term 'cordon' refers only to a stem with small side shoots (flowering spurs). In this system, it is difficult to contain strong cultivars and create poor cordons. Similarly, cultivars that bear fruit at the tips of their branches (tip bearing) are difficult to train because the tips have to be cut off and produce very few fruits. Strong 2-3-year-old trees are used for planting in the cordon system. Trees are planted at an angle of 45 degrees, 5 wire trellises are placed 180 cm apart at 8 m intervals. The top wire is a few centimeters below the top of the post. The first trellis extends 45 cm above ground level and the others are fixed at 30 cm intervals. In the cordon system, trees are planted at a 45° angle which results in the movement of auxin to the basal parts of the tree thus making the lateral buds productive. Apple rootstock M.27 (ultra dwarf) or M.9 (very dwarf) are most suitable for this training system where the soil is fertile. Pears are best transplanted onto Quince C or Quince A for very poor soils.

Pruning

Pruning is employed to improve both the fruit crop and tree appearance, but it is a practice frequently neglected and misunderstood. Pruning is defined as the judicious removal of parts to obtain good and qualitative yield. The ultimate aim of pruning is to develop a canopy, which intercepts maximum light.

Basic Objectives of Pruning

- To restore root/shoot ratio
- To obtain specific form of the plant
- To develop specific geometry within the plant and its overall topology
- To maintain optimum balance between vegetative and reproductive parts
- Pruning for dwarfing and promotion maximum shoots/fruiting points

General Rules for Pruning of Fruit Trees

- ☆ Prune young tree very lightly, follow rules of minimal pruning suggested by Robinson (2003) for first four year
- ☆ Prune mature trees more heavily, especially if they have shown little growth.
- ☆ Prune the top portion of the tree more heavily than the lower portion.
- ☆ Thin out more shoots towards the end of a well-pruned branch in a mature tree. This will increase fruit size and quality on the remaining shoots.
- ☆ To reduce the height of an excessively tall tree, cut whole limbs at the top, making cuts flush with the bark of a lower limb. Stubs left in the top of a tree would not heal.
- ☆ Following the 3: 1 pruning rule which is advised by Cook and Strydom (2000) the leader should be 3 times the diameter of any of the lateral branches in the upper part of the tree. They have suggested that limbs that are larger than the 3: 1 rule should be removed early on to preserve a hierarchy of branch and leader diameter.

Dormant Pruning

Pruning reduces trunk size and canopy volume as compared to the same tree that is not pruned. Pruning has a dwarfing effect that produces new growth in the following year. Reduction in canopy volume by 21 per cent when peach trees were severely pruned (Miller and Byers, 2000). Dormant hedging or shearing has shown success in controlling tree size, but often results in a proliferation of shoots over the periphery of the tree resulting in very dense tree canopies.

Summer Pruning

Summer pruning combined with passive pruning reduced light interception by 15 per cent, but summer pruning increased photosynthetic photon flux density (PPFD) within the tree canopy. Summer pruning increased light penetration within tree canopies, but levels may decrease during the following year (Marini and Barden, 1987). Summer pruning improves fruit colour in climates where colour develops less than usual in variegated varieties.

Root Pruning

Root pruning reduces the uptake of water and nutrients and creates an imbalance in plant hormones that adversely affects the growth of fruit plant shoots. Commercial use of root harvesting in temperate fruit crops has increased due to tractor-mounted root harvesting and loss of daminozide. Root pruning of young and mature apple trees increased shoot growth, controlled tree size light interception, yield efficiency, fruit color, firmness and soluble solids content and reduced preharvest fruit drop and biennial bearing patterns (Ferree and Rhodus, 1993; Mitre *et al.*, 2012).

Rejuvenation

Rejuvenation is a procedure that involves removing exhausted trees with the aim of restoring production capacity of old, unproductive, and senile orchards to a height of 1-1.5m above ground level or from secondary branches for new shoots. This technology also helps to maintain tree height with an open architecture and healthy shoot canopy that promotes outward growth and light utilization.

Bending

Reduced plant growth is achieved by bent shoots because bending reduces the production of gibberellins, which are also antagonists of flowering (Forshey *et al.,* 1992). Fruit quality is also improved as a result of higher light penetration and more fruit shedding on erect limbs rather than rubbing against the branch. In 'Golden Delicious'/M.9, inclining the trunk 45° or 60° from vertical reduced the rate of shoot elongation by 40 per cent and trunk circumference growth by 17 and 27 per cent, respectively. Bending is used in intense training systems, such as the slender spindle or hybrid tree cone orchard system (HYTEC), to weaken the leader and promote fruiting in the lower scaffolds.

Scoring, Girdling, Bark Inversion

Shoot and trunk growth was inhibited by scoring and girdling. These techniques effectively regulate shoot growth in the year of treatment but they may also alter growth one or more years afterwards (Hoying and Robinson, 14). Girdling of mature trees with Macspur McIntosh resulted in an 18 per cent reduction in average shoot length and a 20 per cent reduction in shoot numbers.

Pinching and Notching

Pinching (removal of immature, apical leaves) increases the number of branches on current-season growth. By interrupting the downward movement of auxin from the shoot tip, notching stimulates lateral branching from buds that would not normally grow. Removal of a horizontal band of bark (2mm wide and one-third leader circumference) down to the secondary xylem directly above each lateral bud in a zone 35 to 65 cm (35-65 cm) above the graft union resulted in more branching in 1-year old 'Empire' apple.

Use of Dwarf and Semi-Dwarf Rootstock

In temperate fruit species, there are two key elements that determine tree design and overall tree size. The first factor is genotype, which affects branching density, the proportion of short vs. long shoots, and the blooming pattern (lateral vs. terminal flowering). The root system is a second aspect that has been used to control tree size as a successful but empirical strategy, with different effects on flowering (Lauri *et al.,* 2006). The root system, which ranges from "dwarf" to "vigorous," has a considerable influence on overall tree size, with own-rooted trees being equal to or larger than trees of the same genotype grafted on the most vigorous clonal rootstocks or on seedling rootstocks (Quamme and Brownlee,

1993). The main features used to characterize the influence of the root system on tree growth are trunk cross-sectional area (TCSA) and branch cross-sectional area (BCSA). Dwarfing rootstock reduces tree size, increases dry matter partitioning to fruit production, and results in early flowering. Rootstock influences tree size both directly and indirectly through increased crop load. With the advent of more dwarfing rootstocks, the temperate fruit industry went through a revolution in training systems especially in apple. The most dwarfing *i.e.* M.27, V.3, G *etc.*, dwarf *i.e.* M.9, P.22, P.2, B.9, B.146, B.49, Mark, MAC.39, G.16, G.65, OAR1, Bemani, 0.3, V.2, J.9, F *etc.* and semi-dwarf M.7, MM.106, B.490, G.13, G.30, 0.8, H *etc.* are available for commercial high density planting in apple. Generally dwarfing rootstocks, M.8, M.9, M.27, C.6, B.9 and Mark –are used as inter-stocks to control tree size. The most common dwarfing rootstocks for pear are Old Home x Farmingdale (OH x F) and quince (*Cydonia oblonga* L.) selections that result in trees about 50 to 70 per cent of a standard size pear tree. Trees on quince rootstocks are reduced in vigor, precocious and productive in cropping and, in most cases, bear fruits of good size and quality. Pear growers in Western Europe use trees propagated on Quince C, Adams, A or BA.29 clone of quince rootstock. Cherry rootstock Mahaleb is about 75-90 per cent of Mazzard size and is often called "semi-dwarf", simply because it is smaller than Mazzard. The introduction of the interspecific hybrid Gisela (Gi) rootstocks, largely based on crosses between *P. canescens* and *P. cerasus*, provided new important traits of precocious flowering, reduced vigor, and high productivity. Gisela rootstocks now available commercially include Gi3 (very dwarfing), Gi5 (dwarfing), Gi6 (semi-vigorous), and Gi12 (semi-vigorous) (Lang *et al.*, 2014).

Use of Plant Growth Regulators in Canopy Management

Many physiological functions of plant growth and development are controlled by phytohormones. They control cell division, growth, and differentiation, dictating plant morphology. Variable quantities of phytohormones are found in the shoots and leaves of apple trees depending on their canopy design (Watanabe *et al.*, 2008).

Induction of Feathers (Lateral Branching)

Dwarf rootstock, which may be utilized in higher-density plantings, has become more commonly accessible in order to meet increasing demands in deciduous fruit tree nurseries. Feathered plants from the nursery, such as apple, pear, and sweet cherry trees, increase total yield earlier in the life A physiological phenomenon known as "apical dominance" promotes the growth of lateral branches in actively growing fruit trees in nursery. Auxin inhibits lateral bud development, whereas cytokinin increases the growth of axillary buds, thus these hormones regulate each other (Leyser, 2009). Plant growth regulators such as Cytokinin, 6-benzyladenine (BA), and gibberellic acid (GA4+7), as well as a combination of these two- (a mixture of 50 per cent BA and 50 per cent GA4+7), that promote the formation of lateral branches. They have been used successfully on a variety of fruit crops overcoming apical dominance and improving feathering in apple nursery trees (Muller and Leyser, 2011), but GA's primary function is to lengthen feathers. BA is also used to encourage proleptic shoot production and to weaken acrotony in young trees newly planted in orchards. Cyclanilide has recently been studied for growth-related effects on nursery apple, pear, sweet cherry, and plum trees and found to be promising.

Control of Vegetative Growth through Chemical

Dormant and summer pruning is done to regulate the growth of the tree. Root pruning and girdling are being used in extreme cases. These methods are very costly and/or carry a high risk of failure and loss of part of the tree assimilates. Plant growth retardants, on the other hand, can improve the balance between vegetative growth and fruit production. Daminozide has historically been used most successfully in apples for growth control. Growth retardants such as chloromiquat chloride, especially used in pears, have been registered in several countries. It is now banned due to intolerable residues from excessive use. Instead, only some countries have approved paclobutrazol for use in controlling fruit tree growth. The difficulties with this substance make correct treatment rather challenging due to its great persistence and significant acropetal movement in treated plants. Ethylene, which is released by ethephon, and chemicals that block the synthesis of growth-active gibberellins (GAs) might be alternatives for controlling shoot growth. Ethephon must be applied at dosages that, in addition to reducing growth, may cause fruit loss, premature ripening, or a reduction in alternate bearing. Prohexadione-calcium (calcium 3-hydroxy-5-oxo-4-propionyl-cyclohex-3-enecarboxylate, Pro-Ca) has recently been introduced (trade names: Apogee® and Regalis®) in a number of countries as a plant bioregulator for pome fruit trees and may offer a desirable new solution for the chemical control of excessive shoot growth. Several investigations on the effect of Pro-Ca on vegetative development and fruit formation (Byers *et al.,* 2004). Pruning costs are reduced by about 30 per cent in treated trees. Larger yields are frequently the result of increased fruit set and the availability of more assimilates for fruit growth (Rademacher, 2004). Surprisingly, Pro-Ca-treated plants were less sensitive to fire blight (*Erwinia amylovora*) and apple scab (*Venturia inaequalis*) (Costa *et al.,* 2004).

Deficient Irrigation

Plant vegetative growth can be reduced without affecting yield by deficient irrigation. Irrigation of pears and peaches at specific times during the growing season has been demonstrated to have no effect on plant growth. The apple tree responds to deficit irrigation, but not in the same way as the peach tree.

Precise Application of Fertilizer

Soil fertility is crucial for successful cropping. Fertilization disproportionately increased root and shoot growth as a result the root to shoot ratio decreased in apple trees Application of selective fertilizer to control tree growth is further made difficult by the potential reservoir of minerals stored in the perennating portions of the tree. Growth of peach trees in spring drew strongly from stored nitrogen, independent of nitrogen supplied in the fall. Likewise, root uptake of nitrogen by apple trees was negligible in early spring as trees used more of the stored nitrogen. Desired tree size is very difficult to maintain due to spatial and temporal variation in nutrient availability in young and mature trees but carefully managed fertilization should help reduce undesirable vegetative growth. Engineered trees with altered architectural characteristics should enable substantial advancements in productivity, and alter their growth habit to induce dwarfness. Over-expression of rol (root loci) genes in plant has demonstrated a drastic effect on plant height, internode and routing number in apple rootstock) and kiwifruit. Cytokinin-over-expressing transgenic peach plants containing the *Agrobacterium tumefaciens* are dwarf in stature and produce more branches Correspondingly, introduction of tumour-inducing (T)-DNA from *Agrobacterium rhizogenes* in Colt rootstock of cherry produced trees with significant reductions in plant height, internode length, leaf area and petiole length (Gutierrez-Pesce*et al.*, 1998). Likewise, rice *phytochrome*A (*phy*A) gene had dwarfing effect on Colt rootstock when over-expressed in transgenic. Introduced FT gene from *Populustrichocarpa* in European plum (*Prunus domestica* L.) cv. Blue Byrd. They reported that FT plum lines expressed pleiotropic phenotypes that rendered them different from normal plum trees. They did not grow with a typical upright tree growth habit, but instead became highly branched resulting in a shrub-type growth habit. Shrub-like growth habit of FT plum potentially makes them suitable for novel production systems.

Orchard Nutrition Management

Integrated nutritional management is the need under sustainable and safe fruit production system. In principle it is based on judicious and balanced application of need based organic manures and inorganic fertilizers at appropriate proportions at specified time schedule depending upon the age of the tree. Manures and fertilizer application programme of any orchard should be based on leaf and soil analysis for essential nutrient elements and also on visual symptoms expressed on vegetative growth and fruits. Leaf analysis is critical for nutritional scheduling in any orchard as leaf is the principle site of metabolism and any change in nutritional balance in plant system are promptly expressed in the leaves as symptoms. For areas experiencing continuous drought application of fertilizer schedule need to be recalculated and modified. In heavy bearing orchards, to regulate next year cropping 1 per cent Urea spray may be beneficial immediately after harvest or at petal fall stage. To improve storability and other calcium related disorders. Application of Calcium Chloride (0.5%) 45 and 30 days before harvest is beneficial. Incorporation of organic manures in the soil and bio-fertilizers is very important for sustainable orchard nutrition. Combination of organic manures, bio-fertilizers and fertilizers ensures balanced orchard nutrition package for better plant performance. In temperate fruits number of recommendation have been made at different places based on filed response trial but their application on the basis of leaf and soil analysis yet to be taken up.

Orchard Floor Management

Orchard floor management is another important component. It involves orchard hygiene, soil aeration, soil moisture conservation, weed growth restriction, soil drainage, which facilitates better interaction among soil micro organisms, nutrient pool and organic matter. Clean cultivation involves regular weeding. However barren soil surface is very much prone to excessive moisture loss and lack of humus. Under sloppy terrain soil erosion and loss of soil fertility is also a problem. To maintain humus and activity of soil microorganisms in the soil, addition of organic matters to the soil is necessary. Sod culture, intercropping and cover crop are other important practices under floor management which need to be worked out for different situations for sustenance of soil biological activities and their impact on productivity.

Orchard Irrigation Management

There is a need to improve the water efficiency, its application at critical stages and develop techniques to harvest, use and conserve every drop of water for good plant growth and crop yield. Due to the prevailing topography, water in the temperate region is a main limiting factor. It provides irrigation only to 15-20 per cent of the fruit area and as a result productivity of fruit crops is very low. Therefore, more efforts are needed to cover additional area with the existing resources to enhance the productivity through increased water productivity using water harvesting and micro irrigation (drip/sprinkler) techniques and moisture conservation techniques like mulching *etc.* The most critical periods of water requirement are April to July, and peak water requirement is after fruit set. Drip irrigation has been found to be adventitious for efficient water use by 2.6-2.9 times than conventional systems but this system is yet to be exploited fully in temperate fruits. The experiments conducted at CITH, Srinagar indicated that on an average during the season 3,840 litres/tree through conventional system and 1600-2000 through drip system is required for apple. On an average 80-90 litres/ tree per irrigation at 5 days' interval by drip is required. Fertigation with water soluble fertilizers like ammonium nitrate, calcium nitrate, urea, potassium chloride, potassium nitrate, and potassium sulfate has become important as it saves lot of nutrients. Urea and ammonium nitrate are highly soluble and clogging is very less. Fertilizers which raise water pH (>7.5) are not desirable as such water reduces micronutrient availability. Use of black polythene/dry grass mulch for moisture conservation in apple also found effective for increasing water productivity and gave maximum fruit set (62.5%) and fruit yield (63.4 kg/tree) besides saving moisture from 40-70 per cent. In sweet cherry, as much as 20 per cent water and 1/3rd dose of fertilizers was saved following drip irrigation and fertigation. The productivity increased up by 45 per cent and fruit cracking reduced drastically to 2.72 per cent. Today, there is a growing awareness for use of drip irrigation for increasing quality fruit production. But to achieve wider adoption of this technology, research on the minimal and optimal fraction of the soil to be wetted under varying weather and soil conditions, critical stages of application of nutrients *etc.* needs to be worked out in most of the temperate fruits including apple. Standardization of rain water harvesting and suitable moisture conservation techniques is an important area for harvesting good yields particularly in rainfed cultivation.

Precision Farming in Temperate Fruits for Increasing Efficiency and Productivity

In traditional orchard system larger plant spacing, vigorous trees, improper training and pruning lead wastage of manpower, insecticides/pesticides and irrigation water. While high density planting, plant architectural engineering and canopy management, precise application of water and nutrients through drip irrigation *etc.* increases productivity and fruit quality; shortens juvenile phase and gives high early returns. These high precision technologies make efficient use of land, light, water and nutrients and also facilitate ease of harvest. Further the success of high density precision farming is depended on availability of suitable rootstocks, scion varieties, canopy, water and nutrient management and tall these precision factors are now being identified and standardized for optimizing yield and quality and getting higher returns.

Organic Farming

The fruits and their products produced through use of synthetic materials are not attracting the world market, because of health and other environmental problems. Consequently, the demand for safe or organically produced products is increasing which emphasis the use of biological materials in combination with agronomic methods for sustainable production by avoiding the use of synthetic materials. This organic approach is regarded as the best solution to restore our natural resources and to safe guard our environment and biodiversity.

Since the global trade in organic foods increased manifold India could be a bigger beneficiary of this boom. The temperate region has a tremendous potential to grow fruits, nuts, vegetables, aromatic/medicinal herbs and spices like kalazeera and saffron which can fetch premium price of 100 to 200 per cent more than conventionally produced. The rainfed farming and use of very low chemicals inputs in raising walnut, cherry, apricots, almonds, kiwi, kalazeera and saffron in this temperate region provide great scope have potential under organic farming regime for earning higher income and export promotion. In temperate hills, walnuts are almost grown without application of synthetic chemicals and so are the cherries and apricots. The advantage of walnut, cherry and apricot need to be propagated and popularized in the region. For this development of organic package involving various components like maintenance and enhancement of soil fertility through biological means, addition of organic manure, use of soil micro organisms and utilization of biomass and crop residues, management of pests and diseases through use of botanicals, biological control agents, cultivation of inter crops, use of crops rotations, use of *Vesicular Arbuscular Mycorrhyza* (VAM) and *Trichoderma etc.* and other methods of biological control of pests and disease.is the priority and need to be standardized for important crops along with guidelines for organic product processing as per IFOAM. The hill states taking the advantage of topography and natural farming can exploit the opportunity on scientific lines to achieve sustainable production and livelihood security through promotion of organic fruit culture.

Rejuvenation Technology

Technology for rejuvenation of old apple orchards has been standardized for three apple varieties involving severe pruning, balanced dose of fertilizers and combination of insecticides and fungicides. The Pruning was done in February and application of Chaubattia paste on the cut pruned surface and spraying of Caustic soda (1.0%) during dormancy sevin (0.1%) spray once in growing season (May) and fungicides (Carbendazim as well as Bayleton) sprayed twice in the month of May and June and application of fertilizers $N_{2:}$ P_2O_5: K_2O @ 450: 450: 450 : g per tree, before fall controlled insect, pest and disease resulted in good growth and yield.

REFERENCES

Ahmed N, Kumawat KL and Lal C, 2016. Canopy management and Plant Architectural Engineering in temperate Fruits and Nuts *In:* Temperate fruits and Nuts: A way Forward for Enhancing Productivity and Quality. Chadha, K.L, Ahmed Nazir and Singh S.K (Eds.) daya publishing house New Delhi-110 002

Awasthi RP and Mehta K, 2004. Strategies for canopy management in tree fruit crops. Proceedings of the First.*Indian Horticultural Congress,*320-330.

Barritt BH, Rom CR, Konishi B and Dilley MA, 1991. Light level influences spur quality and canopy development and light interception influences fruit production in apple. *Hort. Science,* **26**: 993–999.

Byers RE, Carbaugh DH and Combs LD 2004. The influence of prohexadione calcium sprays on apple tree growth, chemical fruit thinning, and return bloom. *Journal of the American Pomological Society,* **58**: 111-117.

Chalmer DJ and van den Ende B, 1975. The 'Tatura Trellis', a new design for high yielding orchards. *Journal of Agriculture Victoria,* **73**: 473–476.

Chalmer DJ, van den Ende B and Van Heek L, 1978. Productivity and mechanization of the Tatura trellis orchard. *HortScience,* **13**: 517–521.

Cook NC and Strydom DK, 2000. The South African high density system. *Acta Horticulturae,* **513**: 321–327.

Costa G, Sabatini E, Spinelli F, Andreotti C, Spada G and Mazzini F, 2004. Prohexadione-Ca controls vegetative growth and cropping performance in pear. *Acta Horticulturae,* **653**: 127-132.

Ferree DC and Rhodus WT, 1993. Apple tree performance with mechanical hedging or root pruning in intensive orchards. *Journal of the American Society for Horticultural Science,* **118**: 707–713.

Forshey CG, Elfving DC and Stebbins RL, 1992. *Training and Pruning Apple Trees.* American Society for Horticultural Science, Alexandria, Virginia, p.166.

Gandev S, 2009. Training and pruning of apple tree according to the new system 'Cone'. *Acta Horticulturae,* **825**: 465-470.

Gutierrez-Pesce P, Taylor K, Muleo R and Rugini E, 1998. Somatic embryogenesis and shoot regeneration from transgenic roots of the cherry rootstock Colt (*P. avium* x *P. pseudocerasus*) mediated by pRi 1855 T-DNA of *Agrobacterium rhizogenes. Plant Cell Reports,* **17**: 574–580.

Hoying SA and Robinson TL, 1992. Effects of chain saw girdling and root pruning of apple trees. *Acta Horticulturae,* **322**: 167–172.

Kallow G, Reddy BMC, Singh G and Lal B, 2005. Rejuvenation of old and senile orchard. Pub. CISH, Lucknow. P: 40.

Jackson JE, 1970. Aspects of light climate within apple orchards. Journal of Applied Ecology,7: 207-216.

Ke X, Yoshida H, Hikosaka S and Goto E, 2023. Photosynthetic photon flux density affects fruit biomass radiation-use efficiency of dwarf tomatoes under LED light at the reproductive growth stage. *Front. Plant Sci.***14**: 1-13.

Lakso AN, 1994. Apple. In: B. Schaffer and P.C. Andersen (eds). Environmental physiology of fruit crops; Vol 1: Temperate crops. CPC Press, Boca Raton, FL, pp. 3–42.

Lang GA, Blatt SS, Embree C, Grant J, Hoying S, Ingels C, Neilsen D, Neilsen G and Robinson T, 2014. Developing and Evaluating Intensive Sweet Cherry Orchard Systems: The NC140 Regional Research Trial. *Acta Horticulturae,* **1058**: 113–120.

Lauri PE and Lespinasse JM, 2000. The Vertical Axis and SolAxe systems in France. *Acta Horticulturae,* **513**: 287–296.

Lauri PE, Maguylo K and Trottier C, 2006. Architecture and size relations: an essay on the apple (*Malus domestica,* Rosaceae) tree. *American Journal of Botany,* **93**(3): 357–368.

Lespinasse JM and Delort JF, 1986. Apple tree management in Vertical Axis, appraisal after ten years of experiments. *Acta Horticulturae,* **160**: 139–15.5

Leyser O, 2009. The control of shoot branching: an example of plant information processing. *Plant Cell and Environment,* **32**: 694–703.

Marini RP and Barden JA 1987. Summer pruning of apple and peach trees. *Horticultural Reviews,* **9**: 351–375.

McKenzie DW and Mouat HM, 1963. Semi-intensive planting and orchard management, a new approach to apple growing. *Supplement to the Orchardist of New Zealand,* p. 8

Miller SS and Byers RE, 2000. Response of winter-injured peach trees to pruning. *Hort.Technology*, **10**: 757–765.

Mitre V, Mitre I, Sestras AF and Sestras RE, 2012. Effect of Roots Pruning upon the Growth and Fruiting of Apple Trees in High Density Orchards. Bulletin UASVM Horticulture, **69(1)**: 254–259.

Muller D and Leyser O, 2011. Auxin, cytokinin and the control of shoot branching. *Annals of Botany*, **107(7)**: 1203–1212.

Palmer JW, 1989. In: Manipulation of Fruiting, C.J. Wright (ed.) Canopy manipulation for optimum utilization of light. Butterworths, London, p. 245-262.

Pathak RK, 2009. Canopy management in minor fruit crops. In: Summer School on Management of canopy architecture for higher productivity in sub-tropical fruits. Pub. CIAH, Lucknow, India. pp: 43-48.

Proietti P and Tombesi A, 1996. Translocation of assimilates and source-sink influences on productive characteristic of the olive tree. *Advances in Horticulture Science*, **10**: 11–14.

Quamme HA and Brownlee RT, 1993. Early performance of micropropagated trees of several *Malus* and *Prunus* cultivars on their own roots. *Canadian Journal of Plant Science*, **73**: 847–855.

Rademacher W, 2004. Chemical regulation of shoot growth in fruit trees. *Acta Horticulturae*, **653**: 29–32.

Robinson TL, 2003. Apple orchard systems. In: D.C. Ferree and I.J. Warrington (eds.) *Apples: Physiology, production and uses*. CABI Publishing, Wallingford, United Kingdom, pp: 345–407.

Robinson TL, Hoying SA, Miranda Sazo M, Dominguez LI and Fachinello JC, 2014. Yield, fruit quality and mechanization of the tall spindle apple production system. *Acta Horticulturae*, **1058**: 95–103.

Robinson TL, Lakso AN and Ren Z, 1991. Modifying apple tree canopies for improved production efficiency. *Hort Science*, **26**: 1005-1012.

Tustin S, Hurst P, Warrington I and Stanley J, 1989. Light distribution and fruit quality through multi-layered trellis apple canopies. *Acta Horticulturae*, **243:** 209-212

Verheij EWM and Verwer FLJAW, 1973. Light studies in a spacing trial with apple on a dwarfing and a semi-dwarfing rootstock. *Scientia Horticulturae*, **1**: 25-42.

Wagenmakers PS, 1991. Planting systems for fruit trees in temperate climates. *Critical Reviews' in Plant Sciences*, **10:** 369-385.

Walker CD, 1980. The development of intensive orchards in England, a historical note. *Acta Horticulturae*, **114**: 309–317.

Watanabe M, Bessho H, Suzuki A, Komori S, 2008. Seasonal changes of IAA and cytokinin in shoots of columnar type apple trees. *Acta Horticulturae*, **774**: 75–80.

Wunsche JN and Lakso AN, 2000. Apple Tree Physiology–implications for orchard and tree management. *The Compact Fruit Tree*, **33**(3): 82-88.

Chapter 4

Nursery Production of Unexploited Minor Temperate Fruit Crops

A.S. Sundouri, Saklain Mulani, Sukjai Dhar, Rameet Kour, Zarina Banoo, Amit Kumar and M.K. Sharma

Division of Fruit Science, Faculty of Horticulture, SKUAST-Kashmir, Shalimar, Srinagar–190025, J&K

Nursery management is crucial in successful fruit crop cultivation, ensuring the production of healthy, vigorous fruit trees. This involves selecting the right location, using appropriate propagation methods, providing proper care, managing pests and diseases, and hardening off seedlings. A well-managed nursery produces high-quality trees that thrive in the field, leading to improved productivity and profitability for growers. Careful planning and attention to detail at every stage of nursery management are essential for long-term success in fruit farming.

A nursery is a place where plants are cultivated, cared for, and sold. Growers typically require high-quality saplings or grafts of the correct variety. The primary goal of effective nursery management is to supply planting material of the highest quality for new development areas and replanting efforts. Poor-quality planting material can lead to low yields and additional costs for thinning out weak or under performing plants in the field. Therefore, selecting good planting material and implementing strict culling practices in the nursery are crucial steps. Recognizing the value of high-quality planting material it is essential to known the importance of nursery as follows:

1. Seedlings and grafts are produced in nursery from which the fruit orchards and ornamental can be established with minimum care, cost and maintenance.
2. The nursery planting materials are available at the beginning of the planting season.
3. This saves the time, money and efforts of the farmers to raise seedlings.
4. It assures the production of genetically improved quality planting materials.
5. It provides employment opportunities for technical, skilled, semi-skilled, unskilled labour.
6. Nursery provides the quality planting material for meeting the demands for fruits and processing industries.

Pecanut

Pecan (*Carya illinoinensis*) are a type of nut and tree of the walnut family (*Juglandaceae*). Native to North America, primarily grown for their rich, flavourful seeds. The pecan tree is a deciduous tree that thrives in warm, temperate climates, particularly in the southern United States and parts of Mexico. Pecans are known for their high oil content and rich, buttery flavour, making them a popular ingredient in both sweet and savoury dishes, including pies, snacks, and salads.

Pecans grown from seed are not true to type. This means that offspring or seedlings, do not exhibit the same characteristics or traits as the parent plant. In fact, each seedling is unique and have extremely variable nut quality. Seedlings can be prepared bare root and in containers. Seeds are stratified in moist sand for 70-80 days at low temperature (below 7°C) to improve germination and seedling development. Bare root plants can be transplanted in the containers or may be directly planted in the field. The propagation of pecan can also be done through grafting and budding. These methods are used to propagate a tree of a specific variety because it ensures that the new tree will be a genetic clone of the parent tree, maintaining all of the desired characteristics (such as fruit quality, disease resistance, growth habit) and provides uniformity in the orchard. Also grafting buds or shoots from a parent tree onto a seedling rootstock, growers can also take advantage of the rootstock's beneficial traits, such as disease resistance or adaptation to soil conditions, without compromising the fruiting characteristics of the parent variety. Seedlings can be prepared bare root and in containers. Seeds are stratified in moist sand for 70-80 days at low temperature (below 7°C) to improve germination and seedling development. Bare root plants can be transplanted in the containers or may be directly planted in the field.

Raising of Seedling Root-Stock

Most commercial nurseries use open-pollinated seeds from domestic or native varieties that perform well in a given region. In USA, the most commonly used rootstock seed is from variety 'Elliott' or 'Curtis, while in other countries seeds are used irrespective of variety. Regardless of the variety, make be sure that nuts are from the current season, properly filled and free of insects and disease used for raising rootstocks.

Pecan nuts should be stratified for six to eight weeks prior to planting. This can be done by placing the nuts in a container of alone peat moss, sand, cocopeat or layering them with sand, cocopeat or peat moss. Stratified seed should be kept moist but not wet at a temperature of 35 °F. Stratification hastens germination once the nuts are planted and helps the seedlings to emerge more uniformly, facilitating the ease of weed control in the nursery.

Following stratification, nuts should be soaked for two to three days prior to planting in a container of water to activate enzymes that stimulate growth. A hose left dribbling into the container provides oxygen, which helps to prevent the growth of harmful fungi. Additional measures to protect against infection with seed-rotting pathogens include keeping the soaking nuts in 35-40 °F temperatures and changing the water daily. If nuts are soaked at room temperature, the water should be changed every few hours. Pecan seeds should be planted from mid-March to mid-April for best results. Seeds can be planted in a seedbed or in containers. As each seed from an individual tree is not identical to the parent, there is a considerable amount of variation in rootstock, even from seed arising from the same parent tree. This accounts for much of the variation in tree performance as the grafted tree develops and comes into full production.

Planting in the Seedbed

Seedbeds provide easier sowing and germination. Fertility of the seedbed should be determined before planting based on soil test recommendations. Seeds should be planted 4 to 6 inches apart in the row and 3 to 4 inches in deep.

Figure 4.1: Pecan Nursery Trees growing in a Seed Bed.

Figure 4.2: Pecan Nursery Trees grown on Plastic.

Planting depth should be shallower in clay soils than in sandy soils. Round-shaped nut can be planted easily with a mechanical planter whereas cylindrical or oblong nut varieties tend to hang up in the planter and must be planted by hand. Plastic mulch should be laid in raised beds over the rows prior to planting in order to minimize weed competition. Holes can then be punched into the plastic for planting the nuts.

Weed Management

Weed management is one of the most difficult and challenging job to nursery production of pecans. Following germination, but just before seedling trees break the soil surface (five days to two weeks after planting, depending on weather, *etc.*), apply oryzalin herbicide or paraquat. Paraquat is a contact herbicide without systemic activity and will only damage tissue that it touches, whereas, Oryzalin is a pre-emergence herbicide and may be sprayed over the top of seedlings after emergence to prevent emergence of grasses and small-seeded broadleaf weeds. A hand sprayer with a shielded may be used to direct glyphosate or paraquat application to weeds without damaging small trees.

Potting Mixture

Pecan seedlings also grow well in a variety of container shapes and sizes. A container with an open bottom should be used. Roots should be air pruned by placing the pots on a screen wire bench. By using this method, the roots will grow no farther than the bottom of the pot before being air pruned. This prevents circling of the taproot in the container and promotes a fibrous root system. A potting soil mixture that allows free movement of water through the pot should be utilized. A mix of three parts ground pine bark, one part peat moss/cocopeat and one-part coarse sand works well. All potting mix should be sterile. Amend potting media by incorporating micro-nutrients and lime into the media. This should provide all the required nutrients for the first year and early into the second year. In the second year, additional slow-release fertilizer may be needed, but there is no need for lime or micro-nutrients.

Graft Wood Collection and Storage

Stored graft wood is required for several grafting methods. Unless the wood is collected and stored properly, the grafting endeavour is destined for failure. A variety of problems may occur in the collection and storage process, including drying out or freezing the wood, immature buds, old shoots, and insect and/or disease damage.

Scions should be cut from the previous season's growth of a productive tree. Scions should be straight and smooth, measuring 10-15mm in diameter and 8-12 inches in length having three to four healthy buds and should be insect and disease free. Water sprouts often work well as scion wood. Scions from older wood may be used, but are generally less successful. When whip grafting, graft wood can be used immediately and storage may not be necessary. When using the 4-flap or bark graft techniques, graft wood should be cut when the trees are completely dormant (January-February). Budwood used for spring budding should be collected in early to mid-March. Graft wood can be stored for few months at temperatures between 30 and 38 °F. The wood should not freeze, but it should remain dormant. Pecan buds may swell when stored at 35 °F for a prolonged period of time.

Budding

Patch budding may be used to propagate nursery stock or top-work larger pecan trees. Budding simply involves removing a portion of the bark along with a single dormant bud from scion wood. The bark and bud from the scion wood are fitted into a space cut into the bark of the understock. After the bud begins to grow, the shoot above the bud is pruned away and the grafted bud becomes the primary shoot. Patch budding requires that the bark of both the stock and scion be slipping easily. It is usually done in late July or August, but can also be done in the spring.

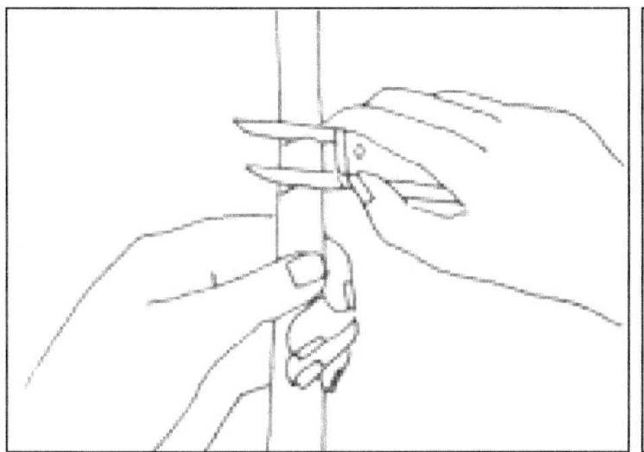

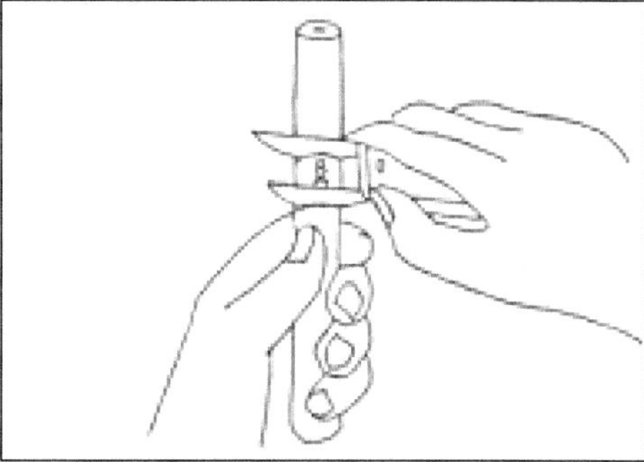

Figure 4.3: Cutting the Stock with

Figure 4.4: Cutting the Scion. a Budding Knife.

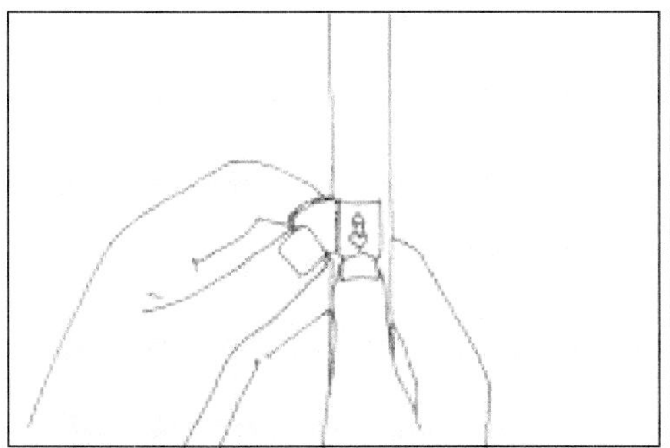

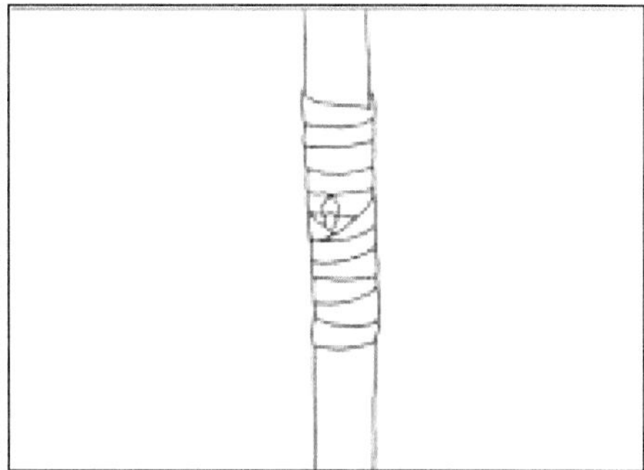

Figure 4.5: Placing Bud Patch onto Stock.

Figure 4.6: The Completed Budding Process.

Bark Grafting

Bark grafting is often used to graft a relatively small scion onto rootstock that is too large for the four-flap graft or whip graft. Bark grafting should be done in the spring, usually late April to early May, about two to three weeks after growth begins in the spring. The primary rule for bark grafting is that the bark on the scions should be tight, while that of the rootstock must slip.

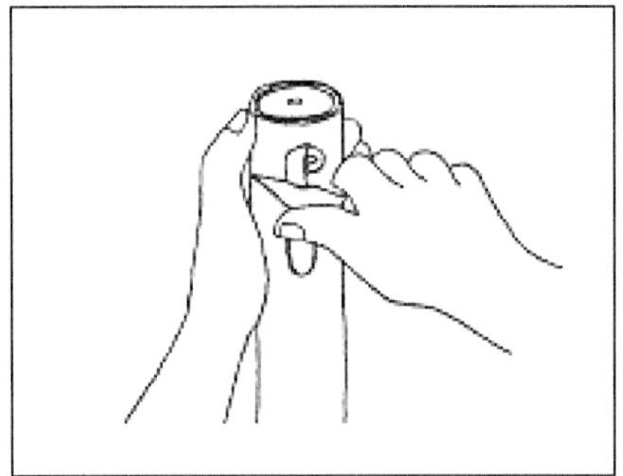

Figure 7: Removing the Rough, Outer Portion of the Bark where the Scion will be Inserted.

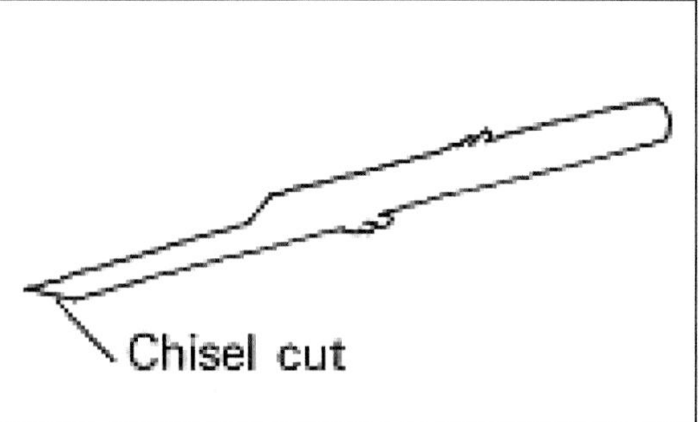

Figure 4.8: Scion Prepared for Bark Grafting.

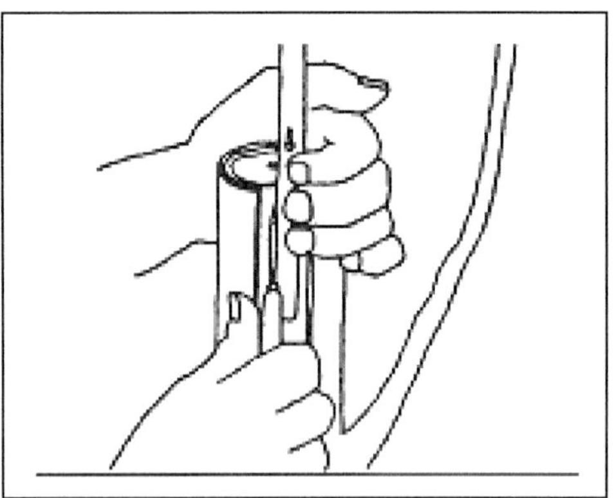

Figure 4.9: Fitting the Scion.

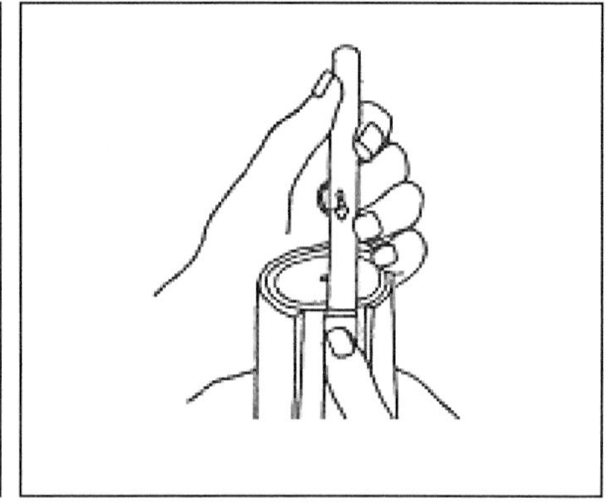

Figure 4.10: Inserting the Scion.

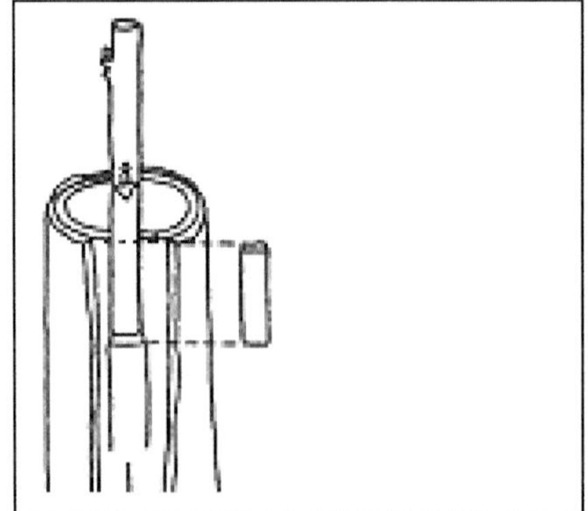

Figure 4.11: Removing the Bark Strip.

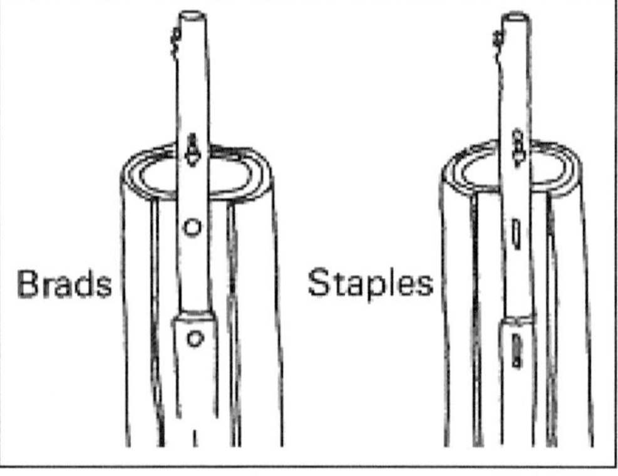

Figure 4.12: Attaching the Scion with Brads (Left) and Staples (Right).

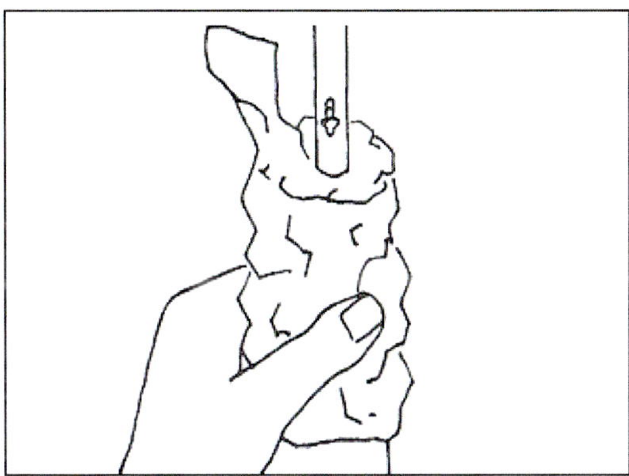

Figure 4.13: Wrapping with Aluminium Foil.

Four Flap Grafting

The four flap grafting is one of the easiest grafts to use, which is ideal for small calliper trees up to 1 inch in diameter. This grafting is most successful when the scion and rootstock are of similar size. For the best fit, the scion should be slightly larger than the stock. The high degree of success with the four-flap graft can be attributed to the large amount of contact between the cambium of the rootstock and scion. Whereas most grafting techniques only allow for the cambiums to connect at two locations, the four flap allows eight opportunities for cambium contact with each graft. Therefore, a precise cut and fit is less critical with the four-flap graft. Scion wood for the desired cultivar should be collected while dormant in late February or early March. Use healthy, 1-year-old wood with well-developed buds for the scion. The four-flap graft can also be used on older trees. When this is done, use limbs less than 1 inch in diameter. The four-flap graft should be made in early spring after the bark begins to slip (April and May). Place the knife blade straight across the cut surface of the stock and slice down through the centre about ¾ in. (Figure 4.14). After this cut is made, make a second cut perpendicular to the first.

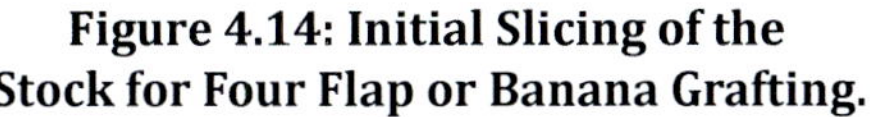

Figure 4.14: Initial Slicing of the Stock for Four Flap or Banana Grafting.

Figure 4.15: Cutting the Scion.

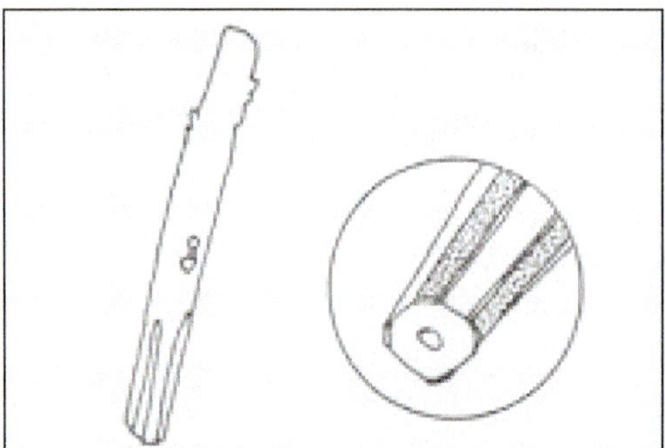

Figure 4.16: Scion Prepared for Four-Flap Graft.

Figure 4.17: Peeling Back of the Four Flaps.

Figure 4.18: Cutting the Stock for Four-flap Grafting.

Figure 4.19: Rubber Band Holding Scion in Place.

Figure 4.20: Wrapped Scion.

Figure 4.21: Completed Four Flap Graft Wrapped with Parafilm.

Whip Grafting

Seedling trees and nursery stock of 1, 2, and 3 years of age with a diameter up to 1 inch may be grafted by the "whip" method. Whip grafting may be done in February and March when the buds are dormant.

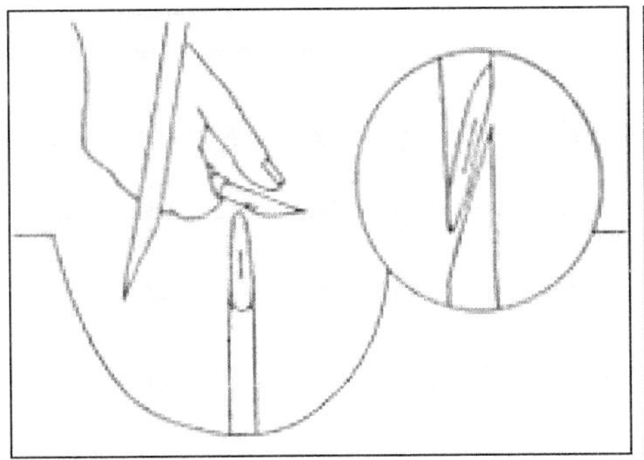

Figure 4.22: Cutting the Seedling Stock for Whip Grafting.

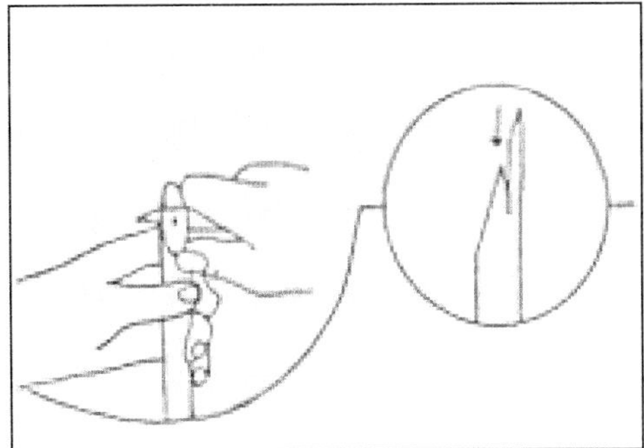

Figure 4.23: Cutting Tongues on the Stock and the Scion.

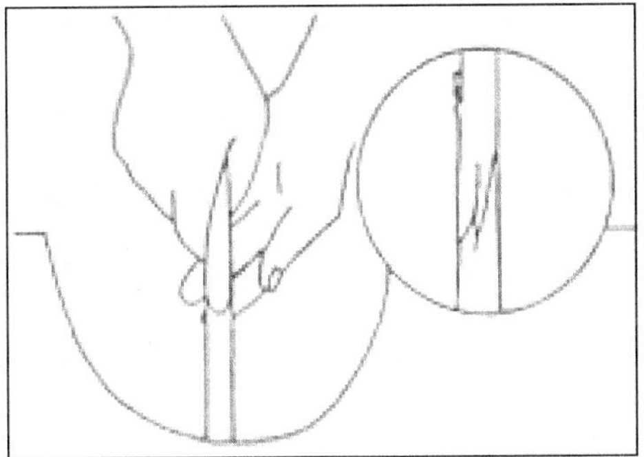

Figure 4.24: Fitting the Scion and Stock.

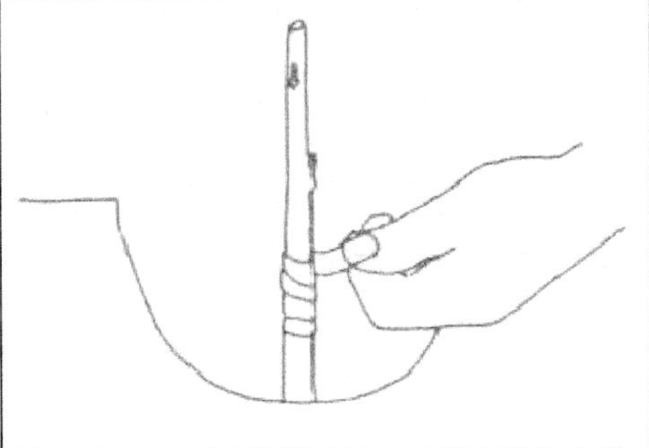

Figure 4.25: Taping the Graft.

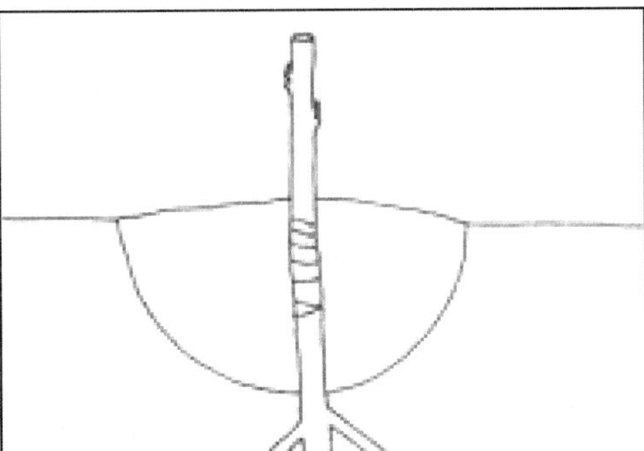

Figure 4.26: The Completed Whip Graft.

Fig

Figs (*Ficus carica* L) is a group of flowering plant in the *Moraceae* family, widely cultivated for their sweet and nutritious fruit. Fig has its center of origin, diversity, and domestication in the Middle East and Western Asia. It has been cultivated for thousands of years, with evidence of fig cultivation back to ancient civilizations in Egypt, Greece, and Rome. However, due to its adaptability, it is cultivated in the sub-tropical, tropical, and temperate regions of the world. Figs are consumed fresh or dried, and are also processed into products such as bread, jelly, jams, candy, and fig-flavoured coffee.

Seed Propagation

Plants grown from seed are not guaranteed to have the same traits as either parent. Additionally, only some seeds will produce female trees with edible fruit while the others will produce male fig trees with small inedible fruit. Figs contain many small fruits and seeds. Viable seeds can be separated from inviable seeds by floating them in water. Viable seed will sink and can be germinated. Seeds that do not sink to the bottom are inviable and should be discarded. Fertile seeds can be germinated by spreading them on flats of moist soil mix or sphagnum moss.

Vegetative Propagation

Fig is most commonly propagated by cuttings to ensure that new trees are female and "true to type", with the same traits as the tree from which cuttings were collected. Shield- or patch budding, or cleft- or bark-grafting techniques can also be used to topwork an existing orchard.

Cuttings

Fig plants are commonly propagated by cuttings. Dormant cuttings of 2- or 3-year-old wood, or basal parts of vigorous first year shoots with a heel of two-year branch at the base, should be used for propagation. Cuttings should be ½ to ¾ inches in diameter and eight to twelve inches long (Morton, 1987). The best cuttings should be prepared in early spring well before bud break. Grow cuttings for one or two seasons (12-15 months) in the nursery before transplanting to a permanent location. Occasionally, two cuttings can be set in one location to ensure establishment if one does not survive. In India, Propagation studies of fig show success by quick dip treatment of cutting with 1500 ppm IBA under mist chamber. In North India, cuttings are made during January-February at the time of pruning, while in South India cuttings are usually made and planted in the rainy season. Pre-girdling of fig cane 30 days before taking cuttings results in quick and increased rooting.

Grafting and Budding

The fig can be budded onto existing rootstock to change fruit varieties. Fig can be topworked by inserting buds into vigorous 1-year-old shoots on heavily pruned trees by wedge grafting, or patch-budded onto older shoots. Cleft- or bark-grafting can also be used to topwork existing varieties.

Layering

Although commercial plantings should utilize nursery stock, propagation by ground or air layering can be successful in figs. Layering should be done in early spring using one-year old branches to ensure sufficient rooting my midsummer. Fig can be air layered successfully during monsoon. One year old branches if layered in June can be planted in the field by August-September, using sphagnum moss as medium, moistened with humic acid (4%).

Micropropagation

Fig shoot tips and nodes have been successfully micro propagated using several techniques. In general, roots can be propagated from shoot tips by placing them on semi-solid propagation material with IBA. Nodal explants can be propagated by placing them on propagation material with BA (Bapat and Mhatre, 2005). Each fig variety may require different concentrations of growth hormones for successful propagation.

Scion Selection

There are two primary types of commercial fig, the "Smyrna" and the "common" fig. The common fig produces fruit without pollination (parthenocarpic fruit). Popular varieties of common figs grown include Mission (a dark purple fig), Kadota (a white fig) used for canning and drying; Adriatic, and Conadria. Adriatic is a white fig once commonly used in bars and paste which has now been largely replaced by Conadria in fig paste production. Calimyrna is the most popular Smyrna cultivar. The San Pedro cultivar produces the first crop parthenocarpically but must be pollinated to produce a second crop.

Rootstock Selection

Figs are frequently propagated by cuttings. As a result, rootstocks are not commonly used in fig production. Exotic fig cultivars like Deanna, Excel and Conadria perform better on Brown Turkey rootstock than on their own roots, as this imparts vigour to these scions. While using pre-cured scions, tip cuttings are preferred to others to do wedge grafting. If scion material is scarce, chip budding is the best alternative method. When in situ grafting/budding is done, success rate is more.

Chestnut

Choosing the Right Chestnut Seeds

Chestnut trees are a type of deciduous tree that produces edible nuts. They are native to Europe, Asia, and North America. There are many different varieties of chestnut trees, each with its own unique characteristics. When selecting the seeds for seedling raising, choose only seeds that are plump and free of cracks or blemishes. Discard the hallow, semi filled, cracked and immature seeds

Preparing the Seeds for Planting

Before sowing chestnut seeds, it is important to prepare seeds properly. This will help to improve the germination rate and ensure that the seedlings grow healthy. To prepare chestnut seeds for planting, it is necessary to soak the seeds in water for 24 hours. This will help to soften the seed coat and make it easier for the seeds to germinate.

i) Scarify the seeds. Scarification is the process of scratching or nicking the seed coat to help the seeds to germinate. Scarify the seeds by rubbing them with sandpaper or using a sharp knife.

ii) Stratify the seeds. Stratification is the process of exposing the seeds to cold temperatures to mimic the natural winter conditions. Stratify the seeds by placing them in a plastic bag with moist peat moss or potting soil and storing them in the refrigerator for 3-4 months.

Chestnut seed requires special care that is different from many other types of seed. After harvesting, chestnut seed requires the following:

1. Keep the seed moist and don't let them to dry. To avoid the chestnut seeds from drying keep in moist peat or sawdust, or keep small quantities in plastic bags, or plant in moist soil. Never let the chestnuts dry out, even for a few days. If the chestnuts are stored in plastic bags, there should be a few pin holes in the bags to admit oxygen.
2. Before they will germinate, chestnuts must be chilled at 30-45°F. This can be done in a refrigerator, in an unheated building, or underground. If chestnuts freeze and are killed at temperatures below 25 F. If they are fall planted, be sure to plant them about 1 to 3 inches deep and mulch to protect them from freezing. In late winter (usually Feb), chestnuts will start germinating in the refrigerator. The root emerges first, then the shoot. Ideally, they should be planted by this time. However, if they do start growing before planting, just plant them carefully to avoid breaking off the sprout.
3. Plant in a well-drained, acidic (pH 5-6) soil or potting medium. chestnut roots are very susceptible to water-logged soil conditions. They need to be moist, but not too wet.
4. Protect seed from animal and pests depredation like mice, birds, bear, deer, and many other animals that love chestnuts.

Quince

Quince (*Cydonia oblonga* Mil.) is a deciduous, multi-stemmed shrub or tree which grows to 8-12 ft. (2.44-3.66 m) and bears pome fruits. Taxonomically, *Cydonia* is a monotypic genus in the family *Rosaceae*, which also includes apple and pear. Some varieties have twisted, rangy branches and can become attractive specimen trees when old. Leaves are long and oval in shape, and covered with pubescence (fine hairs), as are the buds and fruit. Quince is similar to pear and bears a many-seeded pome fruit which is bright yellow when mature. It is high in pectin and was commonly grown for use in the production of jellies before the advent of commercial gelatin. The fruit can be highly astringent and typically must be cooked before consumption, although some cultivars can be eaten fresh.

Sexual Propagation

Where quince seedlings are desired, the quince seed is extracted from mature fruit, cleaned, stratified in sand and stored in a cool place until planting out in late winter - early spring. If the seedlings grow well, they will be large enough to bud with the desired varieties during the following late summer–autumn. Trees on seedling rootstock should produce some fruit by about the fifth year (Campbell, 2001).

Mound Layering

Mound layering is a traditional method of propagation also called 'stool layering'. Quince can be propagated readily by mound layering, which is useful with heavy-stemmed, closely branched plants and appropriate for fruit rootstock production. The original plant may be cut back to the growth of new shoots from the base. During the following spring, a mound of soil containing sphagnum peat moss is pressed around the new growing shoots. Roots from the new shoots grow into the mound of soil. New plants are ready for digging and transplant during the following autumn or spring (Lerner and Dana, 2001).

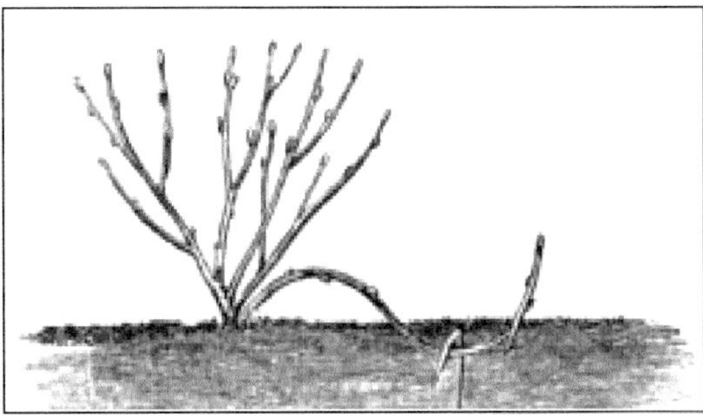

Hardwood Cuttings

Quince roots easily by hardwood cuttings from one-year-old wood. The cuttings, about 25 cm (9.8 in), are taken during the late autumn–early winter. This is a common method of propagation; however, the resultant trees tend to produce suckers, which require removal. Cuttings from two- to three-year-old wood also root easily. After one season, cuttings can usually be transplanted to their permanent location.

Persimmon

Persimmon (*Diospyros kaki* L.), is native to China. Persimmon has a low chilling requirement (less than 100 hours). As a result, the buds may break dormancy after periods of early warm spells, only to be damaged by spring frosts that occur later in the season. Trees do not produce well in regions with high summer heat, like the desert, because bark and fruit will sunburn. Although it can be grown in a wide range of soils, persimmon does not tolerate salinity (Farrar, 1999). The persimmon tree begins to bear fruit after 3 to 5 years and has an average life span of 60 years (Das *et al.,* 2001).

California commercial cultivars 'Hachiya' and 'Fuyu' of *D. kaki* (oriental persimmon) can set seedless fruit without pollination (parthenocarpic) when no other pollinating varieties are nearby (LaRue, 1982). Cross pollination, and seed development within fruits occur when varieties are planted within half a mile of each other (Farrar, 1999). In the eastern and southern US, D. Virginiana (American persimmon) requires pollination to set fruit. Parthenocarpy in persimmon varies by cultivar, region, climate, and nutritional status, and is not well understood (Woodburn and Andersen, 1996).

Seed Propagation

Mature seeds are used to propagate rootstocks. Seedlings may be grown from fresh seeds collected in the fall and stratified at approximately 45°F for 60 to 90 days. After stratification the seeds can be germinated in boxes at 70°F, although germination from seed has a relatively low success rate (25-35%). Seedlings are lined out in the spring when the soil temperatures are 55°F or higher. Plants develop a better lateral root system when they are transplanted at 6 to 8 inches tall. Seedlings can be grafted at the end of one season's growth, when the rootstock and scion are dormant (LaRue *et al.,* 1982).

Grafting

Grafting should be carried out during the dormant period, before vegetative growth begins, on rootstocks at least 1/3-inch diameter. The scion wood should be 1/4 to1/3 inch in diameter and about 3 to 5 inches long with 2 to 4 buds. Scion cuttings should be collected from healthy, well lignified plants during winter dormancy and kept in cold storage until the time of grafting. A whip and tongue graft is used when the diameter of the scion is thin, while a wedge graft is appropriate for scions with thicker diameters (Bellini, 2002).

Cuttings

To propagate rootstocks with cuttings, use softwood cuttings, root with bottom heat under mist, and treat with indole-3-butyric acid (IBA) (Hartman *et al.,* 2010). Rootstocks from cuttings can also be obtained by removing large sized roots from young plants. These cuttings are put in a greenhouse under mist and moved to a nursery in the spring. Self-rooted cuttings can support a graft within one year (Bellini, 2002).

Micropropagation

Multiple plantlets are grown from shoot tissue taken from a single plant, resulting in identical clones. This method is preferred for rootstock development.

Micropropagation is a method used to produce a large number of identical plants from a small amount of plant tissue. This method is advantageous when there is a limited source of plant tissue; however, it is labour intensive and expensive, requiring sterile laboratory facilities. The performance of micro propagated trees varies by cultivar. For example, micro propagated trees of cultivars 'Jiro' and 'Hiratanashi' grow easily and more quickly than grafted trees and showed less shock at transplant compared to grafted trees (Tetsumura *et al.,* 1999).

Rootstock Selection

Seeds from mature trees or cuttings are used for rootstocks. Three common rootstocks are: *Diospyros lotus, D. kaki*, and *D. virginiana*. *D. lotus* is the most common rootstock and is compatible with most commercial cultivars. It tolerates a wide range of soils, produces uniform seedlings, grows vigorously, resists drought, and produces a fibrous root system that transplants easily. *D. lotus* is tolerant to Armillaria, but susceptible to crown gall and Verticillium and does not tolerate poorly drained soils (Hartman *et al.,* 2010). *D. kaki* produces long taproots with few fibrous laterals. Seeds from this variety produce uniform and vigorous seedlings, which is desirable for rootstock propagation (LaRue *et al.,* 1982). *D. kaki* is resistant to Agrobacterium and Armillaria but susceptible to *Verticillium* (Hartman *et al.,* 2010). *D. virginiana* has a fibrous root system tolerant of both drought and excess moisture, however trees propagated on this rootstock are not uniform and are prone to suckering (LaRue *et al.,* 1982).

Scion Selection

Persimmon fruits are classified into two groups based on their astringency at harvest: astringent or non-astringent. Astringent cultivars develop water soluble tannins throughout the flesh of the fruit until the fruit is harvested. Non-astringent cultivars stop accumulating tannins at an early stage of fruit development. In cultivars and regions where pollination is required, persimmon fruits are further classified by flesh colour in response to pollination (Yamada *et al.,* 2012). The flesh colour of pollination-constant cultivars is not influenced by pollination, whereas pollination-variant cultivars have dark flesh around the seeds when pollinated. Persimmon fruits are commonly classified, based on astringency and flesh colour in response to pollination, using the following four groups:

- Pollination constant non-astringent (PCNA)
- Pollination variant non-astringent (PVNA)

- ✰ Pollination constant astringent (PCA)
- ✰ Pollination variant astringent (PVA)

Pollination Variant Astringent Cultivar 'Hachiya'

Fruit is oblong and conical in shape with a deep orange to red color when ripe. The flesh of unripe fruit is soft and astringent due to high concentrations of tannins. When it is fully ripe, 'Hachiya' fruit loses this astringency and becomes sweet and edible (Morton, 1987). 'Hachiya' trees grow better in cooler areas, are typically heavy bearing and should be planted in solid blocks without pollinators to produce seedless fruit.

Pollination Constant Non-Astringent Cultivar 'Fuyu'

Fruit is flat and squarish-round with an orange to red color. The flesh is firm and non-astringent when the fruit is ripe. 'Fuyu' fruit keeps well and is excellent for packing and shipping (Morton, 1987). 'Fuyu' trees produce only female flowers and fruits without pollination. However, if another persimmon variety with male flowers is present, cross pollination may occur, producing fruit with seeds. 'Fuyu' trees grow better in warmer areas, and have an upright and vigorous growth habit which is easy to train and maintain.

REFERENCES

Bassil NV, Postman JD, Hummer KE, Mota J, Sugar D and Williams R, 2011. Quince (*Cydonia oblonga*) Genetic Relationships determined using microsatellite markers. *Acta Hort.* (ISHS) **909**: 75-83.

Bellini E, 2002. Cultural practices for persimmon production. First Mediterranean symposium on persimmon 51: 39–52. Retrieved March 10, 2013: Brun, C. A. Persimmons: an overview of cultivars, production, harvesting and marketing.

Campbell Jill, 2001. Quince Growing. New South Wales Department of Primary Industries. *AgFact* H4.1.3. Retrieved April 16, 2014.

Carroll D, 2013. Persimmon pest management in San Joaquin Valley. Retrieved on March 5, 2013: http: //aaie.net/wp-content/uploads/2012/06/Persimmon-Pest-Management-in-the-San-Joaquin-Valley.pdf

Das S, Shillington L and Hammett T, 2001. Fact Sheet 13: Persimmon. Non-Timber Forest Products. Retrieved Feb. 16, 2012: http: //www.sfp.forprod.vt.edu/factsheets/persimmon.pdf

Farrar K, 1999. Crop Profile for Persimmons in California. Retrieved on March 5, 2013: http: //www.ipmcenters.org/cropprofiles/docs/capersimmons.pdf

Hartman HT, Kester DE, Davies FT and Geneve RL, 2010. Hartmann and Kester's Plant Propagation: Principles and Practices. 8th ed. Chapter 19 Propagation Methods and Rootstocks for Fruit and Nut Species. p. 728-29. Retrieved Oct. 20, 2011: http: //aggie-horticulture.tamu.edu/faculty/davies/index.html

LaRue JH, Opitz KW and Beutel JA, 1982. Growing persimmons in California. Division of Agricultural sciences University of California leaflet 21277.

Lerner B Rosie and Michael N Dana, 2001. New Plants From Layering. Purdue Univ. Coop. Ext. Service. Hort. Publ. HO-1-W. 3pp.

Morton J, 1987. Japanese Persimmon in: J. Morton, Fruits of Warm Climates. Miami, FL: 411-416. Retrieved on Feb. 14, 2012: http: //www.hort.purdue.edu/newcrop/morton/Japanese _persimmon.html

Ryugo K, Schroeder C, Sugiura A and Yonemori K, 1988. Persimmons for California. California Agriculture July-Aug. p. 7-9.

Tetsumura T, Tao R and Yukinaga H, 1999. Orchard growth, flowering and fruiting of micro-propagated Japanese persimmon trees. *J. Horticultural Science and Biotechnology,* **65**: 121-130.

Yamada M, Giordani E and Yonemori K, 2012. Persimmon.pp 663-693 In: Fruit Breeding. Badenes M.L. and Byrne D. (eds). Handbook of plant breeding.ISBN 978-1-4419-0762-2. Springer New York Dordrecht Heidelberg London.

Chapter 5

Orchard and Canopy Management in Unexploited Minor Temperate Fruits

A.S. Sundouri, Shivali Sharma, Suraya Hussan, Rameet Kour and Amit Kumar

Division of Fruit Science, Faculty of Horticulture, SKUAST-Kashmir, Shalimar, Srinagar–190025, J&K

Orchard and canopy management pertains to the careful management of the orchard soil in such a way that the soil is maintained in a good condition suitable to the needs of the tree with least expenses. This involves maintenance of the physical condition of the soil, its moisture and nutrient content. A good system of orchard cultivation should ensure weed control and saving in moisture and nutrients, little disturbance to soil and preventing soil erosion and reduced cost of cultivation, whereas, Orchard management refers to the management of the orchard soil in such a manner that the fruit trees give higher yield of quality fruits in successive years for sustainable economic returns. Management practices for orchard management should be executed in a timely manner during the lifetime of the orchard. The decision about how the orchard will be managed should always be taken before an orchard is planted. An efficient orchard management program always ensures higher return to the grower.

There are several systems available for managing the orchard, each having its own advantages and disadvantages which are to be consider before choosing a particular system for an orchard. However, no single management practice can be recommended for all orchards. A particular program is followed depending on factors like climate, location of orchard, topography, tree spacing, planting system/ orchard design *etc.* Therefore, the system should be combined with cultivation practices like pruning, nutrient and water management, provision for pollination, fruit thinning and disease and pest management in order to obtain desirable tree growth and vigour with the aim to promote precocity with high productivity having objectives:

1. To provide optimum soil moisture during the critical plant growth stages.
2. Prevent or reduce soil erosion.
3. Increase organic matter and improve nutrient status of soil.
4. Control weeds and strengthen the plant growth.
6. Improve soil structure and loosen it for good aeration and water percolation.
7. Enhance and improve microbial activity

Orchard Management

Soil is the basis for the growth of fruit trees and is a basic component of the orchard ecosystem, which directly provides water and mineral nutrients for fruit trees. The physical and chemical properties of soil affect the abilities of the soil to fix fertilizer and of the trees to absorb nutrients. The relatively stable and healthy development of the soil environment is conducive to the growth and development of fruit trees and the sustainability of orchards (Li *et al.,* 2019).

Clean Cultivation

Clean cultivation is the most widely used soil management method used in orchards globally. However, there is a general lack of effective soil management measures in fruit orchards. The disadvantages of clean cultivation are being gradually exposed after the long-term management mode of clean cultivation. Clean cultivation causes a series of ecological and environmental problems, such as the decline in soil quality, the decrease in fertility, and the aggravation of pests and diseases, and seriously affects the yield, quality, and economic and social benefits of an orchard. The clean culture which is extensively followed and involves regular ploughing and removal of weeds, bushes and thachis leads to following disadvantages.

1. Humus will be completely depleted rapidly due to frequent cultivation.
2. Frequent cultivation causes injury to the feeding roots, the trees may be short lived or stunted in growth.
3. Clean cultivation aids in more aeration leading to the depletion of nitrogen.
4. Hard pan is created in the soil.
5. Frequent cultivation causes more soil erosion.

The above mentioned imperfection in clean cultivation can be minimized by avoiding deep and frequent cultivation in minor fruit crops.

Figure 5.1: Clean Cultivation in Fruit Crops.

Clean Culture with Cover Crops

Soil management involves raising of a cover crop or green manure after removing the weeds. If clean cultivation is attempted during the rains, considerable erosion is almost sure to occur. It is probably best to plant a green manure crop between the trees early in the spring and plough it into the soil towards maturity. In minor fruit trees, green manure crops like Sunhemp, Cowpea, Daincha, Lupins *etc.* are more commonly used. Legume cover cropping in loquat, persimmon, fig, and other fruit crops is becoming a common practice in the management of orchards. Cowpea and French beans grow well under loquat and fig tree.

Figure 5.2: Crops grown for Green Manuring.

In some places to prevent soil erosion, certain permanent cover crops like *Calapogonium muconoides, Centrosema pubescens* and *Peuraria phaseoloides* are raised in the alley spaces. They are leguminous crops, establish in a short period, dry up during summer to conserve moisture. With summer showers they come up again because of their profuse seeding habit and spread themselves as a vegetative mat by the time the heavy monsoon starts pouring in (Jiao *et al.,* 2017)

Inter-row planting of legume can improve the soil micro-ecological environment of orchards and increase the soil organic matter and nutrient content of orchards. After grass cultivation, the contents of total nitrogen and total phosphorus in the 0–40 cm soil layers of the orchard were significantly higher than those of the clean cultivation control (Huo *et al.,* 2011). Moreover, natural grass between rows in orchards has a significant effect on soil organic matter and a large number of nutrient elements (Wu *et al*; 2013). Grass cultivation reduced the nitrogen content of leaves, but increased the phosphorus content of leaves (Yang *et al.,* 2020).

Mulching

Mulching is the process or practice of covering the soil/ground in order to improve plant growth, development, and crop production efficiency. Mulch is a scientific word that means "soil coating." Mulch is most likely derived from the German word "molsch," which means "soft to rot," and apparently refers to gardeners spreading straw and leaves over the ground as mulch (Jacks *et al.,* 1955). Mulching prevents runoff and soil degradation, decreases weed infestation, and controls water evaporation, both of which help to prevent soil deterioration. It improves the physical, chemical, and biological properties of soil, as well as the growth and yield, by retaining soil moisture and assisting in the regulation of temperature fluctuations. It also adds nutrients to the soil (Patil *et al.,* 2013). Mulches have been found to be helpful in increasing water retention in soil and its availability by reducing evaporation losses and alleviating the harmful effects of water stress on soil and plants (Singh *et al.,* 2010). Mulches not only preserve soil moisture but also have a variety of other benefits, such as reducing water loss by evaporation and suppressing severe temperature fluctuations, resulting in more retained soil moisture and soil fertility maintenance. The main objectives of mulching are to prevent water loss by evaporation, to prevent soil erosion, to control weeds, to minimize fertilizer leaching, to encourage soil fertility, and to improve the yield and quality fruit crops. As a result, mulching is beneficial to the preservation of our underground resources. For sustainable crop production, water resources, soil, and the environment must all be considered (Kaur *et al.,* 2017). There are two types of mulches basically organic which can be decomposed later and enhance the soil properties and inorganic which cannot be decomposed but can be used for more period of time. Organic mulches include hay, straw, tree leaves, bark shreds, saw dust, compost, biodegradable paper, manure, groundnut hulls *etc.* They have the ability to easily decay due to the presence of insects, slugs, and cutworms that eat them, allowing them to degrade quickly while also adding organic matter and nutrients to the soil. Inorganic mulches are rocks and gravel, polyethylene, LDPE,

HDPE, black polyethylene mulch, crushed stone, marble chips, pebbles, ground rubber *etc.* They perform various actions like weed control, moisture conservation, soil conservation, soil temperature, soil solarization (with transparent plastic mulch controls disease pest) *etc.*

Mulch helps to provide essential nutrients needed for good plant growth. Mulches keeps the fruit clean, protect plants from the cold when applied in the fall, provide a clean and pleasant surface for caring for plants and harvesting fruit. Mulching can be done at almost any time of the year. The greatest effect for improving plant growth is achieved when mulch is applied in May-June after the soil starts to warm up. Once mulch starts to break down, nitrogen, potassium and some minor elements are made available for plant growth. Mulching for winter protection should be done in November, after several frosts have occurred. Organic matter such as straw, hay, wood chips, ground bark, sawdust, leaves, grass clippings, and pine needles are common mulching materials. These should be applied to a depth of 2 to 6 inches and cover the ground around a plant out to the drip line. Black plastic is often used in vegetable production. Mulch provides a very favorable environment for plants to grow in, plants may grow late into the fall, making them susceptible to winter injury. Mulch also provides an almost perfect environment for mice to live in. Damage caused by mice feeding on roots and bark is a common occurrence on mulched plants. To help counteract these problems, much should be pulled away from trees in late August and starting September. Tree growth will be slowed and mice frequently move to other areas where they can be protected. Mulch having a high carbon to nitrogen ratio (sawdust, wood chips) may cause nitrogen deficiency on plants. Bacteria that break these materials down use nitrogen, thus depriving plants of this vital element. The best solution is to compost these materials until they start to break down. This may require a year or two for sawdust. Alternatively, make supplemental applications of a water soluble fertilizer high in nitrogen (Li, 2013)

Figure 5.3: Use of Weed Mat.

Sod

In this method, permanent cover of grass is raised in the orchard and no tillage is given. This type of orchard cultivation is followed in USA and Europe in different kinds of temperate fruits. Orchard grass is an effective method used to maintain soil fertility and increase fertilizer fixing and yield in orchards. It can create a good micro-environment for fruit trees, increase the soil organic matter and nutrients, improve the soil quality in orchards, repair the ecological environment of orchards, improve the fruit yield and quality, improve the economic benefits of planting, and solve the problem of harm caused by long-term clean cultivation (Bal *et al.,* 2015). This may be useful in slopy lands for preventing soil erosion. But they compete for soil moisture and available nitrogen. The drawbacks of this system are the need for increased manuring and water application. Hence sod may be useful with deep rooted trees because soil moisture will be very low on the top layers however they are harmful to shallow rooted trees.

Sod Mulch

This is similar to sod with the only difference is that the vegetation is cut frequently and the cut material is allowed to remain on the ground. This is slightly better than the previous one, as the moisture loss is not so great as in sod. In both sod and sod mulch, more nitrogen should be applied to the fruit trees than usual application because the vegetation utilises more soil nitrogen.

Intercropping

In young orchards, the question of how best one can use the soil between the trees with long juvenile period. If the trees are properly spaced there is considerable land which will not be used by the permanent trees for several years. It naturally appeals to the grower to get some return from this vacant land especially when he is getting no return in the early periods. The practice of growing any economic crop in alley spaces of the fruit trees in the first few years or in the' unoccupied spaces of the long duration crop in the early periods is referred as intercropping. Intercropping is the practice of growing short-term crops, like vegetables, in the early stages of establishing perennial fruit trees. It can be beneficial for fruit crops, and can increase productivity and income also can have a positive impact on the overall performance of plants, but this may depend on the season in which the cropping is done. Intercropping can also have other benefits, such as: improved soil quality, increased biodiversity, weed control, and improved product quality. The following important principles should be observed while growing intercrops.

1. Intercrops should not occupy the area where the roots of the fruit trees are concentrated.
2. Soil fertility should be maintained or improved when intercrops are grown.
3. Water requirements of the intercrops should not clash with those of the main fruit trees. The intercrop may require an irrigation at a time when it would be detrimental to the trees.
4. Intercrops should be selected with reference to their effect on soil moisture. Grain crops remove excessive moisture to the detriment of fruit trees. The intercrops selected should not exhaust the soil water and

nutrients and should not demand more water than is allowed for fruit trees.

Vegetables are the best inter crops when compared to any other field crops. But whatever may be the intercrop grown, it should be kept well away from the main fruit trees and irrigated independently. The intercropping should be stopped when trees occupy the entire orchard space. Thereafter, green manuring or cover cropping should be only practiced.

Many growers prefer some quick growing fruit trees to grow as intercrops. The satisfactory fruits are available for this purpose. In temperate minor fruits crop regions peaches, nectarine and olives are often grown between trees. Similarly, in properly spaced orchard, trees can be planted to bear in two or three years and will produce a number *of* crops before it is necessary to remove them. Such short-lived trees are known as 'fillers'. The danger in using fillers is when they are allowed to remain in the orchard for too long periods. As normally root system makes a faster growth than the branches, the roots *of* the permanent trees come in contact with the roots of the fillers before there is any crowding above ground. Therefore, the fillers should be removed after a few years usually immediately after the main fruit trees have commenced bearing.

Organic Farming

It relies on fertilizers of organic origin such as compost, green leaf manure, FYM, bone meal, *etc.*, and emphasis on techniques such as crop rotation, companion cropping/planting, biological pest control, mixed cropping, and the fostering of insect predators and/or the product developed by organic method. In general, organic farming is designed to allow the use of naturally occurring substances while prohibiting or strictly limiting synthetics. Unexploited fruit crops such as Loquat, Persimmon, Fig, Chestnut, Pomegranate *etc.*, are generally cultivated through a natural farming method by poor farmers, but they can also be cultivated successfully in large scale by adopting the eco-farming/natural farming fruit cultivation methods. Organic farming methods combine scientific knowledge of ecology and modern technology with traditional farming practices based on naturally occurring biological processes. These unexploited fruit crops are by default cultivated in natural habitat under the temperate conditions of India and the system is known as natural farming which has following essentialities over the other system of cultivation such as:

1. National Farming (NF) may have beneficial effect on soil life as they do supply humus and don't have an adverse effect on physical, chemical and biological properties of soil.
2. No cost involve.
3. Natural Farming helps in maintaining the available natural flora and fauna of the particular region.

4. The continuous use of pesticides is ecologically unsustainable as pests acquire resistance.
5. There is a premium for the horticultural produces which are raised under natural farming.

Canopy and its Management

The cultivation of minor fruit species in temperate regions presents unique challenges and opportunities. While traditional crops such as apples and cherries dominate the temperate landscape, there exists a diverse array of unexploited minor fruits such as loquat, hazelnut, pecan nut, chestnut, persimmon and fig —that hold potential for both commercial viability and ecological benefits. The cultivation of these temperate minor fruits presents an exciting opportunity for diversification in fruit production. While these crops are often overshadowed by more widely cultivated species, they offer unique flavors, nutritional benefits, and ecological advantages. Effective canopy management like training and pruning are fundamental horticultural practices that shape the growth of trees, improve their productivity, and ensure tree health. While the general principles of pruning and training apply across many fruit and nut trees, each species has its own unique requirements based on its growth habit, fruiting patterns, and environmental adaptations. In this chapter, we will discuss the specifics of canopy management of temperate minor fruit crops *viz.*, loquat (*Eriobotrya japonica*), hazelnut (*Corylus avellana*), pecan (*Carya illinoinensis*), persimmon (*Diospyros spp.*) and fig (*Ficus carica*) providing insights into how these practices contribute to optimal growth, fruit production, and disease management.

This chapter delves into the principles of canopy management specifically tailored to temperate unexploited minor fruits, highlighting practices that can enhance their productivity while maintaining ecological integrity.

Principles of Canopy Management

- ✰ Optimal utilization of light.
- ✰ Avoidance of microclimates conducive to disease and pest proliferation.
- ✰ Ease in executing cultural traditions.
- ✰ Optimal efficiency with high-quality output.
- ✰ Efficiency in acquiring the necessary canopy structure.

Understanding Canopy Dynamics

Growth Habits of Minor Fruits

The canopy of fruit-bearing plants comprises various layers that can influence light interception, air circulation, and microclimate conditions. In minor fruits, understanding the unique growth habits—whether they are shrub-like or small tree forms—is essential for developing appropriate management strategies. Each of the selected minor fruits has distinct growth patterns and canopy structures. Understanding these characteristics is essential for effective management:

Loquat (*Eriobotrya japonica*)

A fruit tree that typically grows as a small to medium-sized tree. The canopy is generally rounded and dense, with a broad spread. Loquat produces clusters of small, yellow to orange fruit that grow on racemes. The fruiting typically occurs in late spring to early summer, after flowering in the fall and winter. The branches grow outwards and upwards, creating a full, bushy appearance. However, if left unmanaged, they can become crowded, leading to reduced fruit quality. Regular pruning can help manage its height and shape, promote better fruit production, and improve overall health by removing dead or overcrowded branches.

Overall, the loquat's growth habit, characterized by its evergreen foliage, rounded canopy, and fruiting clusters, makes it a unique and valuable addition to temperate gardens and orchards.

Hazelnut (*Corylus avellana*)

A multi-stemmed shrub or small tree that can be managed for both nut production and ornamental value. They have a bushy appearance, with a dense canopy that can reach heights of 10 to 20 feet (3 to 6 meters) and tend to sucker freely from the base. The branching pattern is critical for light penetration and airflow. They produce nuts on catkins that develop in late summer and bloom in early spring. The canopy structure influences light interception, which is vital for photosynthesis and fruit development. Proper management ensures that lower leaves receive adequate light, enhancing overall productivity. The challenge with hazelnuts is managing the suckers while ensuring productive shoots receive adequate sunlight and air.

Pecan (*Carya illinoinensis*)

Pecan trees are large, deciduous trees that grow slowly in their early years but can become quite vigorous as they mature. They have an upright growth habit and produce nuts on wood that is one year or older. Pecans usually have a single central leader with well-developed lateral branches. The branches tend to grow horizontally, creating a dense canopy that can provide significant shade. Pecan trees are monoecious, meaning they have separate male and female flowers on the same tree. The male flowers (catkins) bloom in the spring, while the female flowers are small and appear at the ends of new shoots.

Chestnut (*Castanea* spp.)

Chestnut is a tree that can grow tall with a wide canopy, producing large nuts and requiring thoughtful pruning for optimal light exposure. Chestnut trees typically grow as large deciduous trees, reaching heights of 30 to 100 feet (9 to 30 meters). They often have a straight trunk and a broad, rounded canopy, which can become quite expansive as they mature. Chestnuts usually exhibit a well-defined central leader with horizontal branching that creates a dense canopy. The branches can be quite sturdy, allowing for heavy fruit loads. Chestnut trees are monoecious, meaning they have both male and female flowers on the same tree. Male flowers appear as long catkins, while female flowers are smaller and found at the tips of shoots.

Persimmon (*Diospyros* spp.)

Persimmons trees can vary from small trees to larger specimens, with a dense canopy that can benefit from management to enhance fruit visibility and ripening. Persimmons are deciduous trees that can vary in size from small to large, depending on the species and cultivar. They typically have an upright, spreading habit and bear fruits on the current season's growth.

Fig (*Ficus carica*)

Fig trees typically grow as small to medium-sized deciduous trees or large shrubs, reaching heights of 10 to 30 feet (3 to 9 meters). They have a wide, spreading canopy that can become quite bushy with age. Figs develop a strong central leader, though multiple trunks can also be common, especially in older trees. The branches tend to grow horizontally, allowing for an open canopy structure that supports fruit development. Figs are unique in that they produce flowers inside their fruit, known as syconium. The flowers are pollinated by tiny wasps in some varieties, while others are self-pollinating. Figs typically flower in the spring and may produce a second crop in late summer or early fall, depending on the climate and variety.

Light Penetration and Photosynthesis

Light is a critical component in photosynthesis, and the ability of lower leaves to access sunlight directly affects fruit development. In minor fruit species, where the growth habit may result in dense foliage, managing the canopy to enhance light penetration can significantly improve yields.

Canopy Structure and Function

The structure of a tree's canopy plays a critical role in photosynthesis, nutrient allocation, and fruit development. Effective canopy management ensures that lower leaves receive adequate light while promoting airflow to reduce disease pressure.

Training Systems

Benefits of Training

Training trees to achieve an optimal canopy structure can enhance light exposure, improve air circulation, and facilitate harvesting.

Training Techniques

Loquat

A central leader system can be beneficial for loquat trees, promoting upward growth while allowing for the removal of competing branches. Loquats are typically trained to a central leader or modified central leader system during their early years. The goal is to develop a well-balanced scaffold of branches that allows light penetration and air circulation, which reduces disease pressure and promotes uniform fruit ripening.

- **Central Leader System**: Start by selecting a strong central shoot and pruning the young tree back to encourage the growth of lateral branches. Maintain a dominant central leader while promoting scaffold branches spaced evenly around the trunk.
- **Modified central leader**: This system is a variation of the central leader, where the leader is cut back to reduce its dominance and encourage more spreading of the scaffold branches.

Hazelnut

Training as a multi-stemmed shrub can help manage height and improve accessibility for harvesting. Hazelnuts are often trained either as a single-stemmed small tree or as a multi-stemmed bush, depending on the grower's preference and the cultivar.

- **Bush System**: This is the most common training method for hazelnuts. In this system, multiple stems are allowed to grow from the base, and the plant takes on a natural, bush-like form. To maintain health and productivity, the bush should be kept open in the center, with older stems regularly removed to promote new, productive growth.
- **Tree System**: This system involves training hazelnuts to a single trunk. Once the central leader is established, scaffold branches are selected and the tree is pruned to develop an open canopy. This method facilitates easier harvesting but requires diligent removal of suckers.

Pecan

Use a modified central leader approach to encourage vertical growth and manage lateral branches, facilitating better nut exposure.Pecans are typically trained to a strong central leader system in order to create a tall, upright tree that can support a large canopy for nut production.

- **Central leader system**: Train the tree to have a dominant central leader with well-spaced lateral branches. In the first few years, encourage the development of 5-7 strong scaffold branches at wide angles to the trunk. This will create a strong structure capable of supporting the weight of nuts and withstand wind stress.

Chestnut

Train young trees to develop a strong central leader with well-spaced lateral branches to create a balanced canopy.

- **Central leader system:** Encourage a single, dominant central leader that will serve as the main trunk. Remove competing leaders to maintain this structure. Scaffold branches should be evenly spaced around the trunk.
- **Open center system:** Maintain an open center by pruning back any inward-growing branches. This encourages an airy canopy, reducing disease pressure and promoting healthy fruit development.

Persimmon

An open center training system can be effective, promoting multiple scaffold branches that allow light to penetrate deep into the canopy. Persimmons are often trained to an open-center (vase-shaped) system or a modified central leader system to encourage sunlight penetration and airflow.

- **Open center system**: This system promotes a wide, open canopy that facilitates light penetration, essential for fruit quality and ripening. After the first year, select 3-5 strong branches to form the primary scaffolds, removing the central leader to keep the center open.
- **Modified central leader**: Similar to loquats, a modified central leader system can be used for persimmons, where the tree is allowed to grow with a slightly dominant central leader and well-spaced scaffold branches.

Fig

Training fig trees is essential for maximizing fruit production, improving air circulation, and facilitating harvesting.

- **Open center (Vase) training**: Develop a wide, open canopy that allows sunlight to reach all parts of the tree. Remove inward-growing branches to maintain this shape.
- **Single leader system:** Encourage a single central leader for a more traditional structure. This helps with tree stability and makes harvesting easier.
- **Espalier and fan training**: For limited spaces, train the tree to grow flat against a wall or trellis. This method maximizes sunlight exposure and fruit production while maintaining an attractive appearance.

Pruning

Objectives of pruning: It aims to:

- Improve light penetration.
- Enhance air circulation to reduce disease risk.
- Promote healthy growth and fruiting.

Types of Pruning

- **Renewal pruning:** Cut back older stems to encourage new shoot growth. This helps maintain productivity and vigor.
- **Thinning:** Remove crowded branches to open the canopy. This increases light exposure for lower leaves and improves airflow.
- **Heading back:** Shorten long branches to encourage branching and improve fruiting potential.

Importance of Pruning

It is a fundamental aspect of canopy management, helping to shape the tree, improve light penetration, and enhance fruit quality. Each species requires specific pruning techniques tailored to its growth habit.

Timing and Techniques of Pruning

The timing of pruning is crucial for maximizing the effectiveness of these techniques. Late winter or early spring, before bud break, is generally ideal for most of these species to avoid disrupting flowering and fruit set.

Loquat

Prune in late winter to remove dead or crowded branches. Focus on opening the center of the tree to allow light to reach inner leaves.

Pruning Techniques for Loquat

- **Initial Pruning (First 3 Years)**: During the first two years, focus on shaping the tree by selecting 3-5 primary scaffold branches spaced evenly around the trunk. Prune off any competing shoots to prevent crowding.
- **Maintenance pruning**: After the third year, prune annually to remove any dead, damaged, or crossing branches. Thin out crowded areas to allow light penetration and remove suckers from the base. Cut back vigorous shoots to control tree height and encourage fruit-bearing wood.
- **Fruit-bearing pruning**: Since loquats fruit on new wood, it is important to prune after fruit harvest to stimulate new growth for the next season's fruiting. Avoid heavy pruning in late summer or fall, as this can promote late-season growth that is vulnerable to frost.

Hazelnut

Use a renewal pruning method to manage growth, encouraging new shoots while maintaining an open canopy for light penetration and air circulation.

Pruning Techniques for Hazelnut

- **Initial Pruning**: In the first few years, prune to establish the desired form, whether a bush or tree. Remove excess suckers regularly.

- **Maintenance Pruning**: Once the bush or tree is established, prune in late winter or early spring to remove dead or weak wood, as well as any branches that are crossing or shading productive shoots. In the bush system, remove about one-third of the oldest stems at ground level every year to promote vigorous new growth. In the tree system, thin out dense growth to maintain air circulation.
- **Sucker Control**: Hazelnuts are prone to suckering, which can reduce productivity if not managed. Regularly remove suckers to ensure that the plant's energy is directed towards nut production.

Pecan Nut

Implement a modified central leader system to promote a strong, upright growth while managing lateral branches for better nut distribution and access.

Pruning Techniques for Pecan Nut

- **Initial pruning**: In the first few years, prune pecan trees lightly to develop a strong central leader and well-spaced scaffolds. Remove any competing leaders or weak branches. Avoid heavy pruning, as this can slow down the establishment phase.
- **Maintenance pruning**: Mature pecan trees do not require extensive pruning but should be pruned to remove dead, diseased, or broken limbs. If necessary, thin out branches in the canopy to improve light penetration and reduce wind resistance. Pecans tend to produce a large number of small, unproductive shoots (known as "water sprouts"); these should be removed regularly.
- **Height control**: In orchards, pecans may need to be topped to keep the trees at a manageable height for harvesting. This should be done cautiously and incrementally, as heavy topping can reduce yield.

Chestnut

Prune to maintain a balanced structure, removing lower branches to allow for upward growth and maximizing sunlight exposure.

Pruning Techniques for Chestnut

- **Maintenance Pruning:** Remove dead, damaged, or diseased branches. Inspect the tree regularly. Cut back any unhealthy branches to healthy wood or remove them entirely to improve overall tree health.
- **Thinning:** Improve air circulation and light penetration. Selectively remove crowded or competing branches. Aim to create a balanced structure with well-spaced branches, enhancing airflow and reducing humidity in the canopy.

- ☆ **Heading Back:** Control height and encourage lateral branching. Cut back long, vigorous branches to a desired length, just above a lateral bud. This stimulates side shoot growth, creating a fuller canopy.

Persimmon

Prune during dormancy to manage tree height and encourage a more open canopy that allows light to reach developing fruit.

Pruning Techniques for Persimmon

- ☆ **Initial pruning**: In the first few years, focus on establishing a strong structure by selecting 3-5 main scaffold branches. Remove any weak or competing branches.
- ☆ **Maintenance pruning**: Once the tree is mature, prune annually to remove dead, diseased, or damaged wood, as well as any branches that cross or crowd the canopy. Persimmons benefit from light thinning of the interior branches to improve air circulation and reduce disease pressure.
- ☆ **Fruit-bearing pruning**: Since persimmons fruit on new growth, it is important to prune after harvest to encourage the development of new fruiting wood. Remove overly vigorous shoots and shorten lateral branches to control size and encourage fruiting.

Fig

Prune during the dormant season (late winter to early spring) before new growth begins to minimize stress on the tree and reduce disease risk.

Pruning Techniques for Fig

- ☆ **Maintenance Pruning**: Remove dead, damaged, or diseased branches to improve overall tree health.
- ☆ **Thinning**: Selectively remove crowded branches to increase light penetration and air circulation. Aim for an open center to reduce humidity and disease pressure.
- ☆ **Heading Back**: Shorten long branches to promote lateral growth, which can lead to increased fruiting.

Conclusion

Orchard and Canopy management are the critical component of successful cultivation for temperate minor fruits like loquat, hazelnut, pecan, chestnut, persimmon and fig. By understanding the unique growth habits and requirements of these species, growers can implement targeted pruning, training, nutrient management, and pest control strategies. As interest in diverse and nutritious fruits continues to grow, effective orchard and canopy management will play a pivotal role in transforming these minor fruits into valuable and sustainable crops.

Continued innovation in management practices will further enhance the potential of these underutilized species, contributing to agricultural diversity and resilience.

REFERENCES

Bal L, Chen X, Wang Z, Lv DG and Zhao B, 2015. Effects of alfalfa and Elymus sibiricus on soil nitrogen mineralization in apple orchards. *Hortic. J.,* **42**: 2469–2477.

Huo Y, Zhang J, Wang MC and Yao YC, 2011. Effects of intercropping grass on soil organic matter and mineral elements and their relationships in pear orchard. *China Agric. Sci.,* **44**: 1415–1424.

Jacks GV, Brind WD and Smith R, 1955. Mulching. Tech. Comm. No. 49. *Commonwealth Bull. of Soil Sci.*

Jiang LL, Gong QT, Wu HB, Sheng FJ and Sun RH, 2019. Effects of different grass treatments on soil microbial communities in apple orchards. *Appl. Ecol.,* **30**: 3482–3490.

Jiao RA, Zhang S.H, Li Y, Li CZ, Wang JP and Jiao J, 2017. Research progress on the effect of grass on the growth and development of fruit trees and orchard environment. *J. Fruit Trees,* **34**: 1610–1623.

Kaur J and Bons HK, 2017. Mulching: A viable option to increase productivity of field and fruit crops. *Journal of Applied and Natural Science,* **9**(2): 974-982.

Li SW, Yang WQ, Zhao R, Han MY and Kou JC, 2016. Effects of intercropping grass on soil fertility in apple orchards. *Grassl. J.,* **24**: 895–900.

Li TT, 2013. Effects of Long-Term Different Fertilization on Soil Fertility and Fruit Tree Growth and Development in Apple Orchard under Grass Coverage. Master's thesis, Northwest University of Agriculture and Forestry, Xianyang, China.

Li XY, Ren LH, Ji BQ and Ma XT, 2019. Effects of grass coverage on soil physical properties in mountain orchards. *Appl. Technol. Soil Water Conserv* **190**: 7–9.

Patil PS, Kelkar TS and Bhaleroa SA, 2013. Mulching: A Soil and Water Conservation Practice. *Research Journal of Agriculture and Forestry Sciences,* **1**(3): 26-29.

Singh AK, Singh S and Rao VV 2010. Meshram, D.T. Effect of mulching on soil properties, growth and yield of NA 7 aonla (*Emblica officinalis*) in semi-arid ecosystem. **78**(3): 193-197.

Wu YS, Zhang YM, Ji XH, Zhang R, Liu DL, Zhang ZY, Li WY and Chen XS, 2013. Effects of natural grass on soil nutrients, enzyme activities and fruit quality in pear orchards in the Yellow River Delta. *Chin. Agric. Sci.,* **46**: 99–108.

Yang L, Mao YF, Hu YL, Wang YY, Zhang LL, Yin YJ, Pang HL, Su XF, Liu YP and Shen X, 2020. The effect of grass on improving soil fertility and apple tree nutrition in orchards. *Plant Nutr. Fertil. J.,* **26**: 325–337.

Chapter 6

Propagation, Orchard and Canopy Management in High Altitude/Cold Dessert Fruit Crops

Amit Kumar, Saklain Mulani, Devesh Attri, Bhavesh and Rajmeet Singh

Division of Fruit Science, SKUAST-Kashmir, Srinagar, J&K

A cold desert, any large arid area of land is a unique type of desert characterized by frigid temperatures, particularly in winter, and low precipitation, often in the form of snow. Unlike hot deserts, which are known for scorching temperatures, cold deserts experience long, harsh winters with temperatures frequently dropping below freezing, while summers can be relatively mild. Cold deserts are found typically in temperate regions at high elevations (such as the Gobi Desert in Mongolia), on plateaus, or in mountainous areas; however, they also occur in polar regions like Antarctica. Life in cold deserts is sparse but resilient, with plants and animals that have adapted to survive extreme temperature fluctuations, strong winds, and minimal water. Despite the challenges, cold deserts play a vital role in the planet's biodiversity and climate systems.

In India cold desert is primarily located in the high-altitude regions of Ladakh and Spiti Valley of Himachal Pradesh.This cold desert of the Himalayan region spans 186,200 square kilometers, where the landscape is marked by barren, rugged terrain and extreme climatic conditions. These areas experience long, severe winters with temperatures plunging below -30°C, while summersare generally harsh due to intense sunlight at high altitudes. Precipitation is minimal, mostly occurring as snow, and water is a scarce resource. The vegetation is sparse, consisting of hardy shrubs and grasses adapted to the cold and arid environment.

Union Territory of Ladakh lies in the northernmost area of India, extending from 32°152 to 36° N latitude and from 75°152 to 80°152 E longitude. The total area is about 98,000 square kilometers. It is bordered on the north by the eastern part of the Karakoram Mountains and on the south by the western edge of the main Himalayas. The elevation varies from approximately 3,000 meters in the lower Indus and Nubra valleys to 7,600 meters in the Zanskar and Karakoram ranges (Kala *et al.,* 2011). In Ladakh, different ranges of soils are found like;sandy loam in nature with low organic carbon and nitrogen content. The soils in this region have low fertility and are susceptible to erosion (Gupta *et al.,* 2016). The high levels of sunlight in the area enhance the distinctive taste and color of fruits of this area, making the region renowned for its quality produce (Banoo *et al.,* 2020). Propagation, orchard, and canopy management are essentialpractices for successfully cultivating fruit crops in high-altitude or cold desert regions. In these harsh environments, where low temperatures, short growing seasons, and limited water availability present significant challenges, proper orchard planning and canopy management are crucial for optimizing fruit production. Propagation involves selecting cold-hardy rootstocks and using grafting techniques to establish resilient fruit trees. Orchard management focuses on efficient use of space, soil conservation, and protecting trees from extreme climatic conditions, while canopy management ensures that the trees' growth is controlled to maximize sunlight exposure, improve air circulation, and enhance fruit quality. Together, these practices help overcome the environmental limitations of high-altitude cold deserts and improve crop productivity and sustainability. The region's unique environment, provides an ideal climate for apricot cultivation. Apricots (*Prunus armeniaca L.*) are the most significant fruit, followed by apples as the principal fruit crop in Ladakh. Grapes, almond, walnut, cherries, peach, prunes, pear, chilgoza, plum are also grown in these cold arid areas but not on a commercial scale. Seabuckthorn, a shrub is a most important crops of the cold arid region called asgold mine of cold desertswhich has so many health benefits.

In Ladakh, apricots (*Prunus armeniaca* L.) are the most significant fruit, followed by apples as the principal fruit crop. The high levels of sunlight in the area enhance the distinctive taste and color of these fruits, making the region renowned for its quality produce. In Ladakh, apricots (*Prunus armeniaca L.*) are the most significant fruit, followed by apples as the principal fruit crop. The high levels of sunlight in the area enhance the distinctive taste and color of these fruits, making the region renowned for its quality produce. In Ladakh, apricots (*Prunus armeniaca L.*) are the most significant fruit, followed by apples as the principal fruit crop. The high levels of sunlight in the area enhance the distinctive taste and color of these fruits, making the region renowned for its quality produce.

Apricot "*Prunus armeniaca* L." is considered as the most significant fruit yield of Ladakh followed by Apple being the principal fruit crops grown to grow a special taste and colour as a result of high intensity of sunlight the area receives, and the region is well known for its quality fruits

Apricot

Grown in the high-altitude, cold desert climate, Ladakhi apricots are known for their unique sweetness, flavor, and high nutritional value. These apricots thrive in the harsh environment due to their adaptability to the extreme cold, dry conditions, and short growing season. Ladakh region comprises of two districts (Leh and Kargil) in Union territory of Ladakh and extends over a geographical area of 96,701 km^2 accounting for 43 per cent area of state and 75 per cent of cold arid regions of India. It is situated between 34.5° to 34.7° North latitude, 76.2° to 76.5° East longitude and at an altitude of 2,500-2,750 meters amsl. Apricot (*Prunus armeniaca* L.) is widely spread over the whole Ladakh with abundance in district Kargil and lower belts of district Leh. Lower belts include the area from Saspol to Batalik, Nubra Valley and larger portion of district Kargil. Apricots are a major part of Ladakh's cultural and economic landscape, with the fruit being used for local consumption, dried for trade, and processed into oils and other products. The review of the area and the production of the region is becoming important from the aspect of knowing the present status and the future scope as the limited studies were carried out in this region (Ali *et al.*, 2018).

Propagation

The practice of propagation is not only vital for sustaining apricot orchards but also helps preserve indigenous varieties, which are celebrated for their unique flavor and resilience. Apricot or its wild forms are fairly drought resistant, salt tolerant and resistant to pests and diseases compared to other fruit plants. Most of the prevalent local selections are of seedling origin (Sharma *et al.*, 2005). Apricots in Ladakh are traditionally propagated through seed sowing, budding or grafting. Farmers graft apricot scions onto hardy rootstocks adapted to the harsh climatic conditions.

Traditional Methods of Propagation

Seed Propagation

Seed sowing is one of the most common methods to propagate apricot tree. It involves growing new plants from seeds collected from mature apricot fruits. To begin, select healthy and ripe apricots for seed collection. Remove the flesh from the fruit and wash the seeds thoroughly to remove any residue or pulp. Once clean, allow them to dry completely before planting.

Growing apricot seeds in cold dessert region starts with drying the apricot pit. Once dry, carefully crack the pit to extract the seed without damaging it. The seed should then undergo process called stratification, where it's time for germination.

Plant the seed in a pot filled with well-draining soil and keep it moist. Place the pot in a sunny location or under grow lights. As the tree grows, maintain a soil pH of 6-7.5 and provide regular watering, pruning, and fertilization for a healthy apricot tree (Dihar, 2011-2012).

Vegetative Methods of Propagation

Grafting

In cold deserts, apricot propagation can be highly successful using the tongue/cleft grafting method in February, with up to 90 per cent success in open field conditions. Nevertheless, the commercial application of this method is constrained by a significant shortage of graftable scion sticks in the region (Dwivedi *et al.,* 2000).

Budding

BuddbbBB In cold desert regions like Ladakh, there are many mature apricot seedling trees with well-established root systems. However, these trees yield poor-quality, small, non-edible fruits. By top-working these healthy but unproductive trees with promising cultivars, they can be transformed into productive ones. For this, old plants are headed back during March leaving 50 cm stub of side branches from where new shoots can arise. Ten shoots on each tree projecting from the outer periphery should be thinned out. After the shoots attain proper thickness desirable apricot cultivars can be budded in mid-June or early July by adopting shield budding. This generally results in 90-94 per cent success. An orchard on a single tree can be established by budding various cultivars onto different branches of the same tree. This method is an ideal technique for transforming mature, unproductive plants into productive ones (DIhar, 2011-2012).

Raising of Apricot Rootstock

Rootstock is the lower portion of a grafted/budded plant which provide root system to the tree. In many fruits crops rootstock is effectively being used for vigour control, tolerance of biotic stress, fruit quality control, *etc.* However, such investigation has not been done for apricot. Apricot is generally grafted and budded on seedling rootstocks, raised by sowing apricot seeds. Wild apricot (Chulli) and wild peach seedlings are generally used as a rootstock in cold dessert areas. The graft union on wild apricot is good and the trees are more vigorous than on plum and wild peach. One kilogram of apricot seeds contain approximately 700-800 numbers of seeds. Freshly harvested seeds are collected and soaked in water for 8-10 days in the end of February so that the hard seed coat becomes soft. Thereafter, seeds are stratified by putting them between layers of sand in a box/container which is kept in shade and water is applied regularly to keep it sufficiently moist 25-30 days after stratification the seed breaks its dormancy and starts sprouting. These should be lifted from the box and sown in well prepared nursery beds at a distance of 30 cmx 15 cm. In Ladakh first fortnight of April is ideal for sowing of stratified apricot seeds for raising rootstock.

The nursery bed should be given light irrigation to keep soil moist till germination of seedlings. After germination normal cultural practices are adopted for raising of healthy rootstocks. In a properly managed nursery approximately 70-80 per cent stratified seeds germinate and 55-60 percent attains the graftable diameter of pencil thickness by the month of November of the same year. These rootstocks can be used for budding or grafting for commercial propagation of apricot (DIHAR 2011-2012).

Apple

Apple is a traditional fruit crop in the high-altitude trans-Himalayan region of Ladakh, ranking as the second most important fruit after apricot. Several native cultivars are cultivated in the area, with Thra, Mongol, and Karkechu being the most popular (Dolker *et al.,* 2021). These local varieties ripen in August but have poor storage capacity. In contrast, exotic delicious cultivars, which ripen in late October, store well for 4-5 months using traditional storage methods. Due to their late ripening and extended storage life, the demand for delicious cultivars is rising in the region (Angmo *et al.,* 2018). Ladakh offers several natural advantages for apple production, including long daylight hours, high light intensity, and relatively warm days with cool nights and low humidity from May to October, making the region ideal for apple cultivation. However, apples in Ladakh are mostly grown as individual trees or in small clusters using traditional methods, without chemical fertilizers or pesticides. Apple propagation in Ladakh involves techniques adapted to the region's unique high-altitude environment, characterized by cold winters, short growing seasons, and minimal rainfall.

Vegetative Propagation

Grafting

Grafting is the most widely used method of apple propagation in Ladakh. Scions from desirable apple cultivars (like Delicious, Thra, Mongol, or Karkechu) are grafted onto hardy rootstocks (wild apple). Grafting is usually done in late winter or early spring, just before the growing season begins. This ensures that the graft has ample time to heal and establish a connection with the rootstock before the growing season. Tongue and whip grafting methods are commonly practiced. Tongue grafting shows high success rate and are suitable for the climatic conditions in Ladakh and also in Lahaul Spiti area of Himachal Pradesh. (Sharma and Chauhan, 2008).

Budding

In addition to grafting, T-budding, chip budding and ring budding are used for propagating apples in Ladakh. This method involves inserting a bud from a desired apple variety into the rootstock. It is typically done in mid to late summer when the bark separates easily. Chip budding done on 15 March and 31 March were most successfully resulting in highest graft success, scion growth and maximum yield of saleable plants, therefore, it is recommended that second fortnight of March is the best time for vegetative propagation of apple through chip budding in cold arid conditions of Ladakh (Dwivedi and Stobdan, 2009). Budding is a simpler method and is often preferred for producing large numbers of treesin cold dessert areas.

Rootstock Propagation

Seedling Rootstock

Apple rootstocks can be grown from seeds of specific apple varieties, such as crabapple. Seeds are typically sown in spring after stratification, a process in which seeds are exposed to cold conditions to break dormancy and encourage germination.

Clonal Propagation

Vegetative propagation through hardwood cuttings or layering is also used. Hardwood cuttings from established rootstocks are planted in nurseries, where they root and grow into new plants. In layering, branches from existing trees are bent and buried in the soil to induce root formation, after which the rooted sections are separated and planted as individual rootstocks.

Given Ladakh's harsh winters, selecting cold-hardy rootstocks is crucial for the survival of young apple trees. Rootstocks like M9, MM106, and M111 are commonly used as they offer good adaptability to cold climates and provide resistance to various diseases and soil conditions. Drought resistance is also essential due to the region's arid conditions. Rootstocks with good drought tolerance, such as M111, are preferable.

Seabuckthorn

Seabuckthorn is an important crop of cold arid region, known as "Tsermang" in Ladakh, is a hardy shrub that thrives in the region's cold desert climate. It grows well in the harsh, high-altitude environment, making it an important crop for Ladakhi farmers. The plant is valued for its resilience, able to withstand extreme temperatures and poor soils, and it plays a key role in preventing soil erosion along the fragile Himalayan slopes. Beyond its environmental benefits, seabuckthorn is prized for its berries, which are rich in vitamins, antioxidants, and fatty acids, offering significant economic potential through products like juices, cosmetics, and medicines. Its cultivation has become an important livelihood in Ladakh, contributing to both ecological balance and local economies.

Propagation

Seabuckthorn (*Hippophae rhamnoides*) can be propagated through various methods, with the most common being seed propagation and vegetative propagation.

Seed Propagation

The seeds are collected from ripe berries, typically during the late summer or autumn. The seeds need to be cleaned and dried. To enhance germination, seeds can undergo stratification, a process of keeping them in moist, cold conditions (0-5°C) for 1-3 months. Seeds are sown in nurseries or directly in the field in spring or early summer. They require light and should be planted shallowly, about 0.5 to 1 cm deep. Germination takes about 2-4 weeks, depending on conditions. Seedlings should be transplanted after they have developed a strong root system (Husain *et al.,* 2018).

Vegetative Propagation

Stem Cuttings

In late winter or early spring, hardwood cuttings are taken from healthy plants. These are typically 15-20 cm long. They are then planted in a nursery or field where they root and develop into new plants (Dwivedi *et al.,* 2002).

Root Cuttings

Root segments of about 5-10 cm can be used for propagation. These cuttings are buried in the soil and sprout into new shoots.

Layering

In this method, low-growing branches are bent down and covered with soil, allowing them to form roots while still attached to the parent plant. Once rooted, they are separated and transplanted.

Suckers

Sucker propagation is a practical method for expanding Seabuckthorn plantations, especially in regions like Ladakh, where Seabuckthorn thrives naturally (Dwivedi and Ahmad, 2008).

Walnut

The walnut tree (*Juglans regia*), commonly known as Persian or English walnut, is a temperate fruit tree that can adapt well to cold climates, making it suitable for parts of Ladakh. While walnut cultivation is not as widespread as apples or apricots, it is gradually gaining popularity as farmers recognize its economic potential. Walnut propagation in Ladakh relies on both traditional and modern methods. While seed propagation is the more traditional technique, grafting is gaining popularity for its ability to produce trees with desirable traits.

Vegetative Propagation

Whip grafting is mainly done for propagation of walnut. This method is typically done in early spring before the sap begins to flow, while both the scion and rootstock are still dormant. A scion (usually a 1-year-old shoot from the desired walnut variety) is carefully cut with a sloping cut at the bottom, and a matching cut is made on the rootstock. The two pieces are then tightly joined, ensuring that the cambium layers of the scion and rootstock align perfectly for proper fusion.

Raising of Rootstock (Sexual Propagation)

Globally due to non-availability of clonal rootstocks in walnut, the cultivation is still relying upon seedling rootstocks. The various species of walnut used as rootstock in different parts of the world are *J. regia, J. hindsii, J. nigra, J. Microcarpa* (*J. rupestris*), *J. Sieboldiana* (*J. ailantifolia*), *J. californica* and some hybrids like Paradox (*J. hindsii x J. regia*) and Royal (*J. hindsii x J. nigra*). Seeds (walnuts) are collected from healthy, productive trees known for their good quality nuts. It's important to choose seeds from trees that have adapted well to Ladakh's climate, particularly those that are cold-hardy and drought-resistant. Walnut seeds require cold stratification for successful germination. This involves exposing the seeds to a cold environment for 3 to 4 months to break dormancy. In Ladakh, this process can be done naturally by storing the seeds outdoors in moist sand during the winter months. Seeds are sown in early spring (March-April) when the soil begins to thaw. In Ladakh, walnuts are often planted in nurseries first, where the young seedlings can be cared for in a controlled environment before being transplanted to the field. Seeds are planted about 5-6 cm deep in well-drained soil, with enough space to allow the seedlings to develop strong root systems. Once the seedlings are 1-2 years old, they are transplanted into the orchard. Transplanting is typically done in early spring after the last frost. Care must be taken to avoid damaging the taproot, which is essential for the tree's long-term growth.

Grapes

Grape propagation in cold arid regions, such as Ladakh and Lahaul Spiti region of Himachal Pradesh, requires careful selection of techniques and varieties that can withstand the harsh climatic conditions, including extreme cold, low humidity, and poor soil fertility. Choosing grape varieties that are tolerant to cold temperatures and have a short growing season is essential. Varieties such as Concord, Frontenac, and Marquette are known for their cold-hardiness and can thrive in colder climates. While grapes are typically grown in more temperate regions, they can be cultivated in cold arid areas with proper care and suitable propagation methods.

Propagation of Grapes

Hardwood Cuttings

In cold arid regions, grapevines are commonly propagated using hardwood cuttings. These cuttings are taken from dormant, mature vines during late autumn or early winter. Each cutting should be about 20-30 cm long, with at least 3-4 nodes (buds). The cuttings are then stored in cool, moist conditions to prevent them from drying out. In early spring, when the ground begins to thaw, the cuttings are planted directly in the field or in a nursery. They are buried with only the top node above the ground. The cuttings will develop roots and shoots as temperatures rise, and by the end of the growing season, they will be ready for transplanting to their final location. Among many other propagation methods, hardwood stem cuttings are the most widely utilized in commercial viticulture due to their reliability in preserving desirable genetic traits (Keller, 2015).

Grafting

In cold arid regions, using cold-resistant grape rootstocks can significantly improve the survival rate of grapevines. The grafting method involves joining a cold-hardy rootstock with a desired grape variety. The rootstock is typically more resistant to cold and drought, while the scion (the upper part of the vine) bears the desired fruit characteristics. In cold desert and high-altitude regions Cleft grafting is a widely used method due to its simplicity and high success rate. This technique involves making a vertical split in the rootstock and inserting the scion into the cleft. The cleft is then sealed with grafting wax to prevent desiccation and infection. Cleft grafting ensures a tight union between the rootstock and scion, which is crucial for nutrient transfer and stability in extreme conditions (Brown, 2019).It is particularly effective for older, established rootstocks and can be used on thicker rootstock bases (Jones and Walker, 2021).

Pear

Propagation of pears in cold arid regions, such as Ladakh, requires careful consideration of cold-tolerant varieties and appropriate techniques to ensure successful cultivation in the extreme climate. Pears, like other fruit trees, are typically propagated through methods such as grafting, budding, and occasionally by cuttings. These methods help ensure the development of cold-hardy trees capable of withstanding harsh conditions.

Propagation

Grafting (Commercial Method)

In cold arid regions, grafting is the most reliable propagation method. Rootstocks such as *Pyrus ussuriensis* (Ussurian pear) or *Pyrus communis* (European pear) are used due to their ability to tolerate cold temperatures and drought. Grafting is typically done in late winter or early spring. Techniques like whip grafting or cleft grafting are used to join a scion (a cutting from a desired pear variety) onto the cold-resistant rootstock. The grafted plant is then grown in a nursery until it is ready to be transplanted.

Budding

T-budding or chip budding is a relatively simple technique and allows growers to propagate pear trees efficiently. Once the bud has fused with the rootstock, it is ready for transplantation in the following growing season.

Root Suckers

In some cases, pear trees produce suckers from their roots, which can be separated from the parent tree and transplanted in the nursery. However, this method is not commonly used in commercial pear cultivation as the offspring may not retain the desired traits of the parent tree. By employing methods like grafting and budding, and selecting appropriate cold-hardy rootstocks and pear varieties, it is possible to propagate pear trees successfully in cold arid regions. Proper management of frost protection, irrigation, and soil quality is also crucial to ensure the long-term growth and productivity of pear orchards.

Prunes

Propagation of prunes (which come from plum trees, typically varieties of European plum (*Prunus domestica*) in cold arid regions presents challenges due to the extreme temperatures and low moisture levels, but with proper techniques and care, they can be successfully cultivated. Prune trees, like many other fruit trees, can be propagated by a few common methods that ensure that they are well-suited to cold, arid environments.

Propagation

Grafting

Grafting is the preferred method for propagating prunes in cold regions because it allows growers to use cold-resistant rootstocks that can survive harsh climates. Rootstocks such as Myrobalan (*Prunus cerasifera*) or St. Julien A are commonly used for their cold tolerance. The scion is selected from a desired prune variety, such as Stanley or Italian, both of which are known for their prune production and cold hardiness. Grafting is typically performed in late winter or early spring. Techniques like whip grafting or cleft grafting are commonly used. After grafting, the trees are kept in a protected environment (greenhouse or cold frame) to allow the graft union to heal and establish.

Budding

T-budding is mainly practice in prune propagation. This method involves inserting a bud from a desired prune variety into the bark of a rootstock. Budding is typically done in late summer when the bark is slipping (easy to peel). As with grafting, the choice of rootstock is crucial for cold arid regions. Myrobalan and St. Julien are also commonly used in budding to ensure cold and drought resistance(Bhutani and Joshi, 1995).

Chilgoza

Chilgoza, also known as pine nuts or pine seeds, comes from the chilgoza pine (*Pinus gerardiana*), which is found mainly in the mountainous regions of the Himalayas, particularly in Pakistan and India. In India Chilgoza is mainly produced in Kinnaur district of Himachal Pradesh, Doda and Kishtwar area of Jammu and Kashmir and cold dessert areas in Uttarakhand. The seeds are encased in cones and are harvested for their rich flavor and nutritional value. Chilgoza nuts are high in protein, healthy fats, and various vitamins and minerals. They are often eaten raw, roasted, or used in cooking and baking. In addition to their culinary uses, chilgoza nuts are also considered to have various health benefits, including promoting heart health and providing antioxidants.

Propagation

***Seeds*:** The main method of propagation in chilgoza is seed. Chilgoza pine seeds taken from healthy trees are insecticide-treated before being sown in the nursery. During the months of November and December, the seeds are sowed at a depth of 3-4 cm in polybags and thoroughly covered with soil mixture. In general, two seeds are planted in each polybag in the nursery. Polybags are watered to maintain the moist conditions required for seed germination. Polybags containing seeds are typically covered with nets or thatches to keep birds and rats away from eating the seeds. After seeding, Chilgoza pine seeds remain dormant in polybags over the winter season. The seeds start germinating during March-April when the environmental conditions become favourable for seed germination. Germination percentage of seeds varies from 80-90 per cent depending on the quality of seeds and climatic conditions during seed germination. In its initial stages after germination, polybags containing seedlings are covered with shade for proper development of seedlings and thereafter, shade is removed from the nursery beds.

Grafting

Propagation through vegetative methods like tender shoot cleft grafting can also be done in chilgoza but survival chances are low (Singh,1992).

Orchard Management

Orchard management of fruit crops in Ladakh/cold arid region requires careful consideration due to the region's extreme climatic conditions, including cold winters, limited rainfall, and high altitudes. Effective orchard management strategies are essential to ensure healthy growth, productivity, and long-term sustainability of apricot cultivation in this challenging environment.

Site Selection and Land Preparation

Fruit orchards in Ladakh are typically established in well-drained, sunny locations, often on terraced slopes to maximize sunlight exposure and prevent frost damage. The soil in Ladakh is often sandy or rocky, so proper soil preparation is critical. Organic matter like manure or compost is added to improve soil fertility and structure. Terracing and contour planting help in reducing soil erosion and optimizing water retention in the sloped landscape.

Planting

As cold arid areas experience cold winters and minimal rainfall, hardy rootstocks like wild apricot (*Prunus armeniaca*), Myrobalan in prunes, *Pyrus ussuriensis* or *Pyrus communis*in pear are commonly used. These rootstocks are well-adapted to local soil and climate conditions. Trees are planted with adequate spacing to allow for proper air circulation, sunlight penetration, and easy movement for pruning, irrigation, and harvesting.

Irrigation

Fruits orchards in Ladakh and Lahaul and Spiti district of Himachal Pradesh rely heavily on traditional irrigation systems like **kuhls** (water channels) that carry glacial meltwater to the orchards. Given the arid conditions and limited water availability, careful water management is crucial. Hot and dry conditions result in high evaporation rates in these regions. Mulching around the base of trees helps conserve soil moisture and reduce evaporation. Modern techniques like drip irrigation are being introduced in some areas to optimize water use efficiency and ensure regular watering during critical growth stages of the fruit crops.

Training and Pruning

Proper pruning is essential for maintaining the structure of fruit trees and promoting healthy fruit production. Regular pruning, especially in late winter or early spring, helps remove dead or diseased branches, improves air circulation, and encourages the growth of strong, fruit-bearing branches.

Fertilization

Due to the nutrient deficiencies in cold arid soils, a balanced mix of nitrogen (N), phosphorus (P) and potassium (K) is often necessary. This can be achieved through application of organic fertilizer, biofertilizer and chemical fertilizers like urea (for nitrogen), single super phosphate or DAP (for phosphorus), and muriate of potash (for potassium).Fruits crops are being grown in a traditional way without the use of chemical fertilizer and pesticides in Ladakh,there is a need to certify and promote fruits of Ladakh as organic. Given the organic nature of farming in Ladakh, farmers use natural fertilizers like farmyard manure, compost, and biofertilizers to improve soil health and boost tree growth. Horticulture department of UT is also giving subsidy to the farmers to build vermicompost unit encouraging them to use organic fertilizers.

Pest and Disease Management

Cold arid region is known for its low incidence of insect-pests and diseases, which can be partly attributed to low humidity, extreme cold winter, low cropping intensity and diversified cropping system. Aphid is a regular pest of apricots in Ladakh. However, in 2016 there was an outbreak of the pest causing a huge economic loss. Codling moth is a serious insect-pest of apricot in Ladakh. In recent years, severe infestation of yellow tail moth (*Euproctis similis*) on apricot trees was seen in Dah-Hanu and Sham belts of Ladakh (Stobdan *et al.,* 2019). Management of these pest are done by mechanical and organic methods. Chemical methods are not practicing in the UT because of the aim of transforming Leh in to organic region by 2025.

Harvesting and Post-harvest Management

Apricots are hand-harvested at full ripeness, typically in mid-summer (July-August), to ensure the best flavor and quality. Apple are also hand harvested during late September to early October. The timing of the harvest is critical to prevent over-ripening or damage from pests. A significant portion of the apricot harvest in Ladakh is sun-dried due to the region's dry climate. Drying apricots in the open air helps preserve them for long-term use and adds economic value to the produce.

Sustainability Practices

In many Ladakhi orchards, farmers practice agroforestry, where apricot trees are intercropped with vegetables or other crops like barley. This diversifies income sources and maximizes land use efficiency. Ladakh orchards are typically managed without synthetic chemicals, making organic farming a common practice. This is in line with the region's clean environment and focus on sustainable agriculture.

Proper orchard management practices in Ladakh are essential to ensure that apricot trees thrive in this unique high-altitude desert environment. By incorporating traditional knowledge with modern techniques, farmers can maintain the health and productivity of their apricot orchards while adapting to changing climatic conditions.

Canopy Management

Standard cultural practices such a straining and pruning of trees are not performed in Ladakh. Lack of training and pruning resulted in trees reaching heights that hinder fruit harvesting. Besides, the fruits remain small and of poor quality. There is a need for an intense awareness campaign on the need of training and pruning in fruit plants. Farmers need to be taken on exposure tours to show orchards where trees are regularly trained and pruned. Front line demonstrations are also need to be conducted in every fruit growing villages (Stobdan *et al.,* 2021).

Canopy management is one of the crucial methods for fruit crop cultivation in cold desert areas, as it affects tree health, fruit production, and the ability to withstand extreme conditions like temperature fluctuations and limited water availability. There are few methods which should be practiced for managing fruit canopies in cold dessert region.

Training Systems

Open Centre (Vase) System

The open centre or vase shape is widely used for apricots. This system allows good air circulation and sunlight penetration, which is important in cold desert regions to reduce humidity inside the canopy and prevent frost damage. It also makes it easier to manage frost-sensitive buds by improving airflow.

Modified Central Leader

This system is often used when cold desert winds are an issue. It provides a sturdy structure with a central leader and scaffold branches arranged to ensure balanced growth and good light distribution. In apple this system is usually followed.

Central Leader System

The central leader system involves training the pear tree with one dominant vertical trunk (the "leader") and several tiers of horizontal branches spaced evenly around it. This is a common system for cold climates as it allows the tree to grow tall while managing lateral growth

Tatura Trellis or Espalier System

These high-density systems can be employed to increase yield while managing canopy shape in limited space. They also help in cold areas by supporting frost protection measures and facilitating irrigation and other management activities.

Head System (Bush System)

This system is widely used in cold regions where low-growing vines are more protected from cold winds and frost. Vines are trained without trellises, and the shoots are kept low to the ground and reduces the risk of frost damage since the vines remain close to the ground, where temperatures are slightly higher. The vine's trunk is kept short, with fruiting canes or spurs pruned annually. This system is often used in combination with techniques that protect the vines from cold during winter, such as mulching or covering the base of the vine.

Guyot Vine Training System

Guyot Vine Training System is usually followed in grapres.The Guyot training system is a part of a group of systems referred to as a VSP system (vertical shoot positioned; other examples include the lyre system).The system was developed in France in the late 1800s and is named after Dr. Jules Guyot. The training system was designed for low to moderate vigor vineyards as it accommodates fewer buds and vine growth per linear foot of vineyard row than other training systems. Vigorous vines are not well suited to the Guyot system, and will result in high shoot density and inner-canopy shading, reducing leaf surface area exposure. The single or unilateral Guyot refers to a one-cane (or one-cordon) system, and a double or bilateral Guyot refers to the training system with two canes or (cordons) originating from the trunk.

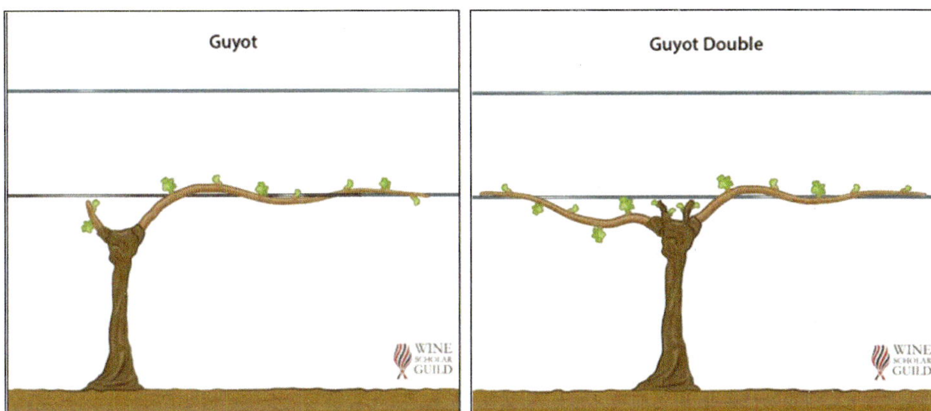

Pruning Techniques

Dormant Pruning

This should be done in late winter before the buds begin to swell but after the risk of severe frost has passed. Dormant pruning helps remove dead or diseased wood, opens up the canopy, and shapes the tree for optimal growth in the coming season.

Summer Pruning

This involves removing excess foliage after the fruit has set to improve light penetration and air circulation. However, in cold desert areas, summer pruning should be minimal to avoid sunburn on exposed branches during hot, dry conditions.

Thinning

Thinning the fruit and branches is necessary to ensure good fruit size and prevent limb breakage under the weight of heavy crops. In cold deserts, thinning helps to balance the tree's energy, particularly in areas where resources like water are limited.

Light Management: Maximizing Sunlight

In cold deserts, fruit crops need adequate sunlight for proper growth and fruit production. A well-pruned open canopy maximizes light exposure, especially for inner branches, leading to better fruit quality and earlier ripening.

Preventing Sunburn

Apricot trees in desert areas can suffer from sunburn on branches and fruit. To mitigate this, avoid over-pruning, especially during the hot summer months. Retain enough foliage to protect branches and fruit from intense sunlight.

Frost Management through Canopy Manipulation

Frost Damage Prevention

Cold desert areas are prone to late spring frosts, which can damage apricot buds and young fruits. Proper canopy management, such as maintaining an open canopy and controlling tree height, allows for the implementation of frost protection methods like wind machines or sprinklers.

Reflective Mulch

Using reflective mulch on the ground can help to increase sunlight exposure on the lower canopy and protect trees from frost during cold periods by retaining heat.

Wind Management

Windbreaks

Fruit trees can be damaged by the strong winds common in desert regions. Strategically placed windbreaks (*e.g.*, hedgerows or netting) around orchards can protect the canopy from wind damage. Windbreaks reduce water loss from trees and help protect against cold winds that exacerbate frost damage.

Water Stress and Canopy Size

Drought Management

In cold desert regions, water conservation is critical. By managing the size of the canopy, you can reduce water stress. Smaller, more compact canopies require less water and are easier to manage with drip or micro-sprinkler irrigation systems.

Deficit Irrigation

This technique involves reducing water supply during less critical growth periods (*e.g.*, after harvest) to manage water resources effectively. A well-managed canopy helps in this practice by reducing the tree's water needs.

Tree Renewal

Older fruit trees can be rejuvenated by cutting back older branches to encourage new growth. This is especially useful in cold desert areas where older trees may struggle to produce due to stress from harsh conditions.

REFERENCES

Angmo P, Chandel JS, Katiyar AK, Targais K, Chaurasia OP and Stobdan T, 2018. Zero energy overwinter storage of apple nursery plants in trans-Himalayan Ladakh, India. *Defence Life Science Journal* **3**: 162-164.

Banoo H and Dar SN, 2020. Impact of horticulture on the economy of cold desert: A study of Kargil, Ladakh.

Bhutani VP and Joshi VK, 1995. Handbook of fruit science and technology- production, composition, storage and processing.

Brown S, 2019. Grafting Techniques for Cold Desert Regions. *American Journal of Enology and Viticulture* **30**(2): 223-234.

Dihar 2011-2012. Multiplication in apricot. Extension Bulletein 18. SKUAST, Srinagar.

Dolker T, Kumar D, Chandel JS, Angmo S, Chaurasia OP and Stobdan T, 2021. Phenological and pomological characteristics of native apple (*Malus domestica* Borkh.) cultivars of trans-Himalayan Ladakh, India. *Defence Life Science Journal* **6**: 64-69.

Dwivedi SK, Singh B and Paljor E, 2000. Studies on vegetative propagation of apricot (*Prunus armeniaca* L) through grafting in Ladakh. *Indian Journal of Horticulture* **57**(1): 39-41.

Dwivedi SK, Paljor E, Attery DP and Singh B, 2002. Propagation of common seabuckthorn through hardwood cuttings in Ladakh. *The Indian Journal of Agricultural Sciences* **72**(4): 228-229.

Dwivedi SK and Ahmed Z, 2008. Seabuckthorn (*Hippophae sp.*) - a Potential Underutilized Fruit Plant for Cold Arid India. *Acta Horticulturae* **1358**: 125-130.

Dwivedi SK and Stobdan T, 2009. Standardisation of optimum time of chip budding in apple under cold arid condition of Ladakh. *Progressive Horticulture* **41**(2).

Gupta RD and Arora S, 2016. Ecology, soil and crop management for livelihoods in Ladakh region: An Overview. *Journal of Soil and Water Conservation,* **15**(2): 178-185.

Husain M, Rathore JP, Rasool A, Parrey AA, Vishwakarma DK and Mahendar K, 2018. Seabuckthorn: A multipurpose shrubs species in Ladakh cold desert. Journal of Entomology and Zoology Studies **6**(2): 1330-1337.

Jones R and Walker D, 2021. Seasonal Timing of Grafting. *International Journal of Horticulture* **12**(5): 412-425.

Kadam PY, Dumal SA and Shinde NN, 1995. Handbook of fruit science and technology- production, composition, storage and processing.

Kala CP, 2011. Floral diversity and distribution in the high-altitude cold desert of Ladakh, India. *Journal of Sustainable Forestry* **30**(5): 360-369.

Keller M, 2015. The Grapevine: A Practical Guide to Understanding Vine Health and Growth. Academic Press.

Sharma DP, Sharma N, Bawa R and Rajesh Kaler, 2005. Potential of Apricot Growing in the Arid-cold Desert Region of North Western Himalayas. *Acta Horticulturae* **696**: 418-426.

Sharma DP and Chauhan JS 2008. Standardization of temperate fruits nursery growing techniques for tribal cold desert areas of north western Himalayas. *Acta Horticulturae,* **699**: 328-335.

Singh NB, 1992. propagation selection and establishment of clonal seed orchard of chilgoza pine (*Pinus gerardiana* Wall.) University of horticulture and forestry, regional research station, Sharbo, Kinnaur (Himachal Pradesh).

Stobdan T, Deen M, Gupta V and Raghuvanshi MS, 2019. Integrated management of yellow tail moth (*Euproctissimilis*) on apricot in Leh Ladakh India. LAHDC Leh, Ladakh. 50.

Stobdan T, Namgial D, Chaurasia OP, Wani M and Phunchok TM, 2021. ZaffarApricot (Prunus armeniaca L.) in Trans- Himalayan Ladakh, India: Current Status and Future Directions. *Journal of Food and Agriculture Research* **1**: 86-105.

Chapter 7

Propagation of Sub-tropical Fruit Crops

***Nripendra Vikram Singh*[1], *Mahesh Kumar Dhakad*[2] *and Vishaw Bandhu Patel*[3]**

[1]*Division of Fruits and Horticultural Technology, ICAR–IARI, New Delhi*
[2]*ICAR–RCER, Ranchi, Jharkhand*
[3]*Horticultural Science Division, Indian Council of Agricultural Research, New Delhi*

Plant propagation is the process of multiplying or growing new plants from a variety of sources, including seeds, cuttings, and other plant parts. It can be done through two main methods: sexual propagation (via seeds) and asexual propagation (via cuttings, layering, grafting, budding, other specialized structure *etc.*). The sexual propagation involves growing plants from seeds, which are formed through the natural reproduction process between pollen and ovules. It results in plants that may vary genetically from the parent in case of cross and often cross-pollinated crops provided there is no apomixis. Asexual propagation is a plant multiplication method that doesn't rely on seeds or sexual reproduction. Instead, plants are multiplied using parts of the parent plant, like stems, leaves, or roots, resulting in saplings that are exact genetic copies or clones of the parent. Vegetative propagation may be further divided into attached method like layering, inarching or detached method like cutting. Propagation of plants can be carried out either on its own roots like cutting, layering, *etc.* or on the root system of other plant to explore the benefits of rootstocks like budding, grafting, *etc.* In this chapter, we will explore various propagation methods used in sub-tropical fruit crops, including seed propagation, vegetative techniques like cutting, grafting, layering and budding. These techniques aim to improve rooting, growth, and success rates for commercial fruit propagation.

Mango

Mango plants can be propagated through both sexual and asexual methods but propagation through vegetative means is highly recommended for commercial production and to avoid genetic variations. Earlier, seed propagation was common method for propagating of both monoembryonic and ployembryonic mangoes; however, nearly all commercial mango varieties are propagated by grafting techniques utilizing mangoes seedlings as rootstocks. There are numerous grafting techniques that have been standardized at various points in time. Important commercial methods of vegetative propagation include inarching, veneer grafting, softwood grafting, and stone/epicotyl grafting. Among the many methods of grafting, approach grafting is one of the old methods of grafting. This procedure often involves grafting the chosen shoot (scion) of a preferred parent tree (mother plant) onto a seedling (rootstock) without separating scion from the mother plant. Despite being laborious and time-intensive, inarching or approach grafting was the predominant method for the commercial propagation of mango plants until 2000, owing to its high success rate (Chua and Yong, 1978; Roy and Sinha, 1994). Veneer grafting technique holds potential for large-scale commercial cultivation. The procedure is straightforward and can be implemented with a high success rate. A downward and inward incision, measuring 30-40 mm in length, is executed in the smooth region of the stock at a height of around 20 cm. At the base of the cut, a shorter incision is made to intersect first, facilitating the removal of the wood and bark. The scion stick is provided with a long slanting cut on one side and a short cut on the opposite side to align with the cuts of the stock. The scion is put into the stock to ensure that the cambium layers make contact on the longer side. The graft union is thereafter wrapped with a polythene strip. Significant success in veneer grafting (> 85%) in mango was seen when scions were grafted at elevated heights (75 and 100 cm) onto seedling rootstock (Kumar *et al.,* 2000). Veneer grafting conducted in July demonstrated to be the most effective for enhanced success and growth of mango grafts (Singh *et al.,* 2012).

Typically, seedlings aged 8 to 10 months are chosen for the soft-wood grafting technique. The grafting is performed on newly emerging shoots. This strategy is advantageous for the in-situ establishment of the orchard. July and August represent the optimal months for softwood grafting. A success rate over 80 per cent was achieved by softwood grafting (Singh and Suryanarayana 1996). Epicotyle grafting applies either the cleft or whip method. The rootstock, aged 12 to 17 days, features coppery-red leaves is used for Epicotyle grafting. The grafts must be situated in a polyhouse that maintains high humidity levels exceeding 80 per cent. Upadhya *et al.*

(2014) observed that in epicotyl grafting of mango the maximum initial (86.76%) and final success (56.74%) of grafts were found when 15 days old seedlings was grafted with 10th day of pre-defoliated scion.

Rootstock has a significant role in regulating the vigour of plants. Reddy *et al.* (2003) observed the influence of rootstocks on growth patterns of 'Alphonso' mango and reported that nucellar seedlings of 'Muvandan' 'Bappakai' and 'Olour' were vigorous rootstocks in decreasing order of vigour, while those of 'Vellaikulamban' imparted dwarfing, in comparison with the scion grafted on its own open-pollinated zygotic seedlings. Dubey *et al.* (2021) studied performance of mango varieties on five mango polyembryonic rootstocks and found that K-2 rootstock imparted dwarfism in Pusa Arunima and Pusa Surya. Pusa Arunima displayed yield superiority with Olour, K-2 and K-3 rootstocks. Rebolledo-Martínez *et al.* (2019) also observed that the rootstock cultivar 'Julie' promoted larger trees, more leaf area and dry matter, and higher fruit production, while 'Thomas' reduced tree size and stimulated lower production of panicles and fruits.

Citrus

In India, citrus trees are propagated using both seeds and vegetative methods. Propagation using seeds is done mainly in acid limes which have high percentage of polyembryony and also for raising rootstocks for budding or grafting purposes (Soost and Roose, 1996). Propagation through air layering is under practice for multiplication of lemon, grapefruit, pummelo and sweet lime (Sharma *et al.*, 2022). Propagation of citrus by mound layering has been reported by Duarte and Medina, 1971. Among Citrus, sweet lime is mainly propagated by hardwood stem cutting (Bajwa *et al.*, 1978). Also, this method has been employed in propagation of lemon cv. Baramasi (Kumar *et al.*, 1995), cv. Pant Lemon (Singh *et al.*, 2013), Sour orange (Ozcan *et al.*, 1993), Citron (Kako Al-Zebari *et al.*, 2014), Kagzi Lime (Bhatt and Bhatt, 2014), Pummelo (Thangjit and Wunnachit, 1994), and Tahiti Lime (Prati *et al*, 1999). The most widely practiced vegetative method is Shield or T-budding, which is typically carried out in spring or September. Shield or T-budding is mostly used for propagating Oranges, Nagpur mandarin, Kinnow mandarin and Sweet oranges. In this method, a bud from a desired variety (scion) is budded onto a rootstock to take advantage of the rootstock's vigorous root system, climate resilience, stress tolerance and to combine the best characteristics of both plants (Table 7.1). A plump bud, ideally from the middle section of the scion shoot and not overly elongated, is selected and excised by making a circular incision around it with a sharp budding knife. On the chosen rootstock, a T-shaped incision is made, just large enough to accommodate the bud. The bud is then inserted into this incision and secured with a strip of plastic film, ensuring that the tip of the bud eye remains exposed. It is preferable to perform budding as low as possible on the rootstock. After 3-4 weeks, the bud should unite with the rootstock. Once the new shoot from the grafted bud reaches about 10 cm, the upper part of the rootstock above the union is cut off, and the plastic film is removed.

Table 7.1: Characteristics of Important Rootstocks Used at Commercial Scale in *Citrus* spp.

Rootstocks	*Features*
Carrizo (and Troyer)-hybrid between Washington navel orange and *P. trifoliata*	Induces precocious bearing, intermediate resistance to *Phytophthora* root rot, nematode and CTV tolerance. Carrizo as rootstock mostly used in Spain, South Africa, Italy, Australia, and California for sweet orange, grapefruit, and most mandarin scion hybrids
Cleopatra-small-fruited mandarin (*Citrus reshni*)	It is cold and drought tolerance, resistance to CTV and some viroids, tolerance to some species of *Phytophthora* root rot. Mostly used as rootstock for mandarin cultivars in USA
Macrophylla (Alemow) earlier considered as a hybrid of citron (*Citrus medica*) and pummelo (*Citrus maxima*) but a recent study revealed it as a hybrid between *Citrus micrantha* (or a close papeda species) and citron (Curk *et al.*, 2016)	Macrophylla is resistant to root rot and tolerant to salt and boron. Macrophylla has been extensively used as a lemon rootstock in California, Spain and Morocco
Poncirus trifoliata (L.) Raf. (trifoliate orange)	Resistant or tolerant to citrus nematode, some *Phytophthora* species, CTV, flooding, and cold. Dwarfing rootstock especially the P. trifoliata cultivar Flying Dragon (Castle *et al.*, 2005; Mademba-Sy *et al.*, 2012). Trifoliate orange rootstock has major impact in Japan, China, Australia, New Zealand, and Argentina, and is significantly used in many other citrus-producing regions
Rangpur lime (*Citrus limonia* Osbeck)	Tolerant to CTV, drought as well as calcareous soils. Used in India
Rough lemon (*Citrus jambhiri* Lush)	Typically, grafted scions on rough lemon produce large crops and large fruit size. Resistance to high pH and drought, and good growth in deep sandy soils. Predominantly used in India and Pakistan, and South Africa, but is currently relatively unimportant in most other countries.
Sour orange (*Citrus aurantium* L.)	Sour orange induces high yields and excellent fruit quality, good tolerance to root rot, high pH, salinity, flooding, cold, and blight.
Sunki (*Citrus sunki* hort ex. Tanaka)	Sunki has tolerance to CTV, xyloporosis, blight, and salinity.
Swingle citrumelo is a hybrid of Duncan grapefruit (*Citrus paradisi*) and trifoliate orange (*P. trifoliata*)	Swingle reported to be tolerant against drought, cold, *P. nicotianae* (but not *P. palmivora*), citrus nematode, blight, and CTV. It is not suitable for highly calcareous soil conditions with salinity, and heavy clay soils.

Source: Bowmana and Joubertb, 2020.

Micro-budding and micrografting techniques are also practices in citrus for multiplication of virus free plants. Shoot tip grafting coupled with chemo and thermo therapy is the most reliable method to recover graft transmissible disease-free saplings. In vitro shoot tip grafting protocols have been standardized in Kinnow mandarin (*Citrus deliciosa*) by Kumar *et al.* (2014). *In-vitro* generated and 2-3 weeks etiolated old seedlings of cleopatra, rough lemon and carrizo were used as rootstocks. The seedlings were grafted by shoot tips measuring 1 mm in length containing apical meristem and one or two leaf primordial which were excised from *in-vitro* generated shoots and grafted on rootstocks. Among different rootstocks used, twelve days old seedlings of Carrizo recorded the maximum success (56.60%). To accelerate the survival per cent of STG, BAP @ 1.0 mg/l and 2, 4-D @ 3.0 mg/l were used. Micro budding is a novel and economic technique which has been used for rapid production of citrus bud-grafts throughout the year and advances the availability of budlings earlier by 9 to 12 months (Vijaykumari and Singh, 2003). It has been used for biological indexing for citrus tatter leaf virus (Skaria, 2000) and enhanced the production of citrus through ultra-high-density planting.

Grapes

Grape propagation for commercial vineyards establishment includes the use of cuttings, budding, layering and grafting methods (Verdegaal, 2009). Propagation through cutting is influenced by factors like cutting type, planting season, and the application of plant growth regulators such as Indole-3-butyric acid (IBA) and 2-4-D (Tofanelli *et al.,* 2014). Rooting media is one of the most important factors for rooted cutting production. It is one of the factors affecting rooting and growth of grape cuttings (Abebe, 2017). The best rooting was usually found in cuttings taken from the basal portion of the shoots (Hartmann and Kester, 1983). This might be due to higher accumulation of carbohydrates and endogenous root promoting substances. Environmental conditions such as temperature and humidity, along with nursery practices like hot water treatment and phytosanitary management, play a crucial role in reducing disease risks and ensuring healthy plant development (Waite and Morton, 2007).

The propagation of grapevines using various rootstocks plays a crucial role in viticulture, influencing growth, disease resistance, and fruit quality. Rootstocks such as *Vitis berlandieri* and *Vitis riparia* are commonly utilized due to their adaptability to diverse soil conditions and resistance to pests and diseases, particularly phylloxera (Fabbri and Lambardi, 1988). For instance, '110R' (*Vitis riparia* × *Vitis rupestris*) is favored for its drought resistance and vigor, making it suitable for arid regions (Goode and Lane, 1983). Other rootstocks like 'SO4' (*Vitis riparia* × *Vitis berlandieri*) are noted for their compatibility with a range of Vitis vinifera cultivars and improved rooting characteristics (Ferrer *et al.,* 1991). Moreover, the selection of rootstocks can enhance fruit quality and yield, as evidenced by studies that demonstrate increased fruit size and improved disease resistance in grafted plants (Galavi *et al.,* 2013; Abou-Rawash *et al.,* 1998). Overall, the choice of rootstock significantly affects the success of grape propagation, with implications for vineyard productivity and sustainability.

Table 7.2: Promising and Recommended Grape Rootstocks in India

Rootstocks	*Features*
1103 Paulsen	Drought tolerant
99Ritcher	Drought tolerant
110Ritcher	Drought tolerant, salt tolerant
1103P	Salt tolerant
SO 4	Lime tolerant
Ramsey	Salt tolerant
St. George	Drought tolerant
Dogridge	Drought and salt tolerant
Saltcreek	Salt tolerant
1613, B2-56	Enhanced bud break
Dogridge, St. George, 34 EM, Ramsey	Enhanced Nitrogen and Potassium uptake
Ramsey, Dogridge, 140 RU, 99 R, 110 R	Can tolerate soil EC more than 2 m mohs/cm and water EC more than 1 m mohs/cm
140 RU, 1613, Ramsey, Dogridge	Can tolerate soil ESP more than 15 per cent and/or water SAR more than 8
Ramsey, Dogridge B, 140 RU. Teleki 5-C	Can tolerate chloride content of water is more than 4 meq/litre
Berlandieri and Rupestris crosses followed by other species like Champinii, Riparia and Rupestris	Water use efficiency, enhanced biochemical properties of berries including proline

Source: https://nrcgrapes.icar.gov.in/Recommended-Varieties.htm; Adsule PG and Upadhyay A. 2004.

Recent innovations, including *in vitro* propagation (Sandoval-Prando *et al.*, 2014), show promise for enhancing growth and sustainability. Meristem culture in combination with various therapies are widely used for production of disease-free planting material and for rapid multiplication of plants. Some disadvantage associated with this technique includes high cost and acclimatization issues (Barlass *et al.*, 1982). However, challenges such as climate change (Trbic *et al.*, 2021) and disease control in nurseries (Rego *et al.*, 2009) require ongoing adaptations to maintain vine quality and productivity.

Litchi

The genetically diverse nature of seedlings produced, long juvenile phase of seed raised plants and tendency of irregular bearing makes propagation through seeds inefficient and uneconomical. Also, viability of litchi seed is extremely short thus, propagation by seed is done only for raising rootstocks.

Air-layering is a preferred method for propagating litchi all over the world (Bhambota *et al.,* 1968). The process involves removing a 2 cm wide bark ring from a healthy one-year-old twig, 2.5-4.0 cm thickness. The girdled area is then surrounded with a mixture of sphagnum moss and soil, wrapped in polyethylene, and sealed. Root formation typically occurs within two months, after which the branch should be separated and potted as a new plant. July to October is the ideal time for air layering. Applying growth hormones like IBA 5,000 ppm paste on the girdled portion significantly improves rooting success (Rahman *et al.,* 2000). The well rooted air layers are detached from the mother plants after 60-75 days after tying the layers. The budding method has also proven highly effective for litchi propagation, with a 70 per cent success rate in Kanjirapally, Kerala during August to November. The resulting budlings are vigorous, with a strong root system and 100 per cent field survival rate (George *et al.,* 2016). In China other than air layering grafting of litchi is also commercially used. Grafting methods provide advantages such as early bearing as scions are mature part of plant with no juvenile phase and higher propagation coefficient, through grafting, rootstock may affect several characters of scion cultivars (tree vigor, flowering time, hardiness), use of seedling rootstock provides better anchoring root structure. Different grafting methods such as Whip grafting, Cleft grafting, Bark grafting can be employed in litchi.

The grafting protocol was standardized at ICAR–RCER, Research Centre, Ranchi. Before scion collection girdling of litchi branches is required for favourable physiological condition of scion, which lead to better union of stock and scion and thereby making the nutrient and carbohydrate supply at desired level. Girdling restricts the flow of carbohydrate in the lower part and reserves it in the scion wood whereas, defoliation leads to higher meristematic activity at the bud level and both combination is the sole cause of better grafting performance. The branches need to be girdled (removing of 3 mm ring from branches of 40-50 mm girth) 1-2 month before the collection of scion sticks for grafting. Defoliation (removal of leaf blades with petioles intact on the shoot) should be performed 7 days before the scion wood collection. Wedge grafting on 1-year rootstock should be performed under 50 per cent shadenet during June-August which will give more than 70 per cent success rate. The union point wrapped with a plastic strip. The exposed part of the scion above the wrapping area need to be covered by a small plastic cap. The grafted plants should be maintained under natural shade net house conditions (Dhakar and Das, 2017). There are some instances on the use of mound Layering or stooling in litchi as they have stiff branches and are capable of producing several shoots each year (Menzel, 1985). In India Raj and Majumder, 1981 reported 84 per cent success against 71 per cent with air layering and survival percent was also high for stooling (82%) while it was 46 per cent in air layering. The advantages of using this method includes simple to perform, quicker and economical.

Micropropagation can be used for regenerating numerous clonal plants from callus culture using young leaf explants (Sarin *et al.,* 2023). Somatic embryogenesis has been proven successful for multiplication in litchi through induction of adventitious buds (Kantharajah *et al.,* 1989).

Pomegranate

Pomegranate cultivation has grown leaps and bounds in India during last two decades generating an accelerated demand of more than 15 million healthy saplings annually (Singh *et al.,* 2022). The requirement of elite bacterial blight free saplings has grown tremendously due to threatening spread of bacterial blight disease to major pomegranate regions of India primarily due to infected planting material. Pomegranate is commercially propagated through air layering, stem cuttings and micropropagation (Levin, 2006; Finetto, 2009; Singh *et al.,* 2021). There are also reports on exploring possibilities of using rootstocks through grafting, budding and other innovative propagation methods in pomegranate (Chandra *et al.,* 2012; Karimi and Farahmand, 2011; Karimi *et al.,* 2011 Singh *et al.,* 2017; Ahire *et al.,* 2017).

Air layering is the most widely practiced pomegranate propagation method in the Deccan Plateau region of India. The method involves 2-3 cm long girdling of upright branches of 8 mm to 15 mm in diameter from healthy tree at spacing of about 20 cm. Rooting hormones like IBA at concentration of 2000 to 3000 ppm or paste of commercial root promoting formulations is used on upper part of the cut (Singh *et al,* 2021; Chandra and Jadhav, 2012; Sharma *et al.,* 2014; Tomar, 2011). The stem is wrapped with moist rooting medium having enough water holding capacity such as sphagnum moss (peat moss) and covered with the help of white polythene strip having 200 to 300 gauze and the ends are tied with coir/jute thread or string (Singh *et al.,* 2021). Girdling encourages roots to form by preventing the downward flow of nutrients, carbohydrates and hormones, which accumulate in the upper part of the cut. The wrap keeps the environment conducive for root development, while the use of IBA, often paired with other compounds, further speeds up root formation, enhancing the success of this propagation method. Etiolation reduces the production of lignins, thus, instead of lignins, phenolic metabolites may be channeled for root initiation (Hudson *et al.,* 2004). Either IBA, alone or in combination with p-hydroxybenzoic acid, is often used to stimulate early roots in air layers for high success (Bhosale *et al.,* 2009; Hore and Sen, 1995; Tomar, 2011).

Globally pomegranate is propagated through stem cuttings and the rooting success for which is mostly affected by maturity of wood (Levin, 2006; Chandra and Babu, 2010). Pomegranate propagation using hardwood cuttings is most effective during or just after the dormancy period of the mother plant. Stems that are 6 to 12 months old, coming from basal parts of the plant and upright in growth are typically choosen and the ideal cuttings are 20 to 25 cm long, with four nodes and a thickness of 6 to12 mm, as these characteristics contribute to a high success rate in rooting and establishing new pomegranate plants. (Chandra and Babu, 2010; Chandra *et al.,* 2014; Singh *et al.,* 2019). Stem cuttings in pomegranate generally face challenges with low root initiation due to factors like hormone levels, nutrient reserves, genetic material, and the age or maturity of the stem.

However, preconditioning techniques such as etiolation, and basal wounding can enhance the chances of successful rooting. These methods improve the conditions for root formation, making propagation through stem cuttings more effective in pomegranate cultivation (Singh *et al.,* 2021; Yesiloglu *et al.,* 1997). Cutting success can be also influenced by plant growth regulators, mostly auxins like IBA, NAA and IAA (Table 7.3). Auxins play a crucial role in the success of pomegranate cuttings by promoting the formation of callus and facilitating vascular differentiation through hydrolytic activity. This process is vital for root development. Both external and endogenous auxins are essential for the formation of adventitious roots on stems, as they stimulate the division of initial root cells. By enhancing these physiological processes, auxins significantly increase the likelihood of successful rooting and overall propagation in pomegranate cuttings (Kumar *et al.,* 2021; Singh 2014; Frick *et al.,* 2018).

Table 7.3: Various Growth Regulators and Duration of Treatments Used for Propagation through Stem Cutting in Pomegranate

Variety and Reference	*Type of Cutting*	*Growth Regulator Used*	*Treatment Type*
Ganesh, Dholka (Tripathi and Shukla, 2004)	Hard wood cutting	1000 ppm PHB + 5000 ppm IBA	Quick dip
Jalore seedless (Saroj *et al.,* 2008)	Hard wood and semi- hard wood cutting	2500 ppm IBA	Quick dip
Bhagawa (Singh *et al.,* 2014)	Hard wood cutting	2500-5000 ppm IBA	Quick dip
Wonderful (Sarrou *et al.,* 2014)	Hard wood cutting	100 ppm IBA + 500 ppm GA_3	Quick dip
Ganesh (Sharma *et al.,* 2009)	Hard wood and Semi hard wood cutting	500 ppm IBA + Borax 1 per cent	15 minutes treatment
Pomegranate-ME12, CR02 and PT08 (Melgarejo *et al.,* 2000)	Hardwood cutting	12000 ppm IBA	Quick dip and basal wounding

Screening and evaluating wild pomegranate germplasm for resistance to various abiotic and soil-borne challenges can be beneficial in identifying promising genotypes. Utilizing these resilient pomegranate genotypes as rootstocks through techniques like grafting and budding can help mitigate the impacts of biotic and abiotic stresses. This approach fosters the development of a climate-resilient pomegranate production system, enhancing overall sustainability and productivity in the face of environmental challenges (Singh *et al.,* 2021; Singh, 2017).

Vegetatively propagated saplings often face a higher risk of carrying infections or latent diseases, as mother plants are exposed to various biotic and abiotic stresses in open field conditions where nurserymen have limited control. In contrast, micropropagation offers a solution by allowing the mass production of high-quality planting material using a small number of mother plants. These mother plants can be maintained under protected environments, reducing exposure to external stresses and ensuring healthier, disease-free saplings for pomegranate cultivation. This method improves both the quality and uniformity of the propagated plants (Singh *et al.,* 2022; Sheela and Nair, 2001).

Avocado

Avocado propagation in India is primarily carried out through seeds which results in non-uniform plants with long juvenile phase and inconsistent quality (Chithiraichelvan *et al.,* 2006). The genetic diversity caused by cross-pollination makes it challenging to produce the homogeneous plants required for commercial orchards. Vegetative propagation of superior avocado clones through grafting or budding is crucial to overcome this issue and promote phytophtohora and salinity tolerant rootstocks. For effective avocado multiplication in the humid sub-tropical regions, cleft grafting during September or March has been identified as the most suitable method (Tripathi *et al.,* 2019).

Duke seedlings are known for their resistance to root rot and cold, while Pollock rootstock helps in mitigating salinity issues. Additionally, both Green and Purple serve as effective rootstocks, with Green offering greater vigor to the scion compared to Purple.

Loquat

Loquat can be propagated by seeds or vegetative methods. Seed propagation is simple, while air-layering, inarching, budding, and grafting are effective vegetative methods. Air-layering is an effective, economical method for loquat propagation. The process involves making a ring cut on a 3-month-old shoot during rainy or spring season, then wrapping the area with sphagnum moss and polyethylene to encourage root growth. Applying growth hormones like 3 per cent NAA or 2,500 ppm IBA to the ringed area can enhance root formation. Inarching and shield budding are common grafting techniques. Budded or grafted plants are preferred over seedlings for true-to-type and earlier bearing trees. Inarching and shield budding are preferred grafting methods, typically performed during rainy or spring seasons. For commercial cultivation, budded or grafted plants are recommended over seedlings due to their true-to-type characteristics and earlier fruit production. Various rootstocks can be used for grafting, including apple, pear, *Mespilus vulgaris,* and quince.

The "franqueamiento" technique is a clonal propagation method for juvenile loquat plants. It involves completely girdling the stem of a young plant while it's still attached to the parent tree. This process stimulates the formation of adventitious roots above the girdled area. After about two months, when sufficient roots have developed, the stem is separated from the parent plant. This method produces a large quantity of evenly distributed roots and has shown promising results in efficiently propagating loquat (Castro *et al.*, 2006).

Mangosteen

Mangosteen is primarily propagated through seeds, that are azygotic and produce offspring similar to the mother plant. The seeds have low viability and are short-lived once removed from the fruit, exhibiting "recalcitrant storage" behavior. Seeds should be sown immediately after extraction, with freshly harvested seeds achieving over 90 per cent germination if sown within five days. In humid conditions at 25°C, seeds remain viable for up to seven weeks. A few seeds display polyembryony, producing 2-4 seedlings. Growth is slow, and even after two years, seedlings reach only about 15 cm in height, which poses a challenge for rapid multiplication.

Vegetative propagation has been explored to enhance growth, shorten the pre-bearing phase, and control canopy size in mangosteen, but methods like cutting, layering, and budding have been largely unsuccessful. However, softwood grafting using mangosteen as both rootstock and scion has shown a 74 per cent survival rate under Konkan agro-climatic conditions, using one-year-old rootstocks and four-month-old scions for successful grafting. Scions for vertical growth should be carefully selected from orthro tropic growing shoot.

Rambutan

Propagation through is common but not recommended for commercial production, as over 50 per cent of seedlings produce only male flowers. Rambutan seeds should be planted fresh immediately after extracting from aril, as they lose viability rapidly. Freshly planted seeds have high germination rates, ranging from 87-95 per cent within two days, but this drops to 50-60 per cent after just one week. To prolong their viability, fresh seeds can be stored in moist materials such as moist sawdust, sphagnum moss, or charcoal for 3-4 weeks. Seedlings generally take around 9-10 months to develop enough for grafting, making timely planting and proper storage crucial for successful propagation. Rambutan seeds should be planted immediately after removing the aril, as they lose viability quickly. Germination rates of 2-day-old seeds are 87-95 per cent, compared to 50-60 per cent in one-week-old seeds. Fresh seeds can be stored in moist sawdust, sphagnum moss, or charcoal for 3-4 weeks. Seedlings take about 9-10 months to become suitable for grafting.

Vegetative propagation is crucial for rambutan, as seedlings take a long time to bear fruit and the ratio of female to male trees is around 4-5: 7. While approach grafting is effective, it is labor-intensive, requiring seedlings to be secured around the mother tree and regularly watered if rainfall is inadequate. Space and shoot availability limit the number of grafts that can be made. Typically, 8 to 12-month-old seedlings are used as rootstocks. An experiment at CHES (Central Horticultural Experiment Station), Chettalli, using 10-month-old seedlings in November and December, achieved 42 per cent success (Tripathi *et al.*, 2014). Patch budding is preferred due to its higher success rate, performed on well-established rootstocks of 8 to 12 months in May or July, with success rates as high as 83.6 per cent reported in other countries. The selection of budwood is critical, with the best results coming from shoots collected 3-4 months after harvesting. A one-year-old budwood with a diameter of 1.5 to 2 cm is ideal, and removing leaves one week before budding improves success. Patch budding has proven the most effective, with a 70 per cent success rate, sprouting in 36.86 days, and a low mortality rate of 6.28 per cent (Bhattacharjee *et al.*, 2022).

Persimmon

Propagation of persimmons is primarily done through grafting, where scions are grafted onto seedling rootstocks. Whip grafting is used for smaller diameter stocks, while cleft and veneer grafting are suitable for larger ones (Chia *et al.*, 1989). It's generally recommended to replant young stock instead of top-working old trees. Scion cultivars can be either grafted or budded, with veneer grafting showing an 80 per cent success rate, particularly when done in September as sap movement begins. Tongue grafting achieves a 60-65 per cent success rate, while T-budding is effective if done in spring or early autumn using mature wood. Timing for vegetative propagation depends on local climate conditions.

Three rootstock species are commonly used for propagating Japanese persimmons: *Diospyros kaki*, *Diospyros lotus*, and *Diospyros virginiana*. *D. lotus*, used in India, is susceptible to crown gall and shows incompatibility with some cultivars like Fuyu. *D. virginiana* is used in Israel and the USA, thriving in heavy, damp soils and cold climates, but often results in uneven tree size and vigor, and is vulnerable to *Cephalosporium* wilt. *D. kaki* is preferred as a rootstock for its uniformity, particularly in Japan, where the Saijo variety produces consistent seedlings.

Aonla

Aonla grows naturally in forests through seeds (Bajpai, 1969). The implementation of systematic vegetative propagation techniques for the mass production of plants is essential in crops such as aonla, as vegetatively produced fruit trees are true-to-type and provide fruit earlier (Hartmann *et al.*, 1993). Aonla is most commonly propagated through budding particularly patch budding (Nand, 1962; Pandy and Prasad, 1980; Pathak *et al.*, 1991). A good success rate can be achieved with patch budding under North Indian conditions between the middle of May and September, provided that the budding is performed on seedling rootstocks that are one year old and have a pencil thickness. Numerous researchers have attempted propagation through patch budding across various agroclimatic conditions and timeframes, yielding differing levels of success due to distinct specific features (Table 7.4). Patch budding requires nearly two years from the raising of rootstock to the planting of budded plants. Recently, the bench grafting procedure was standardized in aonla by Tewari and Bajpai (2005) to facilitate the timely propagation of establishment-friendly grafts. The procedure involves sowing seeds in February-March in polythene bags or seed beds. In the subsequent February, rootstocks are physically uprooted, Cleft grafting is executed at the collar or approximately 3 cm above the collar of the rootstock. Dormant scion shoots of pencil thickness from selected mother plants are taken out prior to grafting and preserved in a separate moist gunny bag. Uprooted plants are stored in a shaded area and covered with a damp cotton bag. Consequently, grafts can be transferred over long distances effortlessly and at minimal expense. *In situ* grafting is recommended for the commercial cultivation of aonla in wastelands by transplanting one-month-old seedlings or rootstocks from poly bags and performing chip budding in the last week of June the following year (Negi *et al.*, 2013).

Table 7.4: Different Vegetative Propagation Methods in Aonla and its Success Rate

Sl. No.	*Type of Propagation*	*Month of Propagation*	*Success Rate (%)*	*Special Attributes*	*Reference*
1.	Patch Budding	July	>90	Use of perforated polybags	Saroj *et al.*, 2000
2.	Patch Budding	Jun	100	Patch budding at 25 cm budding height	Yadav and Deen, 2024
3.	Patch Budding	July	85.35	Patch budding found superior over T, Ring and Forkert budding	Kem and Rawat, 2008
4.	Patch Budding	August	52.39	Patch budding performed *in-situ*	Sharma and Mohammad, 2006
5.	Chip Budding	March-April	86.66	The chip budding technique has proven effective in spring.	Sharma *et al.*, 2000
6.	Soft wood grafting	January	86	5 to 12 months old local seedlings rootstock.	Panchbhai *et al.*, 2006
7.	Cleft grafting	July	87	5-6 months old aonla seedlings rootstock	Tewari and Bajpai, 2002

Bael

Traditionally, bael is often propagated using seeds for cultivation (Singh *et al.*, 1976). The viability of seeds is limited and typically susceptible to insect infestation (Purohit and Vyas 2004). Seedlings derived from this sort of seeds exhibit significant variation in physical and biochemical traits due to the heterozygous nature of the plant. Common means of vegetative propagation for bael include budding, grafting, layering, and root suckers (Table 7.5). Currently, patch budding and softwood grafting are being commercially adopted for proliferation of *bael*. *Bael* rootstocks treated with BA (50 ppm) seven days before budding exhibited early bud break, a maximum bud sprouting (100%), increased length of sprouted scion 30 days post-budding, enhanced diameter of sprouted scion, more leaf count, and optimal scion growth initiation in plants. However, the application of plant growth regulators to the scion branch of the mother plant seven days before budding did not yield satisfactory results (Kumar *et al.,* 2017).

Table 7.5: Different Vegetative Propagation Methods in Bael and its Success Rate

Sl. No.	*Type of Propagation*	*Month of Propagation*	*Success Rate (%)*	*Special Attributes*	*Reference*
1.	Patch budding	June	80.25	Propagation under protected structure gave better results than open field conditions	Syamal *et al.,* 2013
2.	Grafting	March	85-95	Under the propagation shed	Ghosh *et al.,* 2012
3.	Patch budding	Mid ofMay to June	90-95	Rootstock consisted of one yearold uniform seedling	Kumar *et al.,* 1994
4.	Whip grafting	June-July	70	Under West Bengal conditions	Maiti *et al.,* 1999

Minor Sub-tropical Fruits

Minor sub-tropical fruits are also propagated asexually to maintain identity and quality. The information is provided in Table 7.6.

Table 7.6: Propagation Methods and Time of Minor Sub-tropical Fruits

Crop	*Method*	*Ideal Time of Propagation*	*Remark*	*Reference*
Passion fruit (*Passiflora edulis*)	Seeds	March- April or Oct		Hand Book of Horticulture Vol. 1
	Softwood and Semi -hardwood cutting	July- August	Commercial	

Crop	*Method*	*Ideal Time of Propagation*	*Remark*	*Reference*
	Cleft graft, whip graft or side wedge graft	July- September	Rootstocks: Yellow passion fruit (*Passiflora edulis* f. *flavicarpa*), *P. alata*, *P. giberti*, *P. caerulea*	Bose *et al.*, 2021
	Serpentine layering	February		Hand Book of Horticulture Vol1
	Micro-propagation		Explant: leaves or internodal segment	
Egg Fruit (*Pouteria campechiana*)	Seed	July- September		Chettri and Singh, 2024
	Side or Veneer grafting	July- September		
Fig (*Ficus carica*)	Cutting	July	Treat cuttings with IBA 3000 ppm	Sivaji *et al.*, 2014
	Semi-hard wood cuttings	August to October February to April	The cuttings with 4-5 nodes having a length of 12-15 cm.	Ara *et al.*, 2023
	Hardwood cutting	-	Cutting with 4-5 nodes Treat cuttings with IBA 4000 ppm by quick dip method for 25 seconds	Patel *et al.*, 2017
Carambola (*Averrhoa carambola* Linn.)	Softwood grafting	June to August	-	Sonawane *et al.*, 2012

Conclusion

It is essential to select the appropriate cultivars and high-quality planting material in order to achieve larger yields and improve the overall quality of the fruits. The planting material should be free of diseases and true to type of plant being planted. It is imperative that the standardized propagation protocol be adhered to in order to ensure the production of planting material on a large scale in fruit crops. The propagation procedure that is currently in use should be followed by the commercial nurseries. When it comes to the proliferation of fruit crops, the government institutions need to provide the nursery operator with training on innovative approaches.

REFERENCES

Abou-Rawash M, El-Wakeel HF, Kassem and Mohamed EA, 1998. Studies on the vegetative propagation of some citrus rootstocks. *Annals of Agricultural Sciences*, **43**: 523-537.

Adsule PG and Updhayay A, 2004. Grape cultivation on rootstock. Nation al Research Centre for Grapes, Technical Bulletin No. 05, 23 p.

Ahire DB, Ranpise SA and Shete MB, 2017. Assessment of Graft Compatibility of Different Rootstocks of Pomegranate (*Punica granatum* L.). *International Journal of Minor Fruits, Medicinal and Aromatic Plants,* **3**(2): 21-27.

Ara R, Khan SAKU, Mannan MA and Rahim MA, 2023. Evaluation of some rooting substrates and cutting types in propagation of fig (*Ficus carica*). *International Journal of Minor Fruits, Medicinal and Aromatic Plants*, **9**(1): 14-24.

Bajpai PN, 1969. Studies on flowering and fruit development in aonla (*Emblica officinalis* Gaertn). *Horticulture Advances*, **7**(2): 38-67.

Bhambota JR, Singh KK and Saini H, 1968. Propagation litchi plant. *Punjab Horticultural Journal*, **8**: 200-203

Bhattacharjee P, Sakthivel T, Chander S and Nimbolkar PK, 2022. Standardization of asexual propagation techniques in rambutan (*Nephelium lappaceum* L.) for humid tropical region of India. *Environment Conservation Journal*, **23**(1 and 2), 31-38.

Bhosale VP, Jadhav RG, Masu MM, Sitapara HH and Patel HC, 2009. Response of different media and plant growth regulators on rooting and survival of air layers in pomegranate (*Punica granatum* L.) cv. Sinduri. Proceedings of 2nd International Symposium on Pomegranate and Minor Including Mediterranean Fruits. University of Agricultural Sciences, Dharwad, India, pp. 72-73.

Bowman KD, Joubert J, 2020. Chapter 6 - citrus rootstocks. In: *The Genus Citrus*. Eds. Talon, M., Caruso, M., Gmitter, F. G. (Woodhead Publishing, Duxford, UK), 105–127.

Castro M, Darrouy N and Iturrieta R, 2006. (April). "Franqueamiento": A new vegetative propagation technique for loquat. In *II International Symposium on Loquat* **750**: 325-330.

Chandra R and Jadhav VT, 2012. Grafting methods and time in pomegranate (*Punica granatum*) under semi-arid agro-climatic condition of Maharashtra. *Indian Journal of Agricultural Sciences,* **82**(11): 990–2.

Chandra R, Pal RK, Rigveda D, Singh NV and Maity A, 2014. Propagation practices in pomegranate: a review. *Indian Journal Arid Horticulture*, **9**: 1-6.

Chithiraichelvan R, Kumar S, Karunakaran G and Sakthivel T, 2006 (June). Standardization of propagation techniques in Avocado (*Persia americana* Mill). In *National Symposium on underutilized horticultural crops*, **68**.

Chua SE and Yong SK, 1978. The use of approach, bud and wedge grafting techniques to propagate durian (*Durio zibethinus* Murr.), rambutan (*Nephelium lappaceum* L.) and mango (*Mangifera indica* L.). *Singapore Journal Primary Industries*, **6**(2): 94-101.

Curk F, Ollitrault F, Garcia-Lor A, Luro F, Navarro L, Ollitrault P, 2016. Phylogenetic origin of limes and lemons revealed by cytoplasmic and nuclear markers. *Annals of Botany* **117**(4): 565–583.

Dhakar MK and Das B, 2017. Standardization of grafting technique in litchi. Indian *Journal of Horticulture*, **74**(01): 16-19.

Dubey AK, Sharma RM and Kumar A, 2021. Long term performance of mango varieties on five polyembryonic rootstocks under sub-tropical conditions: Effect on vigour, yield, fruit quality and nutrient acquisition. *Scientia Horticulturae*, **280**: 109944.

Fabbri A and Lambardi M, 1988. Effects of stock plant CCC treatments on adventitious rooting in '140 Ruggeri' cuttings. *Acta Horticulturae*, **227**: 260-262.

Ferrer M, Guarinoni A and Camussi G, 1991. Effect of the rooting media and date of cuttings extraction on the quantity and quality of rooting in the grape rootstock SO4 (*Vitis riparia* × *Vitis berlandieri*). Boletin - de - Investigacion - Facultad - de - Agronomia, Universidad - de - la - Republica, No.31.

Finetto GA, 2009. Pomegranate industry in Afghanistan: Opportunities and constraints. Proceedings of 2nd International Symposium on Pomegranate and Minor Including Mediterranean Fruits, University of Agricultural Sciences, Dharwad, India, pp 15-21 (Abstract)

Galavi M, Karimian MA and Mousavi SR, 2013. Effects of different auxin (IBA) concentrations and planting beds on rooting grape cuttings (Vitis vinifera). *Annual Review and Research in Biology*, **3**(4), 517-523.

George SV and Alex N, 2016. Patch budding: a sustainable method of propagation in litchi. In *V International Symposium on Lychee, Longan and Other Sapindaceae Fruit*, **1211**: 181-186.

Ghosh SN, Roy S and Bera B, 2012. Studies on propagation of bael (*Aegle marmelos* L.) under Jhargram conditions. *Journal Horticultural Sciences*, **7**(2): 214-216.

Goode D and Lane R, 1983. Rooting of leafy muscadine grape cuttings. *Hortscience*, **18**(6), 944-946.

Hartmann HT, Kester DE and Davies FT, 1993. Plant Propagation Principles and Practices. Fifth Edition, Prentice Hall of India Pvt. Ltd., New Delhi.

Hore JK and Sen SK, 1995. Role of non auxinic compounds and IBA on root regeneration in air layers of root formation in pomegranate (*Punica granatum* L.). *Current Research University Agricultural Science*, **24**: 83-85.

Kanthrajah AS, Dodd WA and McConchie CA, 1992. In vitro embryo culture and induction of multiple shoots in lychee (*Litchi chinenesis* Sonn.). *Annals of Botany*, **70**(2): 153-156.

Karimi R, 2011. Stenting (cutting and grafting): A new technique for propagation pomegranate (*Punica granatum* L.). *Journal Fruit and Ornamental Plant Research*, **19**: 73-79.

Karimi R and Farahmand H, 2011. Study of pomegranate (*Punica granatum* L.) propagation using bench grafting. *Journal Fruit and Ornamental Plant Research* **19**: 67-72.

Kem JC and Rawat SS, 2008. Standardization of time and methods of budding in aonla cultivars under sub-tropical conditions of Garhwal Himalaya. *Progressive Horticulture*, **40**(1): 78-82.

Kumar D, Pathak RK and Ali W, 1994. Studies on effect of duration and methods of budding in bael. *Indian Journal Horticulture*, **51**(2): 150-153.

Kumar S, Prakash S, Kumar A, Dhyani BP, 2021. Effect of IBA, boric acid and wounding treatments on rooting and growth of stem cuttings in pomegranate (*Punica granatum* L.). *The Pharma Innovation Journal*, **10**(3): 480-488

Kumar S, Ram S and Singh CP, 2000. Success of veneer and cleft grafting at different grafting heights of seedling rootstocks in Dashehari mango. *Indian Journal Horticulture*, **57**(3): 212-214.

Kumar Y, Sharma MK, Singh D, Kumar K and Verma AK, 2017. Effect of auxins and cytokinin on budding and growth of saplings of bael (*Aegle marmelos* Correa.)-a medicinal plant. *Journal Agriculture and Ecology*, **3**: 12-18.

Levin GM, 2006. Pomegranate, I[st] Edn, Third Millennium Publishing, East Libra Drive Tempe, AZ, USA.

Maiti CS, Nath A and Sen SK, 1999. Studies on the propagation of bael (*Aegle marmelos* Correa) by different grafting methods in West Bengal. *Journal of Applied Horticulture*, **1**(2): 131-132.

Menzel CM, Watson BJ and Simpson DR, 1990. Longan. In: T.K. Bose and S.K. Mitra (eds.) Fruits: Tropical and Sub-tropical. NayaProkash, Calcutta, India, pp. 522-546.

Nand D, 1962. Budding in aonla (*Phyllunthus emblica*). *Sci. Cult.* **28**: 486.

Negi RS, Baghel BS and Gautam US, 2013. Effect of method of orchard establishment and propagation on growth and development of aonla (*Emblica officinalis* Gaertn.) plants in wastelands. *Journal Applied Horticulture,* **15**(3): 227-231.

Panchbhai DM, Athavale RB, Jogdande ND and Dalal SR, 2006. Soft wood grafting-aonla propagation made easy. *Agricultural Science Digest,* **26**(1): 71-72.

Pandy IC, Prasad KS, 1980. Propagation of aonla by budding. *Progressive Horticulture,* **17**: 345-346.

Patel HR, Patel MG and Singh S, 2017. Effect of different levels of IBA and NAA on rooting of hardwood and semi hardwood cutting in fig. *Int. J Agri. Sci. Res,* **7**(4): 519-523.

Pathak RK, Ojha CM, Dwivedi R, Hari Om, 1991. Studies on the effect of method and duration of budding in aonla. *Indian J. Hort.,* **47**: 207-12.

Patil VN, Chauhan PS, Shivankar RS, Vilhekar SH and Waghmare V, 2001. Effect of plant growth regulators on survival and vegetative growth of grapevine cuttings. *Agricultural Science Digest,* **21**(2): 97-99.

Purohit SS and Vyas SP, 2004. In: Aegle marmelos Correa ex Roxb. (Bael), Medicinal plant cultivation- A scientific approach. *Agrobios,* p. 280-285.

Rahman N, Hussain I, Imran M, Jan T and Awan A, 2000. Effect of different concentrations of IBA on rooting of litchi (*Litchi chinensis*) in air layering. *Pakistan Journal Biological Sciences,* **3**(2): 330-331.

Raja WH, Noomphy M, Bam G, Nabi S, Sharma OC, Mir JI, and GS M, 2021. Potential and prospects of Kiwifruit production in the state of Arunachal Pradesh. *Pharma Innovation Journal,* **10**: 12.

Rebolledo-Martínez A, Peralta-Antonio N, Rebolledo-Martínez L, Becerril-Román EA and Rebolledo-García RL, 2019. Effect of rootstock in tree growth, dry matter, flowering, yield and quality of 'Manila'mango. *Scientia Horticulturae,* **251**: 155-161.

Reddy YTN, Kurian RM, Ramachander PR, Singh G and Kohli RR, 2003. Long-term effects of rootstocks on growth and fruit yielding patterns of 'Alphonso'mango (Mangifera indica L.). *Scientia Horticulturae,* **97**(2): 95-108.

Rego C, Nascimento T, Cabral A, Silva MJ and Oliveira H, 2009. Control of grapevine wood fungi in commercial nurseries. *Phytopathologia Mediterranea,* **48**: 128-135.

Roy S and Sinha AK, 1994. Comparative evaluation of veneer and approach grafting in mango (Mangifera indica L.) propagation. *Journal Applied Biology,* **4**(1/2): 33-34.

Sandoval-Prando MA, Chiavazza P, Faggio A and Contessa C, 2014. Effect of coconut water and growth regulator supplements on *in vitro* propagation of *Corylus avellana* L. *Scientia Horticulturae,* : 91-94.

Sarin NB, Prasad US, Kantharajah AS and Jain SM, 2003. Micropropagation in litchi (*Litchi chinenesis* Sonn.). In: *Micropropagation of Woody Trees and Fruits*, **75**: 721-731.

Saroj PL, Nath V and Vashishtha BB, 2000. Effect of polycontainers on germination, seedling vigour, root characters and budding success in aonla. *Indian Journal Horticulture*, **57**(4): 300-304.

Sharma AK, Sharma K and Badyala SD, 2000. Effect of transplanting time of rootstock and budding technique on rapid propagation of aonla. *Indian Journal Soil Conservation*, **28**(3): 235-237.

Sharma DK and Mohammad S, 2006. Budding in aonla as influenced by season, rootstock raising and cultivar. *Annals Arid Zone*, **45**(1).

Sheela VL and Nair SR, 2001. Growth, flowering and yield potential of tissue culture banana (Musa AAB cv. Nendran). *Journal Tropical Agriculture*, **39**: 1–4.

Singh DB and Suryanarayana MA, 1996. Studies on softwood grafting in mango. *Flora and Fauna*, **21**: 83-84.

Singh KK and Chauhan JS, 2020. A review on vegetative propagation of Grapes (*Vitis vinefera* L.) through cutting. *Global Journal Bio-Science and Biotechnology*, **9**(2): 50-55

Singh NV, Babu KD, Chandra R, Sharma J, Sahu P, Meshram DT and Pal RK, 2017. *Quality planting material production in pomegranate*. In: Pal, R.K. and Singh, N.V (eds), *Pomegranate for nutrition, livelihood security and entrepreneurship development*. Daya Publishing House (A division of Astral International Pvt. Ltd), New Delhi, India, pp. 69-79.

Singh NV, Karimi HR, Sharma J and Babu KD, 2021. Propagation techniques and nursery management. In: The pomegranate: Botany, Production and uses (Ali, S., Yavari, A. and Zamani, Z. eds), CAB International, Wallingford, Oxfordshire, U.K., pp.196-224.

Singh NV, Salunkhe O, Awachare C, Salutgi U, Sharma B, Parashuram S and Marathe RA, 2022. Healthy Saplings – Key to Sustainable Pomegranate Production. *Biological Forum – An International Journal* **14**(4): 1188-1194.

Singh RR, Karuna K, Kumar A and Mankar A, 2012. Studies on the effect of time and methods of grafting on success and growthof mango graft. *Progressive Horticulture*, **44**(1): 153-156.

Singh UR, Pandey IC and Prasad S, 1976. Propagation of Bael (*Aegle marmelos* Corr.). *Punjab Horticulture Journal*, **16**: 57-59.

Sivaji T, Madhavi K and Vani VS, 2014. Effect of Type of Cuttings and IBA Concentrations on the Propagation of Fig (*Ficus carica*) cv. Poona Fig Under Open Conditions. *Trends in Biosciences*, **7**(11): 1087-1089.

Skaria M, 2000. A micro budding technique for biological indexing and ultra-high density planting of citrus. International organization of Citrus Virologists Conference Proceedings (1957-2010): **14**(14).

Sonawane GR, Khandekar RG, Korake GN, Haldankar PM and Mali PC, 2012. Effect of season on softwood grafting in carambola (*Averrhoa carambola* Linn). *Asian Journal Horticulture,* **7**(2): 412-415.

Syamal MM, Maurya VK and Joshi M, 2013. Effect of methods and time of propagation in bael under different growing conditions. *Indian Journal Horticulture,* **70**(1): 127-129.

Tewari RK and Bajpai CK, 2002. Propagation of aonla (*Emblica officinalis*) through grafting in polybags. *The Indian Journal Agricultural Sciences,* **72**(6): 353-354.

Tewari RK and Bajpai CK, 2005. Bench Grafting in Aonla (*Emblica officinalis* Gaertn.) for Easy Transport and Establishment. *Indian J. Agroforestry* **7**(1): 65-67.

Tofanelli MB, Freitas PL and Pereira GE, 2014. 2-4-Dichlorophenoxyacetic acid as an alternative auxin for rooting of vine rootstock cuttings. *Revista Brasileira de Fruticultura,* **36**(3): 664-672.

Tomar KS, 2011. Effect of different concentrations of growth regulators on rooting and survival percentage of pomegranate air layers. *Progressive Agriculture,* **11**: 431-433.

Trbic G, Djurdjevic VI, Mandic MV, Ivanisevic M, Cupac R, Bajic D *et al.,* 2021. The impact of climate change on grapevines in Bosnia and Herzegovina. *Euro-Mediterranean Journal for Environmental Integration,* **6**: 4.

Tripathi PC and Karunakaran G, 2019. Standardization of time and method of propagation in avocado. *Journal Applied Horticulture,* **21**(1): 67-69.

Tripathi PC, Karunakaran G, Sakthivel T, Sankar V and Senthil Kumar R, 2014. Rambutan cultivation in India. *Technical Bulletin,* **1**: 18.

Upadhya B, Baral DB, Gautam DM and Shrestha SM, 2014. Influence of rootstock age and pre-defoliation of scion on the success of epicotyl grafting of mango. *International Journal Research,* **1**(7): 172-182.

Waite H and Morton L, 2007. Hot water treatment, trunk diseases and other critical factors in the production of high-quality grapevine planting material. *Phytopathologia Mediterranea,* **46**: 5-17.

Yadav U and Deen B, 2024. Effect of Budding Height on Success of Patch Budding in Aonla (*Emblica officinalis* Gaertn.). *Journal Advances in Biology and Biotechnology,* **27**(7): 1371-1379.

Chapter 8

Canopy Management in Sub-tropical Fruits

***Kundan Kishore*[1], *Kanchan K. Srivastava*[1], *Dinesh Kumar*[1] *and Pallavi Jha*[2]**

[1]*ICAR–Central Institute for Subtropical Horticulture, Lucknow–226101, U.P.*
[2]*Punjab Agricultural University, Ludhiana, Punjab*

Plant canopy management is a vital component in fruit crops as it influences light distribution, flowering, yield, fruit quality and pest incidence, particularly under HDPS. It also helps to improve photosynthetic activity in plants, which in turn ensures higher productivity. Plant geometry, canopy architecture, pruning and flower induction are vital components of orchard management under high density planting system. Proper spatial arrangement of plants facilitates different cultural operations such as fertilization, weeding, spraying, training and pruning, harvesting, *etc.*, to ensure better crop growth, enhanced flowering, fruit yield and quality. Quantitative and qualitative regulation of branches in early growth stages is prerequisite for developing plant architecture. Moreover, regulation and orientation of primary, secondary and tertiary branches are crucial for development of optimum plant architecture. In mango, jamun and guava open centre training system is invariably followed. However, in recent years, Y-trellis and espalier system are novel approach for getting higher quality yield in mango. Espalier system is followed in guava for quality fruit production. Plant canopy management establishes a relationship among light distribution, photosynthetic activity, flower induction, input use efficiency, dynamics of pests and diseases, yield and fruit quality. Prevalence of irregularity in bearing necessitates use of flower inducer in mango. Among various methods of flower induction, paclobutrazol and girdling are considered as important approaches.

Introduction

In India, most of fruit orchards of sub-tropical region are still under low density plantation. Howeverm some of the fruit orchards of mango and guava are under medium to high density plantation. The main problem with medium- to high-density plantations is the over-crowding of branches affecting the production potential of orchards. Production of mango and guava are highly affected under close spacing plantation. Once the trees start to shade each other, production usually falls due to poor light availability inside the canopy. Attempts to control the growth of the trees with proper management of canopy is often not practiced which results in the loss of production. Canopy management in fruit crops deals with the development and maintenance of plant structure in relation to size and shape for maximizing yield and quality. Reorientation of canopy has been a cornerstone to resolve the issue of low production of quality fruits. Moreover, better understanding of canopy architecture and function, in relation to its environment, appears necessary to design appropriate canopy management practices. Canopy management in evergreen perennial fruit crops such as mango and jamun through pruning was not considered as a common practice, but in recent years annual pruning is considered essential for sustaining yield especially under medium to low density plantation. Demonstration of multiple vegetative flush and distinct pattern of vegetative growth greatly influence flower induction mechanism in mango and jamun which necessitate to develop a balance between source and sink for ensuring regularity in yield in mango. It has been observed that under high density planting system crops bear satisfactorily only up to limited span due to vigorous vegetative growth attributed to the occurrence of multiple vegetative flushes. Under such condition, canopy regulation is a vital component of production technology to enhance economic life of mango trees. In addition, flowering periodicity in commercially important cultivars is a major limitation in mango cultivation. Canopy management needs to take into account the plant geometry, plant architecture, pruning and flower induction. In this chapter plant geometry, recent advances in canopy architecture, canopy management and flower induction are described to signify their importance in sustaining yield particularly under high density planting system.

Plant Geometry

Plant geometry is the spatial arrangement of plants for effective utilization of resources like land, light and soil nutrients. It is basically deals with the system of planting and spacing. Planting system is based on the principal of arranging optimum number of plants in a piece of land with an aim to provide adequate space for proper tree development. Proper arrangement of plants facilitates different cultural operations such as fertilization, weeding, spraying, training and pruning, harvesting, *etc.*, to ensure better crop growth, enhanced flowering, fruit yield and quality. The common planting systems in mango are square, rectangle, triangular and quincunx. These planting systems can be grouped under vertical row planting pattern and alternate row planting pattern. Square and rectangular system of planting are based on vertical row planting pattern, whereas quincunx and triangular system of planting follow alternate row planting pattern. Square system is the most common method of planting in fruit crops. The plants are spaced exactly at right angle to each other. In this system intercultural operations can be done in both directions. Rectangular system is similar to that of the square in its layout except that the spacing between the rows and that of plants in a row are not equal. Rectangular system accommodates more plants in rows. In this system, inter-cultural operations can be carried out with more ease as the distance between two rows are more. This system facilitates better utilization of sunlight for growth and development of plants. This is the most suitable planting system in mango. Triangular system of planting is also practiced in some parts of the region wherein the even numbered rows are kept midway between those in the odd rows. When compared to square system, each tree occupies more area and hence relatively fewer trees are accommodated. On the basis of plant population in unit area, the planting system in mango may be categorized into low, medium, high and ultra high density plantation (Table 8.1). In low density plantation 100-200 plants are accommodated in one hectare area, whereas in medium density, >200-400 plants are accommodated. Under high density planting system >400-1000 plants are accommodated, whereas in ultra high density planting >2000 plants are accommodated.

Table 8.1: Plant population under different planting density

Planting Density	*Planting Distance*	*System of Planting*	*Plant Population (Plants/ha)*	*Crop*
Low density	10m x 10m	Square	100	Mango and jamun
	8m x 8m	Square	156	
Medium density	6m x 7m	Rectangular	238	Mango and jamun
	6m x 6m	Square	277	
	5m x 5m	Square	400	Mango, gauava, jamun
High density	4m x 5m	Rectangular	500	Mango and guava
	3m x 5m	Rectangular	666	
	2.5m x 5m	Rectangular	800	
	3m x 4m	Rectangular	833	
	2.5m x 4m	Rectangular	1000	
Ultra high density	2.5m x 2.5m	Square	1600	Mango and guava
	2m x 3m	Rectangular	1666	
	1.5m x 2.5m	Rectangular	2666	

Plant Architecture

Plant architecture represents the organization of plant components in space, which may change with time. It is a biological trait, which determines growth and fruit bearing capacity. Plant characters like plant height, number of branches, leaf area, respective position of the organs (leaf, stem, flowers and fruits), are genetically controlled. However these traits are highly influenced by environment and human interventions (Tang *et al.,* 2019). Efforts have been made to develop an 'ideotype', ideal plant architecture, to facilitate optimized interception and better distribution of sunlight, to maximize synthesis of photosynthates and to ensure a balance between 'source and sink' which are essential for higher yield and better fruit quality. However it has been successful in annual plants like cereals and vegetables. On the other hand, developing ideal canopy architecture in fruit crops is a tedious job due to its inherent growth behavior.

Usually, plant does not produce quality fruit unless scientific intervention is made. Plants need to be regulated optimally for enhanced production of quality fruit. Canopy architecture is thus one of the most important production factors for ensuring high yield, optimum leaf area index (LAI) and profit per unit area and time. It primarily deals with the regulation, orientation and arrangement of branches at different levels of canopy (Hassan *et al.,* 2010). In many fruit crops, substantial increase in yield (30-50%) with improved fruit quality has been achieved by managing appropriate and efficient canopies particularly under high density plantation system. This intervention is commonly followed in temperate crops due to their growth behaviour and bearing pattern. It has been scientifically validated that plants with short stature harness sunlight more efficiently than larger ones which has direct impact on orchard efficiency. In tropical crop like mango there is an ample scope for canopy architecture ideotype to improve yield, quality and profitability under high density plantation.

An ideal plant architecture (ideotype) not only facilitates better interception and distribution of light, regulation of fruiting terminals and cultural operations but also able to carry crop load. However, profound knowledge on growth behaviour, branching pattern and bearing habit is essential in order to develop an ideotype. While tailoring plants, orientation and angular distance among branches, optimized number of photosynthetically active leaves and fruiting terminals are taken into account. Optimization of plant architecture under high density plantation by regulating number and orientation of branches is key component under high density planting (Mitra 2019). Quantitative and qualitative regulation of branches in early growth stages is prerequisite for developing plant architecture. Three primary branches need to be oriented in such a manner that each should make about crotch angle of 45° - 50° from vertical axis. For better plant architecture there should be 2-3 secondary branches per primary and 2 tertiary branches per secondary should be allowed to develop. Care should be taken to keep the centre of the plant open by removing branches from central portion.

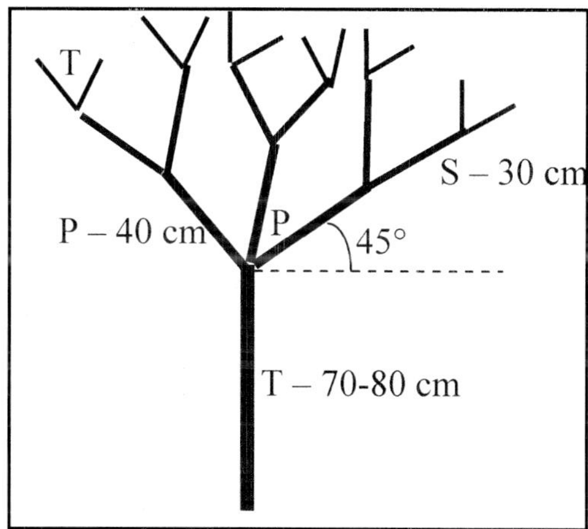

Figure 8.1: Canopy Architecture in Mango.

In terms of light interception and distribution, canopy volume, LAI, yield, fruit quality and pest incidence, a combination of 3 primary, 2 secondary and 2 tertiary branches was found suitable under the plant geometry of 3m x 4m spacing, whereas a combination of 3 primary, 2 secondary and 3 tertiary branches was found to be optimal under 5m x 5m spacing. It was evident from research findings that canopy volume of 20-22 m^3/plant, LAI of 3.3 and light interception of 60 per cent may be considered optimum for an ideal plant architecture for 5m x 5m spacing, whereas canopy volume of 7.5 m^3/plant, LAI of 3.5 and light interception of 58-60 per cent may be considered optimum for an ideal plant architecture for 3m x 4m spacing.

Novel Training System

For the development of an ideal plant architecture training system, orientation of branches and canopy microclimate are vital as they ensure higher yield and better fruit quality. Open centre training system is commonly followed in fruit crops such as mango, jamun, guava, litchi, *etc.*, however novel training systems like Y-trellis, espalier, Open vase, closed vase and palmette system have been assessed for their suitability on enhancing yield and fruit quality of mango (Fivaz and Stassen, 1997; Kishore *et al.*, 2022) Y-shaped trellis system and espalier system are practiced in eastern tropical condition and sub-tropical condition and the efficacy of training systems is compared with the conventional open centre training system.

Y-trellis

Mango and guava can be trained on Y shaped trellis by allowing two opposite arms (primary branches) to grow at right angle from the height of 90 cm. Plants were trained with 6 cordons along the wires on either side of trellis and for that two secondary branches were allowed to develop horizontally in opposite direction from the arms at a distance of 45 cm. From each side of secondary branch, 5 tertiary branches were allowed to develop where in 4 fruiting terminals (cane) were retained. Kishore *et al.* (2022) reported that Y trellis and espalier in mango could significantly promote availability and distribution of photosynthetic active radiation inside the canopy. The training systems exhibited a prominent impact on photosynthetic activities, flower intensity and fruit yield.

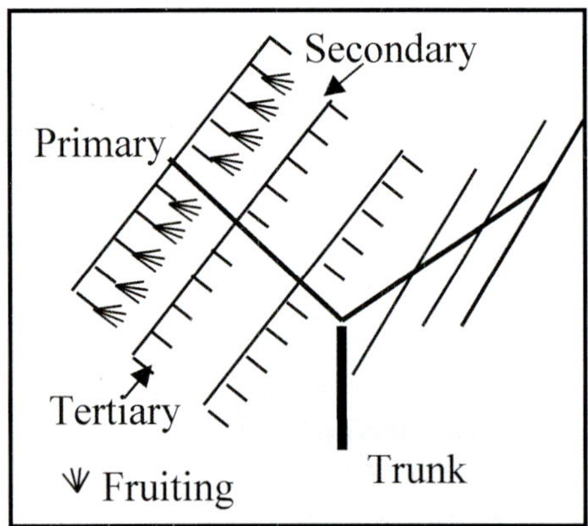

Figure 8.2: Y- trellis.

Figure 8.3: Mango and Guava Trained on Y-trellis.

Espalier System

Mango and guava are trained on vertical trellis by allowing three layers of two opposing arms (primary branches) to grow horizontally at different heights. The first cordon (arm) is allowed to develop at the height of 90 cm, whereas second and third cordons are trained at a height of 1.5m and 2m, respectively. Each cordon has 10 secondary branches (5 on either side) and from each secondary branch 4 tertiary branches were allowed to grow. Fruiting terminals are maintained in every tertiary branch which bears fruits. The suitability of espalier system was assessed in guava under sub-tropical conditions of north India. Srivastava *et al.* (2018) reported that espalier architecture system was a good option for increasing yield and quality

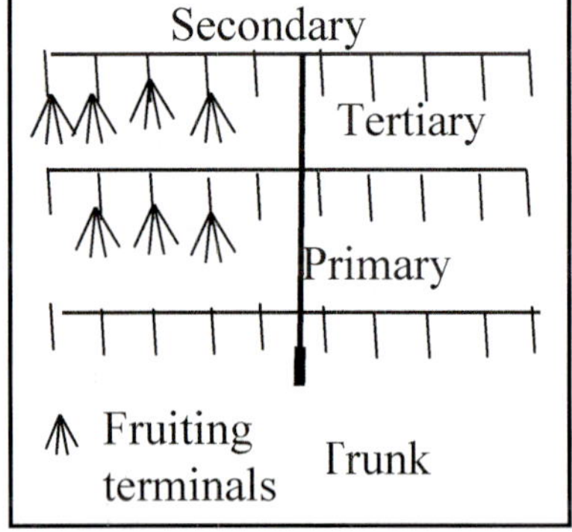

Figure 8.4: Espalier System.

of guava due to maximum exposure of sunlight to scaffolds branches. Moreover it facilitates production of attractive and larger fruits with reduced disease and insect incidence.

Figure 8.4: Mango and Guava Trained on Espalier System.

Canopy Management

Plant canopy management establishes a relationship among synthesis of photosynthates, utilization of light energy, reproductive development, efficient use of nutrients, population dynamics of pests and diseases, fruit yield and fruit quality. Spatial and temporal variations in the radiation regime in different locations of a canopy are vital to integrate the activities of individual leaf and their responses to the natural environment and to evaluate the performance of various plant canopy forms (Wang *et al.,* 2009). Plant with dense canopy has higher proportion of photosynthetically inactive leaves located inside canopy which act as an unproductive sink. This leads to disturbance in source-sink relationships and as a result major proportion of plant canopy remains unfruitful. Moreover, restricted distribution of solar radiation and circulation of air inside the dense canopy, build up the microclimate congenial for pests and diseases (Silva *et al.,* 2014). All these contribute towards low fruit yield and poor quality of the produce. Hence successful fruit cultivation essentially requires efficient harnessing of solar radiation and conversion of light energy into photosynthates for viable sinks like flower and fruits. Fruit crops responds differently to canopy management due to their growth pattern and bearing habit. Guava responds better than mango and jamun to canopy modification due to its fruit bearing habit. Guava bears on the current season emerging shoots. Regular pruning of shoots after fruit harvest encourages development of lateral shoots from which flowering occurs. Optimization of plant density and canopy management techniques is required for higher yield and quality guava production (Mitra *et al.,* 2018). Singh *et al.* (2018). Maintaining a dwarf framework of jamun trees is a tedious task. However open centre training system with annual pruning in September is effective to reduce plant canopy and to improve flowering, fruiting and fruit quality in jamun (Singh *et al.,* 2018).

Approaches for Canopy Management

Approaches to canopy management are continually evolving in response to changes in management practices. It should be aimed to promote mechanization and to reduce cost of cultivation without compromising productivity and quality. This intervention acts a catalyst to refine approaches to canopy management in a low input context. The basic concept in canopy management of fruit crops is to make best use of land and climatic factors for an increased productivity by manipulating the tree vigour and by efficient utilization of available sunlight and temperature, respectively. Canopy management not only supports heavy crop loads but also ensures yield sustainability, fruit quality and minimal load of biotic stress. In order to develop optimized plant canopy, knowledge on tree growth behaviour, branching pattern, flowering and other important processes are needed. It has been established that canopy management is the most important tool not only for yield sustainability but also for efficient utilization of resources. There are different approaches whereby canopy can be managed judiciously and among them rootstock, variety and pruning are vital.

Rootstock

With the intensification of fruit production, the role of rootstocks in commercial fruit production has increased considerably. Rootstocks demonstrate several applications such as improving fruit quality, imparting adaptability to climatic and edaphic conditions, inducing dwarfing, *etc.* the priorities of rootstock selection in the tropics and sub-tropics is focused mainly on canopy management and ensuring regularity n fruiting (Reddy *et al.,* 2003). Rootstock can modify tree size, flowering, and fruit yield and quality of mango and guava. In mango polyembryonic genotypes such as Valiakolumban, Bappakai, Chandrakaran, Kurukan, Muvandan, Mylepelian and Olour are used as rootstocks. Among them only Valiakolumban exhibited its potential in imparting dwarfness to Alphonso. Whereas, Rebolledo-Martínez *et al.* (2019) reported that Esmeralda, Irwin morado, Chauza and Thomas demonstrated smaller canopy area and can be preferred for high density plantation of mango in Mexico. On the other hand the genetic potential of dwarf variety like Amrapali has been exploited in different regions of India However it has been observed that the expression of dwarfing character of genotype is highly influenced by growing environment as under the tropical eastern climatic condition the growth pattern of Amrapali is similar to Langra and Dashehari. In the eastern tropical region, pruning is the most important management practice to sustain yield and quality of mango. *Psidium friedrichsthaliamum*, and Pusa Srijan are considered as dwarf rootstock for guava which can be used for high density plantation.

Variety

For establishment of high density orchard, scion and rootstock play crucial role in imparting vigour to plants. Mango breeding in India resulted in the development of number varieties. However only vary few are genetically dwarf. Iyer (1991) reported that dwarfness of mango is associated with higher content of phenolics in the apical bud and lower stomatal density. Varieties like Amrapali, Manjeera, Arka Aruna and Arunika exhibited dwarfing characters (Dinesh *et al.,* 2016). Mallika was also found to be suitable for growing under high density planting system due to its semi-vigorous growth. Dubey *et al.* (2021) reported that Pusa Arunima and Pusa Surya exhibited dwarfness when grafted on K2 rootstock. Tommy Atkins, Palmer, Osteen and Allison Red exhibit dwarfing character, hence they are cultivated under high density cultivation. However the growth behavior of plants varies with the growing conditions, as dwarf varieties become vigorous under humid tropics.

Pruning

Pruning is an effective tool of canopy management to regulate density and number of photo-synthetically active leaves, balance between source and sink and shoot and root ratio. Recently mature leaves are the major source of photoassimilates which are transported to sinks like buds, developing leaves, flowers, fruits and root and thus a coordination between source and sink is maintained in the plant system. Growth, development and productivity of plant are affected if an imbalance between source and sink is created. Excessive vegetative growth, more number of unproductive and unwanted shoots, high crop load, *etc.* are important factors for creating an imbalance between source and sink. In such case pruning can be an effective tool to restore the balance. Pruning is usually not practiced in evergreen plant like mango. But with the adoption of high density plantation, annual pruning is considered mandatory to contain canopy growth, to minimise the incidence of pests and diseases and to achieve sustainability in productivity. In bearing plants, annual pruning is essential to maintain canopy within the allotted space, to improve the light distribution inside the canopy and to make a balance between source and sink. Unpruned mango plants usually possess dense canopy resulting in excessive vigour, interior canopy shading, poor flowering high incidence of pest and diseases and poor fruit quality. In mango, pruning of one year old shoots after fruit harvest is recommended to maintain the canopy of plants and to ensure regularity in flowering under high density planting. Pruning of old branches should be avoided as it promotes vegetative shoots (Kishore *et al.*,). Mango is primarily terminal bearers and usually mature shoots (4-6 months old) bear flowers. Hence pruning should be carried out in such a manner that shoots should attain requisite age at the time of floral bud differentiation. However in some of the case young shoots are also converted into generative shoots due to the prevalence of low temperature for sufficiently long period. In bearing mango trees, not more than 25 per cent biomass should be removed at a time; otherwise it results in excessive vegetative growth with reduced reproductive shoots. Additionally, diseased or dead branches should also be removed to minimize the pest and disease incidence and wastage of nutrients as they act as a sink. There are different strategies of pruning hence following aspects should be taken into account while practicing pruning. Attempts have also been made to prune the trees to particular shapes to improve light interception and distribution. Many researchers have recorded lower yields in pruned trees compared with unpruned trees. The best responses have been achieved with annual light pruning. Yields are more consistent with this approach compared with the response achieve with less frequent but more severe pruning. Pruning of guava (trained on espalier) at an intensity of 65-75 per cent during mid-May improved yield and quality (Srivastava *et al.,* 2018). It was also observed light and moderate pruning resulted higher number of flower buds and fruits. On the other hand, Samant *et al.* (2016) reported that branch bending in May was more effective than pruning in inducing flowering yield and quality of guava in eastern tropical region of India. Jamun is generally not pruned for flowering and canopy management. However Lawande *et al.* (2014) reported that heading back of vigorous jamun trees at 3.6 m and semi-vigorous trees at 3.6 m height facilitated higher flowering intensity, yield and fruit quality.

Flower Regulation

Crop periodicity is one of the most important limitations for mango as it is markedly evident in most of the commercially important varieties. The 'on' year of mango is characterized by the prominence of reproductive shoots whereas, 'off' year is marked by the dominance of vegetative shoots. There are many factors responsible for inducing crop periodicity however hormones play the central role. Floral induction includes all of the developments necessary for the irreversible commitment by the shoot to produce an inflorescence. Control of floral induction and initiation involves signals from different plant parts which in influenced by environmental stimulus like temperature. Often, interactions between environmental stimuli and endogenous developmental cues particularly hormone exert some control over floral induction and initiation. Some aspects of flowering in mango like juvenility, crop periodicity and variations in the amount of flowering between trees, within trees and between branches on the same tree, make it challenge for physiologists, breeders, and growers. Most of the commercially important varieties of mango (Langra, Dashehari, Himsagar, Bombay Green, *etc.*) suffer from the intricate problem of irregular bearing or cropping periodicity as well as staggered or erratic flowering behavior, leading to considerable loss in production. Flowering in mango depends primarily on low temperature, shoot age, hormonal balance and C/N ratio. Among them plant hormones and temperature play more crucial role in regulation of vegetative and reproductive phases of plant.

Flowering is a highly coordinated physiological phenomenon which is influenced by various factors. Under sub-tropical conditions, mango flowers in response to cool temperatures. A day/night temperature regime of 15/10° and duration of low temperature for 7-15 days are the most important stimulus for flower induction. Under tropical conditions, where cold inductive temperatures are brief and erratic, it is unclear what exactly leads to floral initiation; however, factor like prevalence of day/night temperature regime of 20/15°C improves the likelihood of flowering. Shoot age is considered an important factor for flower induction in mango. Under tropical climate shoot age of 4-6 months are considered optimum for becoming reproductive shoot. It is also proved that pruning that manipulates the timing of flush development and synchronized canopy flushing also favours increase in flowering intensity. However pruning of canopy should be planned in such a manner that new flush should attain sufficient maturity (4-6 months). In eastern tropical region, where flowering occurs in December – January, pruning may be practiced during June-July to facilitate new shoots to attain optimum maturity for flower initiation.

Flower Induction

Prevalence of irregularity in bearing necessitates use of flower inducer in mango. Potassium nitrate, ethrel, girdling and paclobutrazol are commonly used flower inducers, however their efficacy vary with the climatic conditions. The application of KNO_3 on 4-5 months old shoots is suggested for inducing flowering in tropical region. Girdling is also practiced in fruit crops especially in litchi and jamun to induce flowering. Haldankar (2014) reported that girdling in tertiary branch of jamun induced flowering, enhanced flowering intensity, number of flowers and fruits per branchlet, yield and reduced period from flowering to harvesting.

Among various flower inducers, paclobutrazol (PBZ) was found to be the most effective in mango. It is a plant growth retardant which inhibits biosynthesis of gibberellins (Nartvaranant *et al.,* 2000). It has demonstrated its usefulness by regulating flowering and fruit yield in various perennial fruit crops including mango. PBZ induces flowering in mango by reducing gibberellin biosynthesis in plant which in turn increases auxin: gibberellins ratio which is considered as essential for flower induction. Floral induction in mango is considered to be the result of elevated levels of florigenic promoter (FP) and lower content of vegetative promoter (VP), primarily gibberellins, whereas the reverse condition promotes vegetative growth. Paclobutrazol, a gibberellin inhibitor, reduces VP level and thereby increases FP/VP ratio which stimulates flowering shoots and regulates vegetative growth. PBZ effectively regulates tree canopy and induces flowering in mango (Davenport, 2007). The efficacy of PBZ depends on the rate, time of application, pruning intensity, soil texture, soil moisture status, soil organic matter content and residue level.

The dose of PBZ may vary with age, canopy volume and pruning intensity. It has been observed that its higher dose severely affects plant in terms of leaf shortening, compaction of shoot and panicle and yield potential. The optimized dose of PBZ is worked out to be 1ml/m canopy spread (0.25 a. i) in September (Kishore *et al.,* 2019). It can be applied in soil either through collar drench or ring method as both are equally effective, however collar drench method is relatively cost effective. Application of PBZ at optimized dose not only increases flowering by more than 70 per cent but also advance flowering by 10-15 days. The application of optimized dose of PBZ doesn't affect leaf size, shoot length and panicle length. On the other hand, higher doze (3-4 ml/m canopy spread) severely affect the growth of leaves, shoots and panicle (Negi and Sharma 2009).

Paclobutrazol (PBZ) was also found effective in inducing flowering in jamun. Deepika *et al.* (2019) reported that application of paclobutrazol @ 1.0 g a.i. m-1 of canopy diameter induced early flowering and advancement in fruit harvest in south India. However, high dose of PBZ caused reduction in leaf area, compaction in shoot and panicle, and severe reduction in plant growth. It is also evident that in compact panicles fruit set is poor and there is high incidence of pest like webber. Annual pruning of current season shoot (forward pruning) followed by application of PBZ induces flowering without affecting canopy, whereas application of PBZ after severe pruning or heading back affects canopy growth and productivity severely due to shoot and panicle compaction. It is recommended that PBZ should not be used unless canopy of plant is sufficiently developed. Hence in order to increase the production of mango under high density plantation (HDP) judicious use of PBZ may be considered as an important input.

Reckless uses of paclobutrazol in certain mango growing belts of the country have raised concern among scientific community about its implication. Paclobutrazol inhibits gibberellin bio-synthesis in the plant which is responsible for cell elongation and internode extension. If plants are treated with higher dose compressed shoots and panicles are produced due to inhibition in gibberellins biosynthesis. Application of PBZ also causes severe stunting if pruning is anticipated (Davenport, 1993). The stunting effect may sustain for several years as paclobutrazol persists in soil; however it has low potential to contaminate surface water and groundwater. Our efforts to review the available world literature, finds that PBZ is moderately hazardous for human being with remote chance of being genotoxic and carcinogenic (Kishore *et al.,* 2015). The important question is whether the continuous use of paclobutrazol is safe or not? Though paclobutrazol has remote chance of ecological contamination due to its low mobility and high persistence, the risk cannot be completely ruled out. In order to minimize residue threats substantially, it is advisable to use the optimal dose at the right time and avoid using in sloppy land. Data of residue indicated that soil residue level increases with the increase in application rate. Usually its residue sustains for 5-6 months in soil. No residue is detected in fruits even with the application of higher dose of PBZ (3-4ml/m canopy spread). These findings clearly indicate that health concern may not associate with the application of optimized dose of PBZ.

Conclusion

Efforts have been made to control the size of fruit trees under high density planting through effective canopy management which exhibited a prominent impact on photosynthetic activities, flower intensity and fruit yield especially under high density planting system. Novel training system exhibited their potential in improving yield and fruit quality in eastern tropical region of India. Under high density planting system, optimized canopy architecture plays vital role in sustaining yield and improving fruit quality. Research has been conducted to assess whether the growth of the trees can be controlled through better canopy management and the application of growth regulators. There are only a few examples where these strategies have provided good productivity over the long term. Research needs to integrate the main factors affecting the productivity of high-density orchards. Pruning at the wrong time, at the wrong severity or at the wrong interval can inhibit flowering and cropping for one or more seasons. Growth regulators such as paclobutrazol have been used to control tree growth, but there are only a few examples where they have been integrated in the management of high-density plantings. Hence effective canopy management strategies have yet to be developed that can be readily integrated into high-density growing systems.

REFERENCES

Davenport TL, 1993. Floral manipulation in mangos. (In) Proceedings on Conference on Mango in Hawaii. Chia C L, Evans D O (Eds). 9–11 March, University of Hawaii.Honolulu (HI), Hawaii, pp. 54–60.

Davenport TL, 2007. Reproductive physiology of mango. *Braz. Jour. Pl. Physio.* **19**: 363-376.

Deepika V, Rajangam J, Swaminathan V, Venkatesan K and Janaki D, 2019. Effect of paclobutrazol on floral bud initiation, flowering and fruit set of jamun (*Syzygium cuminii* Skeels.). *Jour. Pharmacog. Phytochem.* **8**: 1547-1549.

Dinesh MR, Ravishankar KV and Sangma D, 2016. Mango breeding in India - Past and future. *Jour. Hort. Sci.* **11**: 1-12.

Dubey AK, Sharma RM and Deepak Kumar A, 2021. Long term performance of mango varieties on five polyembryonic rootstocks under sub-tropical conditions: effect on vigour, yield, fruit quality and nutrient acquisition. *Sci. Hort.* **280**: 109944.

Fivaz J and Stassen PJC, 1997. The role of training systems in maintaining higher density mango orchards. *Acta Hort.* **455**: 407–412.

Haldankar PM, Dalvi NV, Parulekar YR and Lawande KE, 2014. Induction of flowering by girdling in Jamun cv. Konkan Bahadol. *Jour. Pl. Stud.* **3**: 46-49.

Hassan HAS, Sarrwy SMA, Mostafa EAM and Ahmed DM, 2010 Influence of training systems on leaf mineral contents, growth, yield and fruit quality of "Anna" apple trees. *Res J Agri Bio Sci* **6**: 443–448.

Iyer CPA, 1991. Recent advances in varietal improvement in mango. *Acta Hort.* **291**: 109–132.

Kishore K, Samant D, Singh HS and Sharma D, 2017. 3 Ps: Mantra for maximizing mango yield under high-density planting system. *Curr. Sci.* **113**: 2078-79

Kishore K, Singh HS and Kurian RM, 2015. Paclobutrazol use in perennial fruit crops and its residual effects: A review. *Ind. Jour. Agri. Sci.* **85**: 863–72.

Kishore K, Singh HS, Sharma D, Rupa TR, Kurain RM and Samant D, 2019. Influence of paclobutrazol on vegetative growth, nutrient content, flowering and yield of mango (*Mangifera indica* L.) and its residual dynamic. *Jour. Agr. Sci. Tech.* **21**: 1557-1567.

Kishore K, Singh HS, Nath V, Baig M, *et al.,* 2022. Influence of canopy architecture on photosynthetic parameters and fruit quality of mango in tropical region of India. *Hort. Env. Biotech.* https: //doi.org/10.1007/s13580-022-00500-z.

Lawande KE, Haldankar PM, Dalvi NV, Parulekar YR, 2014. Effect of pruning on flowering and yield of jamun cv. Konkan Bahadoli. *Jour. Pl. Stud.* **3**: 114-118.

Mitra SK, 2019. High density orcharding and canopy management of some tropical and sub-tropical fruits. *Acta Hort.* **1255**: 127–138.

Mitra SK, Ghosh B and Pathak PK, 2018. High density orcharding and canopy management in guava. *Acta Hort.* **1205**: 955-958.

Nartvaranant P, Subhadrabandhu S and Tongumpai P, 2000. Practical aspects in producing off-season mango in Thailand. *Acta Hort.* **509**: 661-668.

Negi ND and Sharma N, 2009. Effect of paclobutrazol applicationand planting systems on growth and production of peach (*Prunus persica*). *Ind. Jour. Agri. Sci.* **79**: 62-64.

Rebolledo-Martínez A, Peralta-Antonioa N, Rebolledo-Martíneza L, Becerril-Románb EA and Rebolledo-Garcíaa RL, 2019. Effect of rootstock in tree growth, dry matter, flowering, yield and quality of 'Manila' mango. *Sci. Hort.* **251**: 155-161.

Reddy YTN, Kurian RM, Ramachander PR, Singh G and Kohli RR, 2003. Long-term effects of rootstocks on growth and fruit yielding patterns of 'Alphonso' mango (*Mangifera indica* L.). *Sci. Hort.* **97**: 95–108.

Samant D, Kishore K and Singh HS, 2016. Branch bending for crop regulation in guava under hot and humid climate of eastern India. *Jour. Ind. Soc. Coastal Agri. Res.* **34**: 92-96.

Silva DD, Han L, Costes E, 2014. Light interception efficiency of apple trees: a multiscale computational study based on MAppleT. *Eco Model* **290**: 45–53.

Singh S, Singh AK, Mishra DS and Rao VVA, 2018. High-density planting in jamun doubles farmers' income. *Ind. Hort.* **63**: 14-16.

Srivastava KK, Rajan S, Kumar D, Pandey G, Verma A and Pandey S, 2018. Espalier architecture for revolutionizing high quality guava production. *Ind. Hort.* **63**: 18-20.

Tang L, Yin D, Chen C, Yu D, Han W, 2019. Optimal design of plant canopy based on light interception: a case study with loquat. *Front Plant Sci* 10: 364. doi. org/10.3389/fpls.2019.00364.

Wang H, Gu M, Cui J, Shi K, Zhou Y, Yu J, 2009. Effects of light quality on CO2 assimilation, chlorophyll-fuorescence quenching, expression of Calvin cycle genes and carbohydrate accumulation in *Cucumis sativus*. *Jour. Photochem Photobiol. B. Bio.* **96**: 30–37.

Chapter 9

Nursery Production, Orchard and Canopy Management in Unexploited Minor Sub-tropical Fruits

Parshant Bakshi, Reetika Sharma and Rakesh Kumar

Division of Fruit Science, Sher-e-Kashmir University of Agricultural Sciences and Technology of Jammu, Chatha, Jammu–180009, J&K

The world population, currently at 7.87 billion, is growing at an annual rate of 1.03 per cent and is projected to reach 9.6 billion by 2050. India, with its population of 1.38 billion (representing 17.5% of the global population but occupying only 2.4% of the world's surface area), faces immense challenges in feeding its population nutritious and safe food (Anonymous, 2019). This issue is exacerbated by India's ranking of 111 out of 125 countries in the Global Hunger Index, indicating widespread malnutrition, which severely hampers socio-economic progress (Anonymous, 2023). Hunger remains a critical global issue, particularly in India and several African nations, as noted by the World Health Organization (WHO). In response to these concerns, 195 countries have committed to the Sustainable Development Goals (SDGs), a set of initiatives designed to eradicate malnutrition through a holistic approach by 2030 (Anonymous, 2022). One growing trend is consumer demand for high nutrient content fruits, which are recognized for their health benefits, especially in improving immune and metabolic health, a concern that has gained greater urgency in the wake of the COVID-19 pandemic. India's fruit production is dominated by crops like mango, banana, citrus, guava, and apple, which account for 72 per cent of the total area under fruit cultivation.

Despite the high productivity of indigenous crops, they make up only 6.56 per cent of the cultivated area. Climate change, however, poses a significant threat to fruit production, worsening issues like salinity, mineral deficiencies, and pest infestations (Anonymous, 2021). In this context, underutilized fruit crops, which are tolerant to biotic and abiotic stresses, offer promising solutions. These crops, including ber (*Zizyphus* spp.), kair (*Capparis decidua*), bael (*Aegle marmelos*), jamun (*Syzygium* spp.), and aonla (*Emblica officinalis*), are nutrient rich and well-adapted to extreme climatic conditions but they remain underexploited due to insufficient research and lack of governmental support. Recognizing their potential, the Indian government has promoted their cultivation since 2005 under the Mission on Integrated Development of Horticulture (MIDH), particularly to enhance nutrition and socio-economic security in arid regions (Anonymous, 2005). Abiotic stresses, such as high temperatures, salinity, drought, and nutrient deficiencies, contribute to 70 per cent of global yield losses in major fruit crops. These stresses are most prevalent in arid and semi-arid regions, where commercial fruit crops struggle to maintain efficiency. However, underutilized crops like ber (*Zizyphus* spp.), aonla (*Emblica officinalis*), bael (*Aegle marmelos*), jamun (*Syzygium* spp.), and wood apple (*Feronia limonia*) have evolved adaptive traits that allow them to thrive in harsh environments (Choudhary *et al.,* 2019). These crops possess deep root systems, high root-to-shoot ratios and water storage capabilities, enabling them to efficiently extract water and nutrients from deeper soil layers. Their resilience is further enhanced by drought-tolerant traits, such as leaf shedding, thickened stems and mucilaginous sap which help minimize water loss. These fruit crops also demonstrate remarkable biochemical and physiological resilience, accumulating compounds like osmolytes, proline, phenolics and other antioxidants, while synthesizing heat shock proteins and drought-responsive genes to maintain cellular functions under stress conditions. As a result, they can sustain fruit production with minimal yield penalties, even under adverse environmental conditions. By delving further into the genetic basis of these adaptations, these underutilized crops could become pivotal in future breeding programs and sustainable agricultural practices, especially in light of changing climatic conditions. Their cultivation could offer significant economic benefits to farmers in arid regions, providing a stable and resilient source of income (Choudhary *et al.,* 2019; Singh *et al.,* 2022).

Challenges in Cultivation

Low Awareness and Underutilization

Many agricultural practices remain underutilized due to limited awareness among farmers. In rural areas, lack of access to agricultural extension services is a significant barrier to adopting new technologies. While these services aim to disseminate knowledge and promote modern practices, gaps in their effectiveness such as infrequent visits or poor communication hinder the dissemination of vital information. Studies suggest that increased engagement through methods like farm demonstrations and media platforms can help raise awareness and adoption rates for sustainable farming practices.

Lack of Standardized Cultivation Practices

One challenge in agriculture is the absence of standardized cultivation protocols tailored to different regions and crops. This often leads to inconsistent yields and suboptimal resource use. Research shows that while some efforts have been made to develop such guidelines, farmers often lack access to the required training and resources. Strengthening extension services to provide region-specific advice and encouraging the use of modern technologies can help address these inconsistencies.

Market Limitations and Research Gaps

Even when farmers adopt modern practices, they frequently face limitations in marketing their produce, particularly in regions with poor infrastructure or limited market access. Studies indicate that commercialization levels in rural areas remain low, with households selling only a fraction of their production. Additionally, there are research gaps in developing resilient market strategies and improving value chains, particularly for non-traditional and high-value crops. Addressing these gaps through targeted policies and investments in market linkages could significantly improve farmers economic outcomes.

Nursery Production Techniques

Guidelines for Nursery Raising

The timing of sowing and propagule production depends on achieving optimum seedling size (approximately 20 cm tall with good rooting) before the monsoon onset July for the Southwest and October for the Northeast. To calculate the total number of seedlings required from a nursery, use the formula: Total seedlings = W (required plants) + X (mortality in nursery) + Y (transportation/culling loss) + Z (buffer loss). For vegetative propagules, the success percentage must also be considered. Typically, nursery area should be 0.25 to 2.5 per cent of the planting area or about 1 acre for every 30,000 seedlings, with daily water supply of 200 liters per 1,000 seedlings (Kumar and Thakur, 2019).

Site Selection and Preparation for Sub-tropical Minor Fruit Nurseries

Importance of Site Selection Based on Climate and Soil Conditions

There is no standard blueprint for designing a plant nursery, as each nursery's layout depends on specific needs and resources. A well-designed nursery typically includes a water tank or pond, water pump, seed and fertilizer storage, implement shed, germination/mother bed area, potting area, seedling raising area, worker mess, office room, propagation structures, and compost area. The layout usually features a series of beds with pathways in between, and an open area for tasks like soil sieving and container filling. Efficient flow and careful planning of roads and paths are crucial to save labor and time, while maintaining cleanliness and controlling weeds to prevent pest and pathogen issues. Proper site selection is crucial in nursery production as it directly impacts plant health and growth. In sub-tropical regions, optimal sites should avoid extreme conditions such as frost or excessive heat. A moderate climate with ample sunlight is essential, as fruit-bearing plants require adequate photosynthesis. Additionally, soil selection is pivotal. The soil should be well-draining, loamy, and rich in organic matter to support root development. According to research, well-aerated soils with a balanced pH of 6.0 to 6.5 enhance nutrient availability and root penetration. Soil texture and moisture retention also affect irrigation management, a critical factor in nursery success (Yadav and Sharma, 2018)

Technical Considerations for Soil and Climate

Soil testing is an important first step in nursery establishment to evaluate parameters such as pH, organic matter content, and electrical conductivity. These factors are critical for ensuring that the soil can support vigorous plant growth. Soil with high salinity, for instance, can hinder nutrient uptake and adversely affect plant health. Amendments like compost or lime may be necessary to adjust soil properties before planting. Furthermore, climatic considerations include identifying windbreaks to protect young plants from wind damage and frost pockets that could expose crops to temperature extremes.

Ideal Conditions for Sub-tropical Minor Fruit Nurseries

In terms of water management, the nursery should have access to a clean, reliable source of irrigation water, which is essential for maintaining plant moisture levels, especially in drier sub-tropical zones. Water sources like wells, when properly managed, ensure a consistent supply, while open sources such as lakes may require additional screening to prevent contamination by algae, debris, and pathogens. Irrigation systems, whether drip or overhead, should be designed based on the nursery's slope and soil type to prevent waterlogging or nutrient leaching. The topography should ideally feature a slight slope to facilitate natural drainage while preventing erosion. A slope between 1 and 2 per cent is often recommended to maintain water flow without causing excessive runoff, which could lead to soil degradation. Sub-tropical fruit nurseries should also prioritize northwestern aspects to reduce the risk of frost damage during cold months while maintaining adequate warmth during the growing season (Dey and Ghosh, 2020)

Inputs for Nursery

In a nursery, key inputs include containers, media, propagules, water, fertilizers, chemicals, electricity, tools, equipment, and labor. Containers can be made of polythene, clay, or iron, with polythene bags being the most affordable option, while root trainers are user-friendly for handling and transport. The growth medium must be firm enough to support seedlings while providing essential nutrients and water; common media include soil, sand, peat soil, sphagnum moss, vermiculite, and coco peat. Nutrient availability is influenced by the pH, with optimal levels between 5.5 and 6.5. Water is crucial for seedling growth, and its quality impacts seedling health; efficient irrigation systems, like sprinklers, help manage water supply and reduce weeds and pathogens. Labor is essential for quality seedling production, requiring skilled and permanent staff. Mother beds are vital for seedling propagation, with various types including raised, sunken, and level beds, each designed to optimize moisture conservation and germination for different species. The nursery must utilize tools and machinery, such as automated filling machines, to enhance efficiency and reduce costs. Proper management of these inputs ensures healthy seedlings and successful nursery operations.

Propagation Structures

For effective propagation, various framed structures like greenhouses, poly tunnels, culture rooms, hardening chambers, and mist chambers are essential. A greenhouse is a transparent-covered structure that allows for controlled crop growth, with designs including shade net houses, plastic film greenhouses, glasshouses, and natural greenhouses tailored to specific needs and resources. Shade net nurseries, typically measuring 20 m × 10 m, utilize GI pipes and UV-stabilized HDPE shade nets to provide 50-75 per cent shade while preventing insect entry with nylon insect-proof nets. Although glasshouses are efficient in utilizing sunlight, their higher construction costs and limitations in automation have led to decreased usage compared to plastic film greenhouses.

Table 9.1: Various Propagation Structures and their Features for Effective Plant Propagation

Propagation Structure	*Description*	*Features*	*Common Use*
Greenhouse	Transparent-covered structure designed for controlled growth of crops.	Types: Shade net houses, plastic film greenhouses, glasshouses, and natural greenhouses.	General crop growth and propagation.
Shade Net Nursery	Structures covered with shade nets to regulate sunlight and protect plants from pests. Typically 20 m × 10 m in size.	Made with GI pipes and UV-stabilized HDPE nets, offering 50-75 per cent shade. Nylon insect-proof nets are used to prevent pest entry.	Nursery production and hardening.
Plastic Film Greenhouse	Greenhouse covered with plastic film, providing a controlled environment for crops at lower construction costs than glasshouses.	Efficient sunlight use with lower cost and better flexibility in automation compared to glasshouses.	Widely used for crop cultivation.

Propagation Structure	*Description*	*Features*	*Common Use*
Glasshouse	Traditional greenhouse with glass covering, offering optimal light utilization but with higher construction costs.	Optimal sunlight utilization, but high construction cost and limitations in automation.	Limited due to high cost.
Poly Tunnels	Arched structures covered with plastic sheeting used to protect crops from weather and extend the growing season.	Economical, portable, and flexible, providing protection from cold weather and enabling early-season crop production.	Seasonal crop production.
Culture Room	Controlled indoor environments used for tissue culture and micropropagation.	Fully controlled temperature, humidity, and lighting systems for efficient growth of tissue-cultured plants.	Tissue culture propagation.
Hardening Chamber	Structure used to acclimatize tissue-cultured plants from lab conditions to the external environment before field planting.	Temperature and humidity regulated to gradually expose plants to outdoor conditions, improving survival rates post-transplantation.	Tissue culture hardening.
Mist Chamber	Propagation structure where misting systems provide high humidity to support rooting of cuttings and young plants.	Maintains high humidity and consistent moisture, crucial for the rooting of cuttings and propagation of moisture-loving plants.	Rooting of cuttings.

Propagation Methods for Minor Sub-tropical Fruit Crops

Sexual Propagation

For minor fruit crops, sexual propagation remains one of the primary methods of establishing new plants, particularly for crops like Indian jujube (*Ziziphus mauritiana*) and tamarind (*Tamarindus indica*). Seed selection is vital in ensuring the health and viability of the resulting plants. High-quality, disease-free seeds are chosen and often treated with methods such as scarification or stratification. For example, tamarind seeds benefit from scarification, where the hard seed coat is softened to enhance water absorption and promote germination. While sexual propagation generates genetic diversity, this variability can result in non-uniform fruiting, which may not always be desirable in commercial cultivation.

Asexual Propagation

For more consistent and uniform plant characteristics, asexual propagation techniques like grafting, budding, and air layering are employed. These methods are particularly important for propagating minor fruits where maintaining specific traits (like fruit size, disease resistance, or flavour) is critical.

- ***Grafting***: This technique is widely used in minor crops such as sapodilla (*Manilkara zapota*) and Indian gooseberry (*Phyllanthus emblica*). By using a robust rootstock and a high-quality scion, grafting ensures plants that are more resilient and produce consistent fruit. Grafting also allows growers to overcome challenges like soil-borne diseases by selecting disease-resistant rootstocks.
- ***Budding***: In minor fruit crops like bael (*Aegle marmelos*), budding is a common method for propagating improved cultivars. The bud is taken from a selected tree and inserted into the rootstock to encourage rapid growth of genetically superior plants.
- ***Air layering***: This method is used for difficult-to-root crops like litchi (*Litchi chinensis*). Air layering involves wounding a branch and encouraging roots to form while it remains attached to the mother plant. Once roots have developed, the new plant can be separated and transplanted. This technique is particularly useful in situations where grafting is not feasible.

Role of Biotechnology in Propagation

In recent years, biotechnology has played a crucial role in enhancing propagation efficiency for minor fruit crops. Micropropagation is one such advanced technique where small pieces of plant tissue (explants) are cultured in a sterile environment to rapidly produce large numbers of disease-free, uniform plants. This method is especially valuable for crops that are hard to propagate through conventional methods, such as carambola (*Averrhoa carambola*). Micropropagation ensures genetic consistency and allows growers to produce plants on a large scale in a short amount of time, which is essential for meeting market demands. Biotechnological advancements like tissue culture and marker-assisted selection also improve the propagation of minor fruit crops by ensuring disease-free planting material and helping to select traits that are favorable for specific climates or markets. These innovations are transforming the way minor sub-tropical fruit crops are cultivated, offering more sustainable and efficient approaches to meet growing consumer demand. By integrating traditional methods like grafting with modern biotechnological tools, the propagation of minor sub-tropical fruit crops can be optimized for higher productivity, uniformity, and resistance to environmental stresses.

Table 9.2: Micropropagation in Minor Fruit Crops

Fruit Crop	*Scientific Name*	*Origin and Distribution*	*Key Compounds*	*Nutritional and Medicinal Value*	*Propagation Methods*	*Micropropagation Techniques*
Indian Gooseberry	*Emblica officinalis* Gaertn.	Tropical South-Eastern Asia, Central and Southern India, Himalayas	Emlicannin A, Emlicannin B, Puniglucon, Pedunculagin, antioxidants, tannins	Rich in Vitamin C (600 mg/100 g), riboflavin, calcium, iron, phosphorus, tryptophane, lysine, methionine; herbicidal properties due to anti-ascorbic value.	Seed propagation (cross-pollination); vegetative propagation via grafting (inarching)	**Explant Waxing:** Explants wax-sealed to prevent oxidative browning; 80 per cent success with shoots. **Media:** MS media with agar (0.8%), sucrose (3%), kinetin (0.4 mg/l), GA_3 (1 mg/l); pH 5.7, cultures at 25°C, 50-55 per cent RH, and 2000 lux light. August-November optimal for culture. **Proliferation:** GA_3 (4.33 µM) + Kinetin (13.9 µM) + Glutamine (342.11 µM) led to high shoot proliferation.
Karonda	*Carissa carandas* L.	Tropical and Sub-tropical regions	High in Vitamin C (1.6-17.9 mg/100 g), iron (39.1 mg/100 g), calcium, phosphorus	Processed into jam, jelly, pickles, chutneys; roots treat intestinal worms, scabies; wood used for combs, spoons, hedges.	Seed propagation, vegetative propagation via cuttings, live fencing	**Shoot Proliferation:** Explants sprout best in spring (Feb-March); MS media with BA (3.0 mg/l). **Rooting:** Best on ½ MS media with IBA (0.8 mg/l) + NAA (0.2 mg/l). Rooted seedlings successfully acclimatized in soil-sand-vermiculate (1:1:1).
Bael	*Aegle marmelos* Corr.	India (deciduous, medicinal)	Psoralen, marmelosin	Ripe fruit acts as tonic, laxative, good for heart and brain. Processed into candy, sherbet, powder.	Seed propagation (slow), root sucker propagation (cumbersome). Seed viability is short.	**Axillary Bud Multiplication:** MS media with plant growth regulators used for in vitro propagation. **Nucellar Callus:** Employed for genetic studies, not true-to-type multiplication.

Fruit Crop	*Scientific Name*	*Origin and Distribution*	*Key Compounds*	*Nutritional and Medicinal Value*	*Propagation Methods*	*Micropropagation Techniques*
Jamun	*Syzygium cuminii* L.	Indo-Malayan region	Antioxidants, polyphenol, flavonoids, carotenoids, vitamins	Used in folk medicine for diabetes, heart disease, liver health. Seeds reduce sugar in urine.	Seedling propagation, juvenile tissue (micro-propagation)	**Shoot Proliferation:** Single node explants cultured on ½ strength MS with BAP (2 mg/l) + sucrose (3%). **Rooting:** On ¼ strength MS with IBA (2.5 mg/l), sucrose (3%). Successfully hardened in vermiculite.
Jackfruit	*Artocarpus hetero-phyllus* L.	India (poor man's food)	Pectin-rich rind, starch-rich seeds	Rind used for jelly, unripe fruit for chips; seeds used in various cuisines.	Not widely propagated due to high heterozygosis, vegetative propagation.	**Shoot Tip Culture:** MS media with BAP (2 mg/l) enhanced shoot regeneration. **Rooting:** IBA (2 mg/l) enhanced root formation.

Source: Singh *et al.* (2022).

Nursery Management Practices for Minor Fruit Crops: Use of Organic Amendments and Biofertilizers for Healthy Seedling Growth

In nurseries, particularly for minor fruit crops, the application of organic amendments such as compost, vermicompost, and well-rotted farmyard manure enhances the physical properties of the soil, improves nutrient availability, and promotes beneficial microbial activity. Biofertilizers, such as nitrogen-fixing bacteria (*Rhizobium*), phosphate solubilizing bacteria (*Pseudomonas*), and mycorrhizae, further improve nutrient uptake, particularly in nutrient-deficient soils. These practices result in robust root systems and enhanced seedling vigor, which are critical for minor fruit crops such as ber (*Ziziphus mauritiana*), aonla (Indian gooseberry), and karonda (*Carissa carandas*).

Irrigation Techniques: Drip, Misting, and Regular Monitoring of Humidity

Irrigation is vital in managing sub-tropical nurseries, where minor fruit crops may face erratic rainfall and high evapotranspiration rates. Drip irrigation ensures precise water delivery to the root zone, reducing water wastage and improving plant growth. Misting systems, on the other hand, are particularly useful in maintaining humidity levels, preventing desiccation of tender seedlings in crops like bael (*Aegle marmelos*) and custard apple (*Annona squamosa*). Regular monitoring of humidity ensures that optimal microclimatic conditions are maintained, promoting healthy seedling growth.

Pest and Disease Management in Nurseries

Minor fruit crop nurseries are often susceptible to pests and diseases, which can adversely affect seedling health. Integrated pest management (IPM) practices such as crop rotation, use of biocontrol agents (*e.g.*, Trichoderma for fungal diseases), and neem-based insecticides have shown promising results in managing nursery pests like aphids, mites, and fungal infections. Furthermore, adopting hygienic practices such as sterilizing nursery beds and equipment, and removing diseased plants, reduces the incidence of disease outbreaks.

Use of Shade Nets and Polyhouses in Sub-tropical Nurseries

Shade nets and polyhouses are effective in protecting seedlings from extreme temperature fluctuations and heavy rains, common in sub-tropical regions. For minor fruit crops, shade nets (with 30-50% shading) provide filtered sunlight, essential for preventing photoinhibition in seedlings like phalsa (*Grewia asiatica*) and jamun (*Syzygium cumini*). Polyhouses, meanwhile, create controlled environments that protect seedlings from adverse weather and pest invasions, leading to better establishment and growth. By integrating these nursery management practices, especially tailored to the needs of minor fruit crops, farmers can ensure healthy seedling growth, improved crop productivity, and long-term sustainability in fruit production systems.

Plant Hardening in Minor Fruit Crops

Acclimatization Process before Transferring to the Field

Plant hardening is a critical process in preparing nursery seedlings for the transition from a controlled nursery environment to the harsher field conditions. In the case of minor fruit crops like custard apple (*Annona squamosa*), phalsa (*Grewia asiatica*), and ber (*Ziziphus mauritiana*), this acclimatization period typically lasts 10-14 days. During this phase, seedlings are gradually exposed to sunlight, reduced watering, and outdoor temperatures, which helps them adjust to fluctuating conditions and strengthens their physiological responses to field conditions such as wind, temperature extremes, and soil variability. Acclimatization also triggers changes in leaf structure and root development, enhancing the plant's resilience to stress.

Techniques to Improve Transplant Survival

To maximize transplant survival, several techniques can be employed during the hardening process:

- ***Gradual exposure to sunlight***: Initially, seedlings are kept under shade nets or in a partially shaded environment to avoid sudden exposure to direct sunlight. The shading is gradually reduced, allowing seedlings to adapt to increasing light intensity without causing sunburn or dehydration.
- ***Controlled watering***: Reducing irrigation during hardening encourages the development of a more extensive root system, improving water uptake efficiency when plants are transplanted. Care is taken to avoid water stress, which can weaken the plants before transplanting.
- ***Nutritional support***: Foliar sprays containing nutrients like potassium and calcium can help enhance cell wall strength and overall plant robustness. Minor fruit crops such as bael (*Aegle marmelos*) and karonda (*Carissa carandas*) benefit from these applications as they enhance the plant's ability to cope with field conditions.
- ***Wind and temperature conditioning***: Exposing seedlings to controlled airflow or mild outdoor winds helps thicken the plant stems, making them less susceptible to breakage during transplanting. Similarly, gradual exposure to varying day and night temperatures ensures that the seedlings become tolerant to the natural temperature fluctuations in the field.

Use of Bio-stimulants and Mycorrhizae

Incorporating bio-stimulants like seaweed extracts or mycorrhizal fungi in the nursery phase can significantly boost root development and stress tolerance. These amendments improve the establishment of minor fruit crops in the field by enhancing nutrient uptake and boosting the plant's immune system.

Orchard Management in Minor Fruit Crops

Orchard Layout and Spacing

- ***Designing orchards for efficient use of space and light:***An optimized orchard layout is critical to achieving maximum yield and maintaining plant health. For minor fruit crops, strategic alignment of trees ensures efficient utilization of sunlight, air circulation, and facilitates easy management practices like pruning, pest control, and harvesting. A north-south orientation is typically recommended to maximize light interception throughout the day, minimizing shading and ensuring uniform fruit development. The row spacing should accommodate the full development of the canopy and root system while maintaining a compact, yet accessible, layout. Canopy management is an integral part of orchard design. A well-managed canopy increases light penetration to the lower branches, enhancing photosynthetic efficiency. Pruning schedules are adapted to the phenology of each species to maintain tree architecture, promote aeration, and reduce disease incidence. For example, in crops like ber (*Ziziphus mauritiana*), a strategy of selective thinning and heading-back is employed to control the tree's sprawling habit and direct resources towards fruiting wood.
- ***Ideal planting distances for different minor fruit species*:** Planting distances are determined by the species growth characteristics, including canopy spread, root expansion, and the need for air and light. For optimal productivity, trees should be spaced in a manner that avoids inter-tree competition while allowing full development of the plants' physiological potential.

Jamun (*Syzygium cumini*): 10-12 meters spacing in a square or rectangular arrangement is recommended to accommodate the large canopy and deep root system of jamun. This reduces the risk of shading and allows ample space for machinery access for operations like pest management.

Ber (*Ziziphus mauritiana*): With a relatively smaller canopy spread, 6-8 meters spacing is sufficient. This spacing allows for efficient utilization of nutrients and water while maintaining open access for orchard management activities.

Bael (*Aegle marmelos*): A spacing of 8-10 meters is advised to allow full expansion of the large, round canopy. Adequate space also ensures that air circulation prevents disease buildup, particularly in humid sub-tropical climates.

Karonda (*Carissa carandas*): As a smaller, bushy species, 3-4 meters spacing is ideal for optimal sunlight exposure while avoiding overcrowding. This close spacing also facilitates more intensive planting systems in smaller orchards or home gardens.

In all cases, the planting distance is determined by the final size of the mature tree and its canopy spread, as well as by site conditions such as soil fertility, water availability, and climatic factors.

Table 9.3: Ideal Planting Distances for Various Minor Fruit Crops

Fruit Species	*Scientific Name*	*Recommended Planting Distance*	*Spacing Arrangement*	*Notes*
Jamun	*Syzygium cumini*	10-12 meters	Square or rectangular	Accommodates large canopy and deep roots; reduces shading; allows machinery access.
Ber	*Ziziphus mauritiana*	6-8 meters	Square or rectangular	Smaller canopy; efficient nutrient and water utilization; open access for management activities.
Bael	*Aegle marmelos*	8-10 meters	Square or rectangular	Large round canopy; ensures air circulation to prevent disease in humid climates.
Karonda	*Carissa carandas*	3-4 meters	Square or rectangular	Bushy growth; optimal sunlight exposure; suitable for intensive planting systems.
Phalsa	*Grewia asiatica*	3-4 meters	Square or rectangular	Compact growth; allows for good light penetration; suitable for small spaces.
Aonla	*Phyllanthus emblica*	8-10 meters	Square or rectangular	Requires ample space for canopy growth; ensures proper air circulation.
Bael	*Aegle marmelos*	8-10 meters	Square or rectangular	Large canopy; sufficient space helps avoid disease buildup.
Ber	*Ziziphus mauritiana*	6-8 meters	Square or rectangular	Optimal spacing for nutrients and water access; promotes better growth.

Land Preparation, Digging Pits and Soil Amendments

Land Preparation

Effective land preparation is the foundation of a successful orchard. This includes deep plowing to break up compacted layers and improve soil aeration. Secondary tillage operations, such as harrowing or leveling, are performed to create a uniform planting surface, which is particularly important on sloped lands to ensure proper drainage. In sloping or undulating terrains, contour planting or terracing is employed to prevent erosion and manage water runoff effectively.

Digging Pits

The size of the planting pit is critical for the establishment of healthy root systems. For large trees like jamun, pits should be dug to a depth and width of 1m x 1m x 1m, ensuring enough room for root expansion and facilitating nutrient uptake. Smaller species like karonda can be planted in 60cm x 60cm x 60cm pits, which are more appropriate for their shallow root systems. The pits should be prepared at least 2-3 weeks before planting to allow for soil settling and exposure to solar sterilization.

Soil Amendments

Amending the planting pit with organic and inorganic nutrients is essential for establishing vigorous seedlings. A mix of topsoil, compost, and well-decomposed farmyard manure is commonly used to enhance the nutrient profile and biological activity of the soil. For soils with low fertility, phosphorus fertilizers (such as single superphosphate at 100g/pit) can be applied at planting to promote early root development. In addition, incorporating biochar or vermicompost improves the water-holding capacity, aeration, and microbial health of the soil. In acidic soils, lime can be added to correct pH imbalances, creating a more favorable environment for nutrient uptake. These amendments not only support early seedling establishment but also contribute to the long-term productivity of the orchard.

Soil Management

Effective soil management is essential for the sustainable cultivation of sub-tropical minor fruit crops, as it directly influences crop yield, quality, and resilience. A balanced approach combining regular soil testing, nutrient management, and tailored fertilizer programs is crucial. Key macronutrients like nitrogen, phosphorus, and potassium, along with micro-nutrients like zinc and boron, play vital roles in fruit growth and quality, while organic matter enhances soil structure and fertility. Both inorganic and organic fertilizers contribute to nutrient availability, with the latter, including compost and biochar, promoting long-term soil health. Practices such as mulching and green manure improve water retention, reduce erosion, and enhance nutrient cycling, especially for crops like guava, ber, and jamun. These integrated strategies ensure higher yields, better soil health, and sustainable productivity.

Irrigation and Water Management

Efficient water management is crucial for sub-tropical fruit crops due to irregular rainfall and high evapotranspiration rates. Best practices include using drip irrigation and sprinkler systems, which allow precise water delivery directly to the root zone, reducing evaporation and runoff. Drip irrigation, in particular, has been shown to improve water use efficiency by 30-40 per cent, while sprinkler systems offer uniform water distribution across large areas. Soil moisture sensors help monitor irrigation needs, optimizing water application during critical growth stages like flowering and fruit set. Additionally, rainwater harvesting in orchards ensures a sustainable water supply, enhancing soil moisture and reducing reliance on external water sources

Fertilization Strategies

Fertilization strategies for sub-tropical minor fruit crops require an integrated approach that combines organic amendments, precise nutrient management, and foliar application to enhance soil health and optimize nutrient use efficiency. Organic amendments such as compost, vermicompost, and biochar improve soil organic matter content, microbial activity, and cation exchange capacity, leading to better nutrient retention and water-holding capacity. Nutrient management should be based on crop growth stages, with macronutrients like nitrogen and potassium applied during vegetative and reproductive phases to support canopy growth, flowering, and fruit development. Micro-nutrients such as zinc and boron are crucial during critical stages for improving fruit set and quality, often applied through foliar feeding to ensure rapid nutrient uptake and correct deficiencies in the plant canopy. Foliar applications are particularly effective for addressing nutrient deficiencies in minor crops, ensuring efficient translocation of essential elements during high-demand periods. Organic amendments for enhancing soil health.

Pest and Disease Management

Pest and disease management in sub-tropical minor fruit crops necessitates a multi-faceted approach due to the prevalence of pests like fruit flies (*Bactrocera dorsalis*), mealybugs (*Phenacoccus solenopsis*), and diseases such as powdery mildew (*Oidium* spp.) and anthracnose (*Colletotrichum* spp.). Integrated Pest Management (IPM) combines biological control agents, cultural practices, and targeted chemical interventions. Organic strategies include the use of neem-based bio-pesticides (*Azadirachta indica*), pheromone traps for fruit flies, and habitat management to enhance predator populations. Preventive care focuses on disease-resistant varieties, such as guava cultivars with higher resistance to Guava Wilt (*Fusarium oxysporum*), and timely pruning to reduce disease incidence. Implementing cultural practices like sanitation and controlled irrigation further minimizes disease outbreaks, while regular monitoring and early interventions ensure sustainable pest and disease control.

Canopy Management

Importance of Canopy Management

Canopy management is essential for optimizing light penetration, improving air circulation, and enhancing fruit quality in sub-tropical minor fruit crops. Adequate light exposure is critical for photosynthesis, which directly impacts fruit size, color, and sugar accumulation. Improved air circulation within the canopy reduces the humidity levels, which limits the spread of fungal pathogens like powdery mildew (*Oidium* spp.) and anthracnose (*Colletotrichum* spp.), contributing to better disease

Table 9.4: The Sustainable Practices and Systems Applied to Minor Fruit Crops, Showing how each Method can Enhance Productivity and Sustainability in Orchards

Description	*Examples for Minor Fruit Crops*	*Expected Impact*	*Reference*
Management Practices			
Cover crops and green manure	Use of leguminous cover crops in Indian gooseberry and karonda orchards	Provide constant soil cover; Increase soil organic matter (SOM); nutrient mobilization; decrease pests and pathogens	Mishra *et al.* (1997), Rastogi (1993)
Crop rotation	Rotation of bael, jamun, and other minor fruit crops on the same land	Improve soil nutrient balance; avoid over-exploiting the soil ecosystem	Ram (1982), Rao *et al.* (1985)
Reduced tillage	Implement reduced tillage in aonla orchards	Avoid soil degradation; increase carbon sequestration; improve soil density and water holding capacity	Mishra and Srivastava (1999)
Intercropping	Intercropping bael with leguminous crops like pulses	Improve nitrogen fixation and nutrient cycles; soil cover; reduce pests and pathogens	Rao *et al.* (1985), Rai and Misra (2005)
Structural elements	Windbreaks and hedgerows with karonda plants in orchards	Decrease impact of weather; improve light accessibility	Kumar and Singh (1993)
Irrigation management	Drip irrigation systems in bael orchards	Maintain optimal humidity; avoid nutrient leaching; reduce water loss	Mishra *et al.* (2006)
Allowed External Inputs			
Organic fertilizer	Application of farmyard manure in jamun orchards	Supplement with required nutrients; improve soil quality	Mishra and Pathak (2001)
Compost	Aerobically decomposed organic matter applied in jackfruit orchards	Improve soil health; increase nutrient content	Warrier *et al.* (1993)
Vermicompost	Use in minor fruit orchards such as Indian gooseberry	Improve soil structure and fertility	Ajithkumar and Seeni (1998)

Description	*Examples for Minor Fruit Crops*	*Expected Impact*	*Reference*
Biofertilizers	Nitrogen-fixing bacteria for aonla and bael orchards	Enhance plant growth and yield	Chaudhary *et al.* (2013)
Fertigation	Drip irrigation combined with nutrient solutions in jamun orchards	Optimize nutrient and water use efficiency	Hossain *et al.* (1994a)
Sustainable Agriculture Systems			
Conservation Agriculture	Minimal soil disturbance, cover crops, and crop rotation in jamun and bael	Improve soil health, reduce erosion, and enhance biodiversity	Kubola *et al.* (2011)
Crop-Livestock systems	Integration of minor fruit crops like bael with livestock	Enhance soil fertility through animal manure; diversify income	Hossain *et al.* (1994b)
Organic Agriculture	Absence of agrochemicals in organic cultivation of aonla and karonda	Promote ecological processes and biodiversity	Mishra *et al.* (2006)
Agroforestry	Integration of trees like bael, jamun with other crops	Enhance biodiversity, improve soil health, and diversify production	Ravikumar and Ved (2000)

resistance. Proper canopy management also enhances yield by ensuring uniform fruit development and improving the microenvironment around the fruit-bearing zones (Ranjan and Chaudhary, 2020).

Pruning Techniques

Pruning plays a pivotal role in controlling tree shape, size, and vigor in minor fruit crops. It helps balance vegetative growth and reproductive development, ensuring optimal productivity. Seasonal pruning is categorized into summer pruning, which encourages new growth and better light distribution, and winter pruning, aimed at removing old, diseased, or unproductive branches to stimulate flowering and fruiting. In sub-tropical fruits like ber (*Ziziphus mauritiana*), precise pruning has been shown to improve fruit set and size. Pruning tools include loppers, hand shears, and saws, and the choice of technique depends on the age and growth stage of the tree (Natranjan, 2020).

Training Systems

Training systems provide structural support and regulate tree architecture to optimize light interception and space utilization. Common systems for minor sub-tropical fruits include the open-center system, which enhances light penetration into the canopy, the central leader system for maintaining strong tree structure, and the espalier system, often used for compact orchards. Tailoring training systems to specific minor fruit varieties, such as the espalier system for pomegranate (*Punica granatum*) and the open-center system for guava (*Psidium guajava*), improves yield and facilitates efficient management (Ranjan and Chaudhary, 2020)

Growth Regulators in Canopy Management

Plant growth regulators (PGRs) are valuable tools for controlling canopy size and shape, particularly in high-density plantings. Paclobutrazol, for instance, is used to suppress excessive vegetative growth in crops like guava, allowing better energy allocation to reproductive growth. Application of PGRs such as gibberellic acid (GA_3) and cytokinins can enhance flowering, fruit set, and yield by promoting cell division and elongation in fruit-bearing areas. Strategic use of PGRs helps achieve a balance between vegetative and reproductive phases (Chakraborty and Sharma, 2019).

Canopy Manipulation for High-Density Planting

High-density planting (HDP) and ultra-high-density planting (UHDP) involve planting more trees per unit area to increase productivity. Canopy manipulation in HDP is crucial to prevent overcrowding, which can lead to poor light distribution and disease incidence. Techniques such as regular pruning, training, and growth regulator applications are employed to maintain manageable canopy sizes in high-density orchards. For minor fruits like jamun (*Syzygium cumini*) and karonda (*Carissa carandas*), HDP systems improve early bearing, yield, and land-use efficiency, though challenges include higher management costs and the need for specialized knowledge in canopy control (Gupta and Verma, 2020).

Sustainability and Future Prospects

Sustainable Orchard Practices

Sustainable orchard management in sub-tropical regions focuses on integrating organic farming, permaculture, and agroforestry to enhance ecosystem services. Organic farming practices, such as using compost and natural pest control, reduce chemical input while improving soil health. Permaculture and agroforestry systems integrate fruit trees with other crops and livestock to create self-sustaining ecosystems that promote biodiversity and resilience. Biodiversity conservation is vital in sub-tropical orchards, as it enhances pollination, pest control, and genetic diversity, which are essential for long-term orchard productivity and sustainability.

Future Research Needs

Future research must focus on breeding programs to develop improved minor fruit varieties with better yields, disease resistance, and adaptability to changing climates. Climate-resilient orchard practices, such as drought-tolerant rootstocks and water-efficient irrigation systems, are increasingly necessary to address the challenges posed by climate change. Additionally, promoting minor fruits like ber, karonda, and bael in global and local markets through improved value chains and post-harvest technologies can enhance their economic viability and boost their adoption by farmers.

Policy Support and Incentives

Government policies and financial incentives play a critical role in promoting minor fruit crops. Subsidies for organic inputs, research funding for minor fruit breeding, and financial support for adopting sustainable practices are essential. Extension services and farmer awareness programs further enhance the adoption of sustainable practices by educating farmers on best practices, new technologies, and market opportunities. These initiatives ensure the long-term viability of minor fruit cultivation while contributing to food security and environmental sustainability.

Conclusion

Minor sub-tropical fruits hold significant potential in modern agriculture, offering unique flavors and nutritional benefits while enhancing food security and diversifying crop production. Integrating these fruits into mainstream agriculture can provide substantial economic advantages by tapping into niche markets and improving small holder farmers livelihoods. Moreover, promoting unexploited fruits like aonla, karonda, bael and jamun contributes to environmental sustainability through increased biodiversity and reduced dependence on major crops. To mainstream these fruits effectively, a collaborative approach is essential, focusing on research for improved varieties and sustainable practices, supportive policymaking with financial incentives and infrastructure development, and market development efforts to raise awareness and demand. By fostering collaboration across these sectors, we can unlock the full potential of minor sub-tropical fruits and ensure their sustainability in modern agricultural systems.

REFERENCES

Ajithkumar P and Seeni S, 1998. Vermicomposting in Indian Gooseberry Orchards: A Study of Soil Health Improvement. *Journal of Organic Farming*, **12**(1): 15-21.

Anonymous, 2005. Ministry of Agriculture and Farmers Welfare, Government of India. Mission on Integrated Development of Horticulture (MIDH). Government of India. Available at: https: //midh.gov.in

Anonymous, 2019. United Nations, Department of Economic and Social Affairs, Population Division. (2019). World Population Prospects 2019: Highlights. United Nations. Available at: https: //population.un.org/wpp/Publications/

Anonymous, 2021. Food and Agriculture Organization (FAO). The State of Food Security and Nutrition in the World 2021: Transforming Food Systems for Food Security, Improved Nutrition, and Affordable Healthy Diets for All. FAO, IFAD, UNICEF, WFP, and WHO. Available at: https: //www.fao.org/publications/sofi/2021/en/

Anonymous, 2023. Global Hunger Index. Global Hunger Index: India's Global Hunger Index Ranking. Available at: https: //www.globalhungerindex.org

Anonymous, 2022. World Health Organization. Malnutrition. World Health Organization. Available at: https: //www.who.int/health-topics/malnutrition

Chakraborty S and Sharma R, 2019. High-Density Planting in Fruit Crops: Techniques and Benefits. *Horticulture International Journal*, **3**(2): 55-62. doi: 10.15406/hij.2019.03.00143.

Chaudhary S, Gupta PK and Singh RP, 2013. The Role of Nitrogen-Fixing Bacteria in Bael Orchards. *Journal of Agricultural Biotechnology*, **8**(3): 120-128.

Choudhary ML, Sharma V and Singh J, 2019. Abiotic Stress Management in Fruit Crops. Advances in Horticultural Science, **33**(1): 23–45. https: //doi.org/10.1016/j.ahs.2018.07.002

Dey S and Ghosh A, 2020. Vegetative Propagation Techniques for Fruit Nurseries: Principles and Practices. *Horticultural Reviews*, **48**: 189-228.

Gupta V and Verma A, 2020. Canopy Management Strategies for High-Density Orchards: A Review. *Journal of Applied Horticulture*, **22**(1): 1-11. doi: 10.5958/0976-4423.2020.00001.1.

Hossain AM and Singh RP, 1994a. Fertigation in Jamun Orchards: A Water Use Efficiency Study. *Journal of Agroforestry and Environment*, **7**(2): 56-63.

Hossain AM and Singh RP, 1994b. Integrated Crop-Livestock Systems in Minor Fruit Orchards. *Journal of Agroforestry and Sustainable Agriculture*, **9**(2): 112-120.

Kubola J, Siriamornpun S and Meeso N, 2011. Crop Rotation and Conservation Agriculture in Jamun Orchards. *Soil and Crop Science Journal*, **16**(1): 78-86.

Kumar A and Thakur N, 2019. Nursery Management of Fruit Crops. *Indian Journal of Horticulture*, **76**(1): 1-8.

Kumar R and Singh AP, 1993. Structural Elements for Minor Fruit Orchards: Windbreaks and Hedges. *Journal of Indian Horticulture*, **19**(4): 112-118.

Mishra VK and Srivastava AK, 1999. Reduced Tillage for Soil Conservation in Aonla Orchards. *Indian Journal of Horticultural Science*, **28**(2): 50-58.

Mishra VK, Pathak RK and Singh AK, 2006. Sustainable Production of Bael. *Indian Horticulture Journal*, **54**(4): 22-28.

Natarajan S, 2020. Impact of Pruning and Training on Yield and Quality of Minor Fruits. *Indian Horticulture*, **65**(4): 32-35.

Rai PK and Misra RP, 2005. Intercropping in Bael Orchards: Benefits and Techniques. *Journal of Agricultural Research*, **45**(1): 97-105.

Ram S, 1982. Crop Rotation Practices in Minor Fruit Orchards. *Indian Journal of Agricultural Sciences*, **52**(4): 290-294.

Ranjan R and Chaudhary R, 2020. Advances in Canopy Management of Fruit Crops: Strategies and Techniques. *Horticulture International Journal*, **4**(1): 1-8. doi: 10.15406/hij.2020.04.00139

Rao SK, Singh J and Shukla P, 1985. Crop Rotation and Intercropping in Karonda Orchards. *Indian Journal of Agroforestry*, **14**(2): 200-206.

Ravikumar K and Ved DK, 2000. Agroforestry Practices in Indian Minor Fruit Orchards. *Tropical Forest Science Journal*, **12**(3): 230-240.

Sharma R, 2021. *Best* Practices for Establishing Fruit Nurseries in Sub-tropical Regions. *Journal of Agriculture and Rural Development*, **12**(2): 125-132.

Singh AK, Jat MK and Yadav R, 2022. Underutilized Fruits: Potentials and Challenges in Future Food Security. *Journal of Horticulture and Postharvest Research*, **5**(3): 145–162. https: //doi.org/10.22077/jhpr.2021.4482

Singh V, Deen B and Singh S, 2022. Micropropagation of Minor Fruit Crops of India: A Review. *Agricultural Reviews*. doi: 10.18805/ag.R-2569.

Warrier PK and Nambiar VPK, 1993. Composting Techniques for Jackfruit Orchards. *Tropical Horticulture Journal*, **15**(2): 65-72.

Yadav R and Sharma P, 2018. Soil and Water Management in Fruit Production: A Review. *Journal of Applied Horticulture*, **20**(1): 1-10.

Chapter 10

A Paradigm Shift in Citrus Clean Stock Production in Central India through Innovative Propagation Techniques

Vijayakumari Narukulla

Principal Scientist (Retd), ICAR–CCRI, Nagpur–440033, Maharashtra

Status of Citrus Propagation

Citrus crops are grown over an area of 10,97,000 ha with a production of 142, 45,000 MT (Horticulture Statistics at a glance 2021). The area under Citrus has grown @ 2 per cent between 2019-20 to 2020-21 and about 5 per cent of thc area is replanted annually. The main citrus commercial cultivars of India *viz.*, mandarin (*Citrus reticulata* Blanco), sweet orange (*Citrus sinensis* Osbeck) and lime and lemon (*Citrus aurantifolia* Swingle), respectively account for 44, 28 and 25 per cent of all citrus fruits produced in the country.

Commercially mandarin and sweet orange crops are vegetatively propagated in open field nurseries through 'Shield' or ' T budding' where as limes and lemons are mainly propagated through seed. Vegetative propagation facilitates harnessing of desirable characters of both scion and root stock plants, however in the absence of sanitized production processes, it promotes inadvertent spread of diseases.

About 15 viruses and virus like diseases are reported from various citrus growing regions. Virus infected trees do not put on normal healthy growth, become stunted, show die-back of twigs and branches. Viruses are most potent factors leading to citrus extinct due to decline. Some of the local varieties of Citrus (*e.g.* Vadlamudi orange of A.P.) had nearly become extinct.

Citrus decline has been the major challenge of Citrus cultivation for many decades and it causes huge economic losses by (i) reducing the productivity and productive life of the citrus orchards and (ii) increasing the cost of orchard maintenance. The average productivity (13 tons/Ha) is far below than the recorded productivity (30 tons/ha). The average life of orange orchard is about 20 years where as well maintained orchards of more than 50 years are reported.

Since no control measures are available, the role of virus free planting material for starting of healthy orchards has been emphasized from time to time and several schemes were started. Bud wood certification, control of vectors, sanitation and cross protection are some of the approaches attempted earlier to minimize the losses caused by the viral diseases but worthwhile results could not be achieved due to technological and cost limitations. In the past, non-availability of rapid diagnostic techniques to monitor the mother tree health frequently and non-adherence to sanitized production methods had restricted the large scale multiplication of disease free planting stock.

It is proven that the citrus orchards established with the healthy planting material had higher productivity and longer longevity than the orchards established with the infected planting material. The virus free plants may get infected in the field through the vectors but the stage at which the first infection takes place has a definite bearing on the degree of losses. Once a plant is established, it can sustain the losses under a good system of management. On the contrary, an infected plant may not attain even the stage of production.

It is estimated that annually more than 18 million plants are needed for expansion (new plantings) and for replacement. Citrus nurseries have mushroomed to exploit the heavy market demand and these open field nurseries are not paying much attention to the sanitized methods of propagation leading to inadvertent spread of diseases from region to region through sale of infected planting stock and aggravated the decline problem. Though a few states like Haryana, Himachal Pradesh, Uttar Pradesh and Maharashtra have the Nursery Registration Act in force, significant results could not be achieved in regulating the genuineness of the planting material and its health due to ineffective implementation of these Acts.

There are only a few reliable sources of disease free planting material. In view of the high economic returns from well maintained orchards of citrus, the demand for quality planting material is increasing. The farmers are willing to pay higher price for high yielding quality planting material. The state agencies having the required expertise are unable to meet the growing demand. Large scale availability of healthy, elite and certified planting material would lead to (i) establishment of healthy orchards (ii) gradual elimination of infection sources and (iii) sanitized cultivation and (iv)increase in the productivity. The advent of biotechnological/ rapid diagnostic techniques like Shoot Tip Grafting (STG), Elisa together had paved the way for large scale multiplication by facilitating sanitation, quarantine, certification, multiplication.

Development of Production Protocols

The Central Citrus Research Institute (CCRI) i.e erstwhile National Research Centre for Citrus, had started developing protocols for elimination of virus and virus like graft transmissible diseases and multiplication of elite healthy planting material which was a felt need of the time. Under the World Bank aided project (1992-96), a new Tissue Culture Laboratory was established and the research work on Shoot Tip Grafting (STG) of Nagpur Mandarin was started.The protocols for cleaning the elite trees and the sanitized production of disease free bud grafts were developed and these were further refined under the institute project(1998-2002). In view of the promising results, National Horticulture Board and the Department of Biotechnology came forward to fund the development of additional infrastructural facilities, such as insect proof screen houses, net houses, glass houses, required to scale up the planting stock production programme as a pilot project (2003-2009) and to transfer of technology from lab to land. Based on the success of pilot project, "Technology Mission on Citrus" was formulated for producing and supplying quality planting material of citrus on large scale to the farmers.

All aspects of quality planting stock production were studied comprehensively and innovative methods of propagation and nursery practices were developed and dovetailed to facilitate rapid multiplication of quality planting stock of Citrus to meet the huge demand. The Technologies developed are explained and an action plan is suggested to improve the productivity and the profitability of citrus cultivation.

Shoot Tip Grafting *in vitro* (STG)

Morel and Martin (1952) first demonstrated that virus-free plants of Dahlia could be recovered from infected plants through shoot apex culture. Murashige *et al.* (1972) had developed the technique of Shoot Apex Grafting and used it for recovering virus-free citrus clones having no juvenile phase. The apical meristem does not have conducting tissue hence the viruses do not reach the tip of ever multiplying meristems. There after many researchers had adopted and modified this shoot tip grafting technique for cleaning and producing virus free plants. Shoot Tip Grafting technique developed by (Navarro *et al.*, 3), at the University of California, Riverside and Elisa diagnostic techniques (Cambra *et al.*, 1), had paved the way for implementing both citrus sanitation and citrus quarantine programmes.

Considering the overall advantages of STG, it has been selected as the basic technique of the Citrus Varietal Improvement Programmes worldwide.

STG based Sanitation

The recommended procedure for sanitation programs involves the following steps *viz.*, selection of mother trees of local cultivars, indexing of mother trees, recovery of pathogen-free plants by shoot-tip grafting in vitro, indexing of micro-grafted plants, horticultural evaluation of healthy plants and maintenance of healthy plants (Navarro 1993).

Selection of Mother Trees

A large number of Citrus orchards were surveyed and outstanding mother trees were selected on the basis of tree vigour, yield (no of fruits per tree), fruit quality parameters such as fruit weight, TSS and tree health. After observing and confirming the yield, fruit quality parameters over a couple of fruiting seasons, mother trees were finalized for sanitation and further multiplication. Bud wood was collected from selected outstanding trees and propagated on vigorous Rough lemon rootstock seedlings and bud grafts were maintained in screen house.

Indexing of Bud Grafts

These bud grafts were indexed for major pathogens associated with citrus decline in India such as citrus tristeza virus (CTV), ring spot, mosaic virus, citrus exocortis viroids (CEVd), citrus yellow corky vein viroid, and greening bacterium, both biologically and serologically. Biological indexing was done according to standard procedures (Roistacher 1991) and indexing of CTV was done by indirect Das Elisa using CTV specific antibodies (Cambra *et al.,* 1990, Vijayakumari *et al.,* 2006).Indexed healthy bud grafts were labelled and maintained in insect proof screen house as mother plants of healthy scion wood.

Rootstock Preparation

Rootstock seedlings were obtained by seed germination in vitro. Seeds of Troyer Citrange were peeled (removing both seed coats), surface sterilized by immersion for 10min in 0.5-0.7 per cent sodium hypochlorite solution containing 0.1 per cent tween 20 wetting agent and rinsed three times with sterile distilled water. Seeds were individually cultured and incubated at a constant temperature of 27°C in continuous darkness for 2 weeks. Seedlings 3.5 cm tall (Figures 10.1a,b) with a diameter of 1.6 – 1.8 mm at the point of grafting were used as rootstocks. Most of the studies with STG have been conducted using Troyer as rootstock. It has the advantage of trifoliate leaves which serve as morphological markers for identification of adventitious shoots produced by the rootstock.

Scion Preparation

Flushes of 3 cm length (Figure 10.2) were collected from healthy bud grafts, shoots are stripped of larger leaves, 1 cm long shoot tips were surface sterilized by immersing in 0.25 per cent sodium hypochlorite solution containing 0.1 per cent Tween 20 wetting agent and rinsed three times with sterile distilled water.

Figure 10.1a: *In vitro* Seed Inoculation.

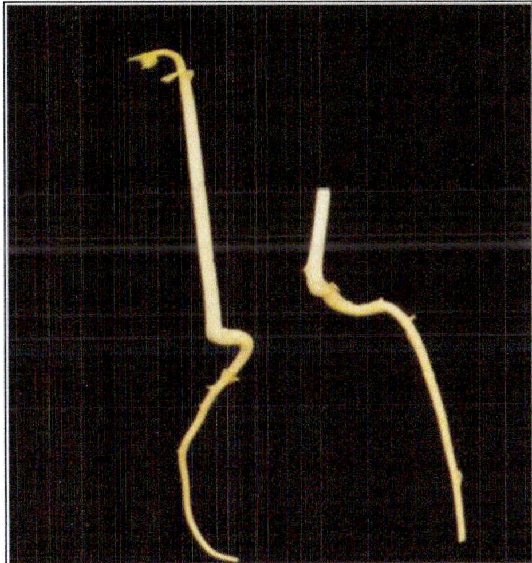

Figure 10.1b: Decapitated Root Stock.

Figure 10.2: Shoot Tip Preparation.

Figure 10.3: Three Week Old Shoot Tip Graft *in vitro*.

Shoot Tip Grafting *in vitro*

The shoot tips comprising of apical meristem and two or three leaf primordia, were excised and with the aid of microscope and transferred to the cortex in apical triangle cut at the top of decapitated rootstock. The grafts were transferred to test tubes having liquid MS media with modified whites vitamins and cultured under 16 hours light, till scion buds had sprouted. Successful grafts (Figure 10.3) were transferred by double grafting (Figure 10.4) on vigorous green house grown Rough lemon, Rangpur and trifoliate orange rootstocks (De lange 1978), (Vijayakumari *et al.*, 1994).

Figure 10.4: Lab to Land Transfer through Double Grafting.

Figure 10.5: Six Year Old STG Plants in Farmers Field at Tumsar in Bhandara District, Maharashtra.

Re-Indexing and Establishment of Field and Protected Foundation Blocks

The double grafts were re-indexed for confirming the elimination of viruses and other pathogens. Only the heathy plants were maintained in protected and field foundation blocks to serve as repositories of disease free mother trees for further multiplication. These plants were re-indexed periodically as required to ascertain health. Each tree in the foundation block was examined several times each year for overall health, fruit quality and symptoms of disease and other abnormalities. Further each tree of the protected foundation block is tested for CTV and Greening once in a year, and for viroids once in 3 years by Sero, bio-diagnostic and molecular techniques.

Multiplication

Multiplication blocks were made under low cost green house. Solarized disinfected potting mixture was only used for filling the poly bags. The rootstocks were grown from disinfected seed and the scion wood collected from virus free foundation blocks was used for raising bud grafts for supply to farmers.

Evaluation of Field Transferred STG Derived Plants

The performance of bud grafts planted in the farmers orchards (1-17 years of age) in different districts of Central India was evaluated periodically in terms of vegetative growth and yield parameters through properly laid out experiments in Randomized Block Design and the data was analysed statistically (Vijayakumari N., 2009).

The planting stock derived from STG had established well in the field and showed higher plant height, stock girth, scion girth and canopy volume compared to conventionally raised grafts procured from private nurseries and planted in the same orchards (Vijayakumari 2012; Vijayakumari and Karihaloo, 2012). The average yield of STG derived quality plants, recorded in farmers field, was 25 tons/ Ha.

STG based Quarantine Programs

Movement of vegetative material of Citrus is indispensable for various purpose like propagation, germplasm exchange *etc.* but there is an inbuilt risk of spreading the diseases. Quarantine programs aim at importing foreign varieties devoid of new pest and diseases that may be carried on the original material (Navarro 1993). CCRI is declared as National Active Collection Center.

Two different quarantine procedures are followed for the safe importations of citrus plant material. The classical method is very slow and consists of propagating imported bud wood in quarantine greenhouses located far away from citrus growing areas. Thereafter, the newly propagated plants may be indexed or submitted directly to shoot tip grafting in vitro or thermotherapy to free them from graft-transmissible pathogens and then indexed to verify their pathogens free status.

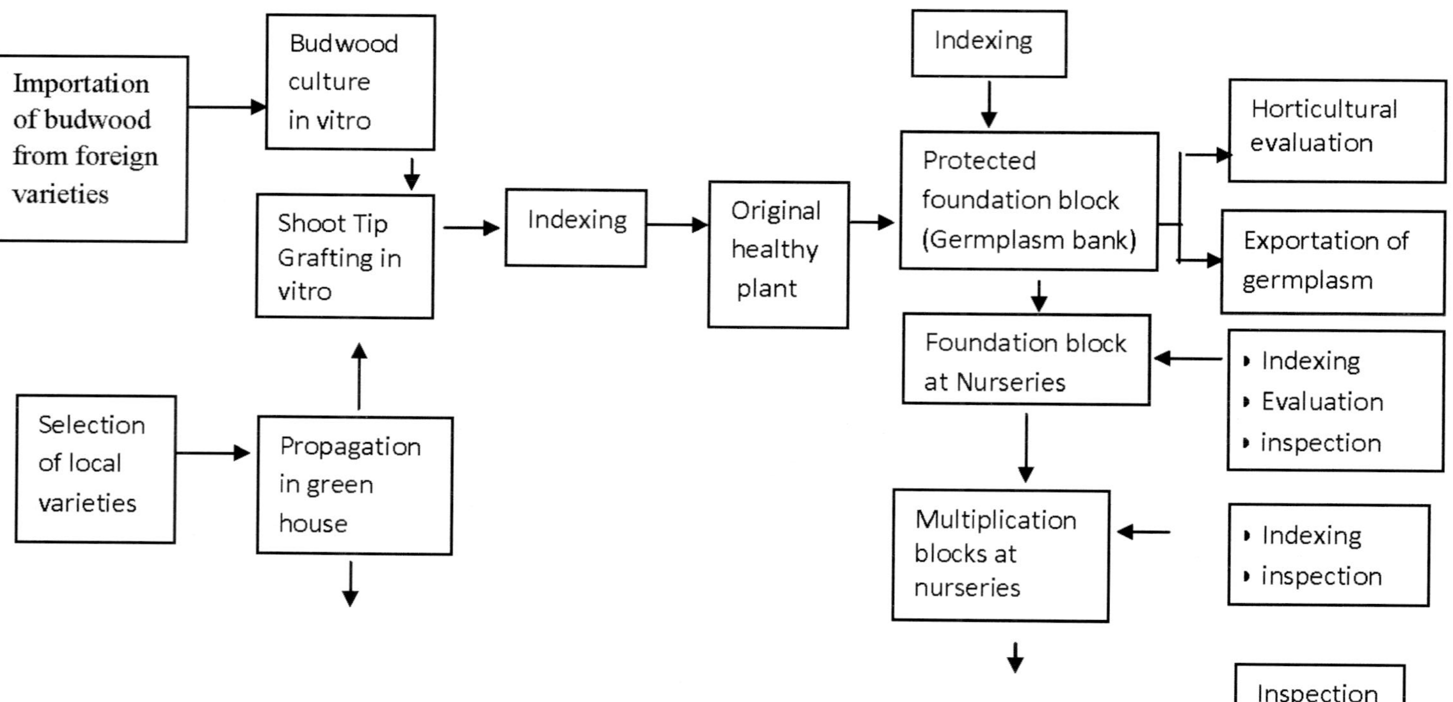

Figure 10.6: Schematic Presentation of Citrus Sanitation, Quarantine and Certification Programs.

This introduction procedure requires the availability of facilities and trained personnel having knowledge of citrus pests, diseases and cultural practices. It is used in countries such as the United States and Australia, having long tradition in quarantine, but it is expensive and difficult to establish for citrus in countries where citrus is minor crop.

An alternative procedure based on STG was developed for the safe introduction of citrus varieties (Navarro *et al.,* 1984) which had proven to be very efficient to exclude citrus pests and diseases (Navarro *et al.,* 1991; Navarro *et al.,* 1984:1988; Vijayakumari *et al.,* 2006, 2008a,b). It consists of culturing *in vitro* the imported budsticks at a constant 32°C or 35°C day and 30°C night, and exposed 16 h daily to 45 p.E/m2Iscc illumination to induce the sprouting of lateral buds and formation of flushes from which shoot tips can be excised and micro-grafted in vitro.

Standardization of STG (shoot tip grafting) based citrus quarantine protocol at C.C.R.I. had paved the way for safe introduction and widening of the gene pool. Using STG as post entry quarantine check, 20 exotic citrus cultivars introduced form USA, France, Australia were cleaned and are being maintained in the Citrus repository. Mandarin introductions (Dhakula local, Corsica No.1 and No.2 Marissol, Clementine Monreal, Frost owari) Sweet orange introductions (Riogrande navel, lane late navel orange, Campbell Valencia, Cutter Valencia, Shamouti orange, Ruby blood orange) grape fruit introductions (Ray Ruby, Flame) and Pummelo (US-145) were cleaned. Out of 21 exotic cultivars cleaned,20 micro-grafted cultivars were recommended for field evaluation. On successful adaptation they can be used for both scientific and commercial purposes. Three exotic collections *viz.,* Cutter Valencia, Flame Grape fruit and US -145 Pummelo were released as varieties for central India. The principal benefit is the availability of bud wood of superior high yielding varieties or clones which meet the fruit quality requirements of export market. The healthy citrus repositories created through in vitro shoot tip grafting are the insurance policies for the future health and economic viability of the Indian Citrus industry (Figure 10.1).

Micro-budding

In Central India, nursery men traditionally raise the root stocks in nursery beds in the open field and after the root stock seedling attain suitable size, budding is done in the winter season from November to January. The conventional shield and 'T' budding is season bound and requires 1-1.5 year old root stock seedlings and easily peelable bud wood with swollen buds ready to sprout. The total time required for production of plantable budgrafts is around 22 months. The budding requirements mentioned restrict the quantum of budgrafts produced in a year.

A simple, economical budding technique named micro-buddding was first developed by Dr. M. Skaria, Texas A and M University, U.S.A. This technique was adopted and modified for propagating Nagpur mandarin for the first time in India. (Vijayakumari and Shyam Singh 2000, 2002 and 2003, Vijayakumari *et al.,* 2008).

The seeds of rough lemon and Rangpur lime rootstock seeds were sown in trays containing sterilized soil. Three month old seedlings of rough lemon and Rangpur lime were transferred to polybags having sterilized soil mixture rich with nutrients to attain luxurious vegetative growth. Delicate young buds are carefully cut from the tender new flush of STG derived scion mother plants and are used for micro budding. Five month old rootstock seedlings are wedge grafted with young buds derived from tender new flush and covered with plastic micropipette tips till sprouting (Figures 10.7 and 10.8). Micro budding was performed under controlled conditions almost throughout the year (except for April/May).

Figure 10.7 Micro-Budded Plant.

Figure 10.8: Large Scale Multiplication through Micro-budding.

Micro budding had shown significant success with various commercial scion and rootstock combinations. At ICAR–CCRI microbudding was successfully performed with various indigenous and exotic citrus scions on 3 commercial citrus rootstocks *viz.*, rough lemon, Rangapur lme and Alemow with success percentage ranging from 40-90 per cent (Vijayakumari, 12).

Impact of Micro-budding Technology

Micro-budding technology had facilitated faster propagation by shortening of citrus nursery phase by 9-10 months and resulted in significant savings in the cost of planting stock production. Further the production programme can be taken up throughout the year. In addition micro-budding can be utilized in biological indexing of virus, viroids, greening bacterium and other disease inoculation and expression studies.

Micro-budding technology had facilitated (i) large scale multiplication by replacing use of young bud wood instead of mature bud wood and young root stocks instead of older root stocks (ii) year round multiplication of bud grafts. This method can easily replace conventional Shield budding/T budding method. The canopy volumes were higher in micro-budded plants than shield budded plants.

Retrofitting the Citrus Nursery Phase for Faster and Economical Production of Quality Planting Stock

The farmers are willing to pay higher costs for quality plants but considering the required quantum of planting stock, it is felt that even a marginal cost reduction would lead to huge savings to the Nursery men/farming community. Further ICAR institutions are mandated to generate 30 per cent of the revenue internally. In this context a study was undertaken to retrofit the nursery phase to reduce (i) the time, cost of production (ii) the extent of root coiling and (iii) improve the growth of field planted bud graft.

In a containerized nursery, the root stock seedlings are raised in primary nursery and secondary nurseries for 12-14 months prior to shield/Inverted T budding. There after budded plants are kept in the nursery for another 6-7 moths. To reduce the nursery phase and save costs the root stock seeds were directly sown in the poly bags. Root stock seedlings of 5 months age were used for micro budding. The time required for raising buddable root stock is reduced by 7-9 months leading to significant savings in root stock seedling raising. The extent of root coiling was reduced significantly because of shorter nursery phase and bypassing the transplanting and this improves the growth of field planted trees.

Research Out Come

About 5.5 lakh disease free quality bud grafts were supplied to citrus growers, nursery men during 2002 to 2023 for raising about 1600 Ha of orchards during 2002 to 2023 to citrus growers/nursery men in various states (Maharashtra, M.P, Gujarat, Rajasthan, Haryana, Punjab, Sikkim, Assam, Orissa, A.P, Tamil Nadu, Karnataka and Kerala). An amount of Rs. 2.93 crore was realized as revenue from the sale of STG derived bud grafts.

Technology Transfers

CCRI has limited capacity to produce the planting stock hence 10 technology transfer Memorandums of Understanding (MoUs) were signed (till 2024) with progressive nursery men and entrepreneurs, who were willing to follow the protocols developed by the CCRI, to scale up the production of quality planting stock.

Dovetailing the Innovative Propagation Methods

The following recommendations were made for effective utilization of innovative propagation techniques *viz.*, STG, Micro-budding and direct seeding methods

1. Clean the promising elite scions using shoot tip grafting technique.
2. Employ double grafting for faster and vigorous growth of scion wood in a protective net house.

3. Reduce the time and cost of root stock raising by directly sowing the seeds in poly bags
4. Use micro-budding technique instead of Shield budding/inverted T budding for faster, economical multiplication of bud grafts in sanitized container nurseries.
5. Introduce mechanism for testing and certification of propagative bud wood by establishing referral labs at CCRI, State Agriculture Universities or/and State Seed testing labs.
6. Implement nursery acts of respective states for ensuring quality of plants produced and supplied.

Conclusion and Way Forward

The protocols of shoot tip grafting, micro-budding and retrofitting developed for faster, economical large scale production of quality bud grafts had proved promising and delivered the desired results. Effective transfer and adoption of this technology by the nursery men would go a long way in improving the productivity and the profitability of Citrus cultivation.

There is a need to implement a sound clonal selection programme to sustain the genetic quality of the stock. A minimum 20 high yielding clones need to be supplied together for overcoming the impending epidemics, if any, in the future.

There is an urgent need to strengthen the inter-institutional cooperation among all the stakeholders to define the roles and coordinate the efforts for boosting the Citrus Industry. ICAR institutes, SAUs need to undertake research works related to varietal improvement, sanitation, establishment and maintenance of protected foundation blocks. Registered nursery men need to establish model nurseries on scientific lines and the line departments need to implement the Nursery Registration Act. The scope of existing seed testing laboratories may be widened and strengthened or referral labs may be established by the ICAR institutes, SAUs to undertake health certification.

REFERENCES

1. Cambra M, Garnsey SM, Permar TA, Henderson CT, Gumpf D and Vela C., 1990. Detection of citrus tristeza virus (CTV) with a mixture of monoclonal antibodies. *Phytopathology* **80**: 103
2. De Lange JH., 1978. Shoot tip grafting a modified procedure. Citrus subtrop. *Fruit J.* **539**: 13-15.
3. Faccioli G and Marani F, 2005. Virus elimination by meristem tip culture and tip micrografting. In: Plant Virus Disease Control. (Eds.): A. Hadidi, R.K. Khetarpal, H. Koganezawa. Second Printing, APS Press, St. Paul. Minn. pp. 346-380

4. Krueger RR, Kahn, TK, Gumpf DJ, Balance PM, Bier OJ and Bash JA, 2003. Shoot tip micro grafting for rescue or therapy of field grown citrus germplasm accessions. In: Proceedings 9th ISC Congress.

5. Morel G. and Martin, 1952. Guerison de Dahlias atteintsd'unemaladie a virus. C. R. Acad. Sci., Paris. **235**: 1324-1325.

6. Murashige T, Bitters WP, Rangan TS, Nauei EM, Roistacher CN and Holliday BP, 1972. A technique of shoot apex grafting and its utilization towards recovering virus free citrus clones. *Hor. Sci.* **7**: 118-119.

7. Navarro L, Roistacher CN and Murashige T, 1975. Improvement of shoot tip grafting in vitro for virus free Citrus. *J. Amer. Soc. Hort. Sci.* **100**: 471-479.

8. Navarro and Jaurez, 1977. Tissue Culture techniques used in spain to recover virus free citrus plants, *Acta Horticulture Tissue culture for horticultural purposes.* **77:** 425-435.

9. Navarro L, 1993. Citrus sanitation, Quarantine and Certification Programs. 12 IOCV Conf. IOCV, Riverside.

10. Shyam Singh, Shivankar VJ, Singh IP, Ghosh DK and Das AK, 2008. Nursery Management for Production of disease free planting material of citrus. National Symposium on Citriculture: Emerging Trends: 24-26 July.

11. Skaria M and Hanagriff RD, 2008. Micro- budded, High Density Citrus planting: Is there an opportunity for HLB control and financial Returns. Proc. of Intl research conf. on Huanglongbing. December. 365-368.

12. Vijayakumari N, Dass HC and Singh A, 1994. Standardization of in vitro shoot tip grafting technique for elimination of virus and virus like diseases from Nagpur mandarin. *Indian Journal. Hort,* **51**: 311-31

13. Vijayakumari N and Singh S, 2000. Shoot tip grafting based citrus quarantine for germplasm exchange. In: Proc. Int. Symp. Citriculture, 227-229.

14. Vijayakumari N and Singh S, 2000. Shoot tip grafting with growth regulators for virus elimination in Nagpur mandarin (Citrus reticulata). *Indian J. Agri Sciences.***70**: 396-397.

15. Vijayakumari N and Singh S, 2002. In vitro shoot tip grafting of citrus exotic germplasm. *Indian J. Hort.* **59(4):** 363-366.

16. Vijayakumari N, Ghosh DK, Das AK, Singh Awatar and Singh S, 2006. Elimination of citrus tristeza virus and greening pathogens from exotic germplasm through in vitro shoot tip grafting in citrus. *Indian Journal of Agricultural Sciences.* **76(3)**: 209-10

17. Vijayakumari N, Ghosh DK, Das AK, Singh S and Singh IP, 2008. Healthy citrus gene bank through STG: A safe insurance policy for sustainable citrus industry. Presented as poster, National symposium on Citriculture: Emerging trends. 30-31.

18. Vijayakumari N, 2009. Superior production performance of STG derived planting stock vs. conventional bud grafts in framers fields of central India. Presented as oral paper, Intl Conf on Hort. Banglalore. India 09-12 November p. 276

19. Vijayakumari N, 2010. Citrus Regeneration systems in Citrus. In Advances Horticulture Biotechnology. Vol I: Fruit crops, plantation crops and spices, Singh H.P., Parthasarathy V.A. Nirmal Babu K, (Ed) Westvile Publishing House, New Delhi. 121 -161

20. Vijayakumari N and Karihaloo J, 2012. Biotechnology in citrus improvement. Souvenir and abstracts. N National dialogue on Citrus improvement, production and utilization. Nagpur, India 27-29 February, p. 33-44.

Chapter 11

Orchard and Canopy Management in Citrus

Thirugnanavel Anbazhagan, Darshan M. Kadam and Surajit Mondal

ICAR–Central Citrus Research Institute, Nagpur–440033, Maharashtra

Citrus (*Citrus* spp.), one of the leading fruit crops in the world are believed to be originated from tropical and sub-tropical regions of Asia and Oceania. Citrus cultivation is practised in diverse climatic and agro-ecological settings, from the tropical regions of Brazil to the sub-temperate regions of China and Japan. The appealing colour, unique flavour, vitamins, minerals, and dietary fibre abundance of citrus fruits make them one of the most prevalent household food commodities across the world. It is grown in 138 countries between 40°N and 40°S latitude and cultivated commercially in 53 countries (Murkute *et al.,* 2023).

Global citrus production has seen a consistent increase over the past thirty years. In 2021, the total area dedicated to various citrus fruits worldwide was 10.22 million hectares, yielding 161.80 million tonnes (FAOSTAT, 2023). Oranges accounted for 46.70 per cent of this total production, followed by tangerines, mandarins, and clementines at 25.93 per cent, lemons and limes at 12.87 per cent, pummelos and grapefruits at 5.91 per cent, and other citrus fruits at 8.59 per cent.

In India, citrus thrives in temperatures ranging from 13.0°C to 37.0°C, with an average annual precipitation of 700 mm. The crop adapts well to a wide variety of soils, with pH levels between 5.0 and 8.0, including sandy loam, alluvial soils, clay loam, deep clay loam, and lateritic/acidic soils (Murkute *et al.*, 2023). Over the past decades, the area and production of citrus have steadily increased. The cultivation area expanded from 0.09 million hectares in 1961 to 1.09 million hectares in 2020-21, while production rose from 0.824 million tonnes to 14.26 million tonnes during the same period. However, this represents only a 1.4-fold increase in productivity, from 9.16 tonnes to 12.66 tonnes per hectare. The main challenges facing the citrus industry include a lack of high-yielding varieties and rootstocks resistant to major environmental stresses, threats from greening, canker, and Phytophthora, rising soil and water salinity, improper orchard and canopy management practices, and a dwindling supply of irrigation water (Kadam *et al.*, 2022).

Mandarins, limes, lemons, and sweet oranges are the most well-known citrus fruits that are farmed for commercial purposes in India. Mandarins account for the largest share of both area and production (approximately 42 per cent each), followed by lemons and limes (31 per cent and 26 per cent, respectively), sweet oranges (19 per cent and 27 per cent, respectively), and other citrus fruits (8 per cent and 5 per cent, respectively), according to the Ministry of Agriculture and Farmers Welfare, Government of India's third advance estimate. However, in terms of productivity, sweet oranges recorded the highest productivity at 18.84 tonnes in 2020-21, followed by mandarins at 13.04 tonnes, lemons and limes at 11.26 tonnes, and other citrus fruits at 7.51 tonnes per hectare.

Climate and Soil

Citrus species are cultivated across various climates, including tropical, sub-tropical, and temperate regions, encompassing humid, semi-arid, and arid conditions. As a warm-climate crop, citrus performs optimally in sub-tropical and Mediterranean climates, which exhibit minimal seasonal changes. Citrus cultivation thrives within a temperature range of 10 to 35°C, with vegetative growth commencing at 12.8°C. Optimal growth conditions for citrus fruits are found between 25 and 30°C, as 30°C is considered ideal for photosynthesis and dry matter production, enhancing tree vigour and productivity (Abobatta, 2021). Growth continues to increase with rising temperatures up to 35°C but begins to decline gradually as temperatures rise further, ceasing almost completely at 49.5°C. Main branches die at 51.5°C. Citrus trees do not tolerate cold or frost well, although they can withstand short periods of low temperatures down to 3°C. Flowers and new shoots are susceptible to damage at -1.7°C, and temperatures below -6°C can result in significant tree damage.

Citrus trees are adaptable to a range of soil types, including clay, loamy, and sandy soils, as evidenced by their cultivation in Egypt. They are tolerant of both high and low pH levels. Despite their adaptability, citrus trees are sensitive to cold or freezing weather and cannot endure prolonged soil flooding, especially in clay soils. Mediterranean climates, however, produce the highest quality citrus fruits. Research has demonstrated the significant impact of soil type on citrus yield and quality in various regions, including Australia, India, Israel, Italy, South Africa,

Western Georgia, and the United States. Notably, the Nagpur mandarin a leading commercial citrus cultivar in central India is predominantly grown in alkaline black soils rich in montmorillonite-based minerals (Srivastava and Singh, 2002; Srivastava, 2015).

Planting

In India, citrus species such as mandarins and sweet oranges are typically established at a spacing of 6 meters by 6 meters, resulting in a planting density of approximately 250-350 plants per hectare. This density can vary based on the specific variety, rootstock, and agro-climatic conditions. In Karnataka, the Coorg mandarin is planted on Trifoliate orange rootstock at a density of 400 plants per hectare and on Rangpur lime rootstock at a density of 275 plants per hectare. In Haryana, Kinnow mandarins are generally planted at a density of 375 plants per hectare using Jatti Khatti rootstock. Additionally, high-density planting of Kinnow mandarins has been successfully achieved using rootstocks such as Troyer citrange, Karna Khatta, and Sohsarkar (Nawaz *et al.*, 2007).

High Density Planting Practices in Citrus

In several countries high-vigour citrus rootstocks like Rough lemon and Volkamer lemon have been utilized in economic citrus production for many years. However, trend is shifting towards high-density planting in modern citriculture (Rabe, 1996). Over the past four decades, Brazilian citriculture has increased average planting density by more than 80 per cent (Girardi *et al.*, 2021a). High-density planting systems represent an innovative strategy designed to enhance yield and net returns, especially during the early stages of orchard development, by accommodating a greater number of plants per unit area compared to traditional planting methods. The primary advantages of intensive cultivation systems include precocity, reduced cost per unit of production, and the potential for increased mechanization with improved input-use-efficiency. In addition, quick return on investment, easier management of cultural operations and improved control of citrus greening are other key rewards. Given the limitations on expanding production space, boosting productivity is essential, and one effective technique to achieve this is through agronomic management, such as high-density planting.

Range of factors regulates size of citrus plants, including rootstock, variety, edaphic and climatic conditions, cultural operations and phytosanitary conditions (Golomb, 1988). Effective control of citrus plant growth can be achieved through various strategies: (a) using dwarfing, semi-dwarfing rootstock and scions with limited growth potential, (b) pruning, (c) biological agents such as viroids and viruses (d) restricting root system growth (e) sloping seedling planting (g) applying plant growth retardants. Among these, the use of dwarfing rootstocks is considered as most effective and proven strategy (Cantuarias-Aviles *et al.*, 2010). In this background, dwarfing rootstocks have garnered significant interest from citrus growers globally due to their suitability for higher planting densities, ease of harvesting, pruning and effectiveness in spraying, all of which contribute to enhanced productivity. The only true dwarf rootstock in citriculture is the Flying Dragon, a size-controlling mutant of trifoliate orange (*Poncirus trifoliata* L. Raft var. *monstruosa*), is noteworthy for significantly decreasing plant size when used as a rootstock (Castle *et al.,* 2007). Consequently, most citrus scions grafted onto Flying Dragon rootstock do not exceed a height of 2.5 meters, with overall growth potentially reduced by 75 per cent compared to those on standard trifoliate orange rootstock (Cantuarias-Aviles *et al.,* 2010). In Brazil the use of tetraploid Swingle citrumelo as rootstock proved dwarfing effect with Valencia sweet orange, decreasing canopy volume > 75 per cent as against vigorous standard (Girardi *et al.,* 2021b). For Tahiti lime, the TS × PT 14 citrandarin exhibited dwarfing characteristics and indicated potential tolerance to HLB (Bettini *et al.,* 2019).

In Brazil, four rootstocks exhibited dwarfing effect, yield efficiency and good fruit quality for Valencia orange these rootstocks include three citrandarins (TSKC × TRFD-003, TSKC × TRFD-006, TSKC × TRFD-007) and one citrumelandarin (TSKC × CTSW-058), hybrids of Sunki mandarin [*Citrus sunki* (Hayata) Hort. ex Tan.] with either Flying Dragon trifoliate orange [*P. trifoliata* var. *monstrosa* (T. Ito) Swing.] or Swingle citrumelo (*C. paradisi* Macfad. × *P. trifoliata*) (Costa *et al.,* 2021).

In Spain, two rootstocks *i.e.* Forner-Alcaide (FA) 418 and FA 517 impart dwarfing traits to Navel orange in comparison to standard rootstock Carrizo citrange. Both rootstocks prompted the heavy production of good quality fruit (Forner-Giner *et al.,* 2014). In case of FA 418 rootstock dwarfing mechanism mediated by improved reproductive growth and fruit development, resulting in compact vegetative development. Consequently, alteration in the pattern of assimilate distribution is one of the main mechanisms of the dwarfing (Lliso *et al.,* 2004).

In China, commonly used semi-dwarfing and dwarfing rootstocks include Xiangcheng orange, Citrange, and Red tangerine, which are considered effective for plant size control (Hayat *et al.,* 2022). In Brazil, high-density citrus orchards typically feature 600–1250 plants per hectare, with row spacing of 4–6 meters and plant-to-plant spacing of 2–3 meters (Hayat *et al.,* 2022). In South-East Brazil, the tree densities of citrus orchards have increased from an average of 337 trees per hectare in the 1980s to over 600 trees per hectare by 2013, eventually stabilizing at 650 trees per hectare (Gonzatto *et al.,* 2022).

With regards to USA, rootstock US-897 citrandarin (Cleopatra mandarin (*C. reshni* Hort. ex Tan.) × Flying Dragon trifoliate orange (*P. trifoliata*)) was advocated in high-density orchards and cultivation under protected structures. Although per plant yield of US-897 is low the potential yield per hectare at the predicted high density spacing can be very high (Bowman and Joubert, 2020). Furthermore, in University of Florida recommended citrandarin UFR-6 (*C. reticulata* 'Changsha' + *P. trifoliata* '50-7') somatic hybrid rootstock for high-density citrus orchards, producing small plants and fruits with high soluble solids content (Bowman and Joubert, 2020).

In India, experiments conducted on Nagpur mandarin to evaluate the impact of different tree densities on production and economics revealed that the yield obtained under 2 m × 2 m spacing (2500 plants per hectare) was twenty-six times higher than the control. However, by the fifth year, the 6 m × 3 m spacing treatment (555 plants per hectare) was found to be economically superior (Ladaniya *et al.*, 2019) (Figure 11.1).

6 x 3 m (Mandarin)

6 x 3 m (Sweet orange)

Figure 11.1: High Density Planting in Citrus.

Another study conducted in central India on acid lime cv. Pramalini with two different spacing, 2.5 m × 2.5 m and 2.5 m × 5 m, suggested that while the yield per area was highest at 2.5 m × 2.5 m, the increased maintenance costs negated the net returns. Conversely, the 2.5 m × 5 m spacing reaped similar profits despite a lower yield compared to ultra-high density planting (Ladaniya *et al.*, 2020) (Figure 11.2). This data collectively suggests that high-density planting is particularly effective in increasing the fruit-bearing volume per hectare. However, irrespective of its prospective to increase productivity in initial years, high density plantation necessitate higher investment and could lead to the lessening of yield owing to excessive tree massing and complications in cultural operations.

Another strategy to reduce tree size is viral dwarfing technique which involves intentional inoculation with the mild strain of citrus exocortis viroid. Technique has been investigated in different countries such as Australia, the USA, and Israel. Bevington and Bacon (1977) demonstrated that inoculating Bellami orange grafted onto rootstocks like Trifoliate orange, Carrizo, Troyer citrange, and Rangpur lime with two mild strains of the CEV virus effectively reduced tree size by 51, 25, 23, and 19 per cent, respectively, compared to non-inoculated controls.

Angle planting also used sometimes to manage tree size; it involves positioning trees at an angle of 30 to 45 degrees or more from the horizontal, with all tree crowns facing the same direction within the row. These rows are typically oriented in a north-to-south direction, and trees are planted in close proximity. This distinctive method, first tested in the early 1960s in Italy and California (Kamatyanatti and Singh, 2019), has been reported to control tree size. Angle-planted orchards exhibit two key responses that necessitate intensive management. Firstly, the inherent gravitropic response causes the trees to grow upright. Secondly, the trees produce numerous vertical shoots along the upper side of the main stem, which can reach unmanageable heights if not regularly topped.

2.5 x 2.5 m 2.5 x 5 m

5 x 5m

Figure 11.2: High Density Planting System in Acid Lime

Consequently, consistent topping is essential for maintaining the desired structure in angle planting. The key findings of different high density planting practices studies performed in citrus are summarized in Table 11.1.

Canopy Management

Canopy management involves adjusting tree canopies and bearing volumes to optimize fruit yield and quality. Generally, larger canopies produce higher yields compared to smaller ones. Effective canopy management in citrus is essential to prevent issues like overcrowding and excessively tall trees, enhance light interception and aeration, and optimize pesticide delivery within the canopy. Additionally, it facilitates equipment operation within the orchard and regulates vegetative growth, which is vital for maintaining healthy and productive citrus trees.

Table 11.1: Impact of High Density Planting Practices on Citrus

Location and Country	*Particulars*	*Key Findings*	*Reference*
Nagpur, India	Scion: Nagpur mandarin Rootstock: Rangpur lime Tree spacing: 6 m × 6 m, 6 m × 3 m, 4 m × 4 m, 4 m × 2 m and 2 m × 2m	The yield obtained under 2 m × 2 m spacing 2500 plant ha^{-1}) treatment was twenty-six times higher than control. In 5th year 6 m × 2 m spacing 2500 plant ha^{-1}) treatment was twenty-six times higher than control. In 5th year 6 m × 3 m spacing treatment (555 plants ha^{-1}) was found economically superior.	Ladaniya *et al.*, 2019
Nagpur, India	Scion: Acid lime cv. Pramalini Rootstock: Nil Tree spacing: 2.5 m × 2.5 m and 2.5 m × 5 m	The yield on area basis at 2.5 m × 2.5 m was highest, it could not be converted in to net returns due to increased maintenance cost. On the contrary 2.5 m × 5 m reaped similar profit in spite of lower yield than ultra-high density planting.	Ladaniya *et al.*, 2020
Indian River citrus district, Florida, USA	Scion: Sweet orange cv. Valencia Rootstock: Kuharske citrange Planting density: 955 trees/ha	Adoption of high tree density, with fertigation augmented fruit yield in citrus greening infected sweet orange orchards.	Ferrarezi *et al.*, 2020
Sao Paulo, Brazil	Scion: Sweet orange cv. Valencia Rootstock: IAC 1697, IAC 1710, Swingle and Swingle 4x Planting density: 513, 696, and 1000 trees ha^{-1} trees/ha	Planting vigorous rootstocks at moderate to high tree densities improves the land-use efficiency of citrus orchards.	Girardi *et al.*, 2021a
Haryana, India	Scion: Kinnow mandarin Rootstock: Rough lemon Tree spacing: 6 m × 6 m, 6 m × 5 m and 6 m × 3 m	Spacing of 6 m × 3 m may be adopted by kinnow growers for getting higher productivity, WUE and economic gain in the initial year of planting up to six years.	Dalal *et al.*, 2013
Pakistan, Faisalabad	Scion: Kinnow Rootstock: Rough lemon Tree spacing: 3.30 m × 6.60 m, 3.30 m × 3.30 m and 6.60 m × 6.60 m	The highest yield was observed at (3.30 m × 6.60 m) spacing density.	Nawaz *et al.*, 2007

In mature groves that are not regularly managed, insufficient light can negatively impact yield and fruit quality. Sunlight interception and distribution are crucial for flowering and fruit set, as well as for improving fruit quality and colour development. Better light penetration within the canopy also promotes faster drying of foliage after rain, potentially reducing fungal pathogen growth (Wani *et al.,* 2021).

Training and pruning, key elements of canopy management, adjust tree form and the ratio of the framework to the fruit-bearing part of the canopy, alter the shoot-to-root ratio, and influence the distribution of carbohydrates and sugars. These practices also affect the amount of sunlight captured by trees, improve aeration, and regulate leaf temperature. Consequently, they regulate vital processes such as photosynthesis, respiration, and the incidence of diseases and pests (Dhillon and Thakur, 2014). An effective training strategy focuses on the strategic arrangement of plant parts to develop an optimized plant architecture that maximizes sunlight utilization and boosts productivity. Light is essential for tree and fruit growth, as leaves capture sunlight to produce carbohydrates and sugars, which are then transported to buds, flowers, and fruits. Planting density and orientation significantly affect light penetration in an orchard. High-density planting can lead to rapid shading, while an east-west row orientation causes more shading compared to north-south alignments. Strong bearing branches are associated with larger fruit production. The initial challenge for a grower is to establish a strong, sturdy, and balanced framework for the trees with well-distributed limbs, followed by equipping them for optimal fruiting (Singh, 2010).

Key principles in citrus canopy management include:

1. Maximizing light utilization
2. Preventing microclimates conducive to disease and pests
3. Ensuring convenience in performing cultural operations
4. Augmenting productivity with high-quality fruit production
5. Achieving desired canopy architecture economically

In citrus, pruning vigorous mature trees can reduce yields in proportion to the quantity of foliage removed and can delay fruiting in young trees. Pruning should be limited to what is necessary for the future development of the canopy and to facilitate efficient cultural and harvesting practices. In Indian conditions, pruning of commercial citrus species, such as mandarin, sweet orange and acid lime, is limited to training for framework development during the juvenile phase and pruning dead, diseased, damaged, and intermingled branches during the production phase. The response of trees to pruning depends on various factors, including species, variety, climatic conditions, tree age, timing of pruning, and orchard management practices. Each situation must be critically analysed before making significant canopy management decisions.

Canopy Management for Young Citrus Trees

Heavy pruning and training of juvenile, non-bearing citrus trees should be avoided as it tends to delay fruit production. Typically, citrus trees do not require pruning during the initial years, except for the removal of sprouts below the bud union and water suckers. Vigorous sprouts should be removed to prevent them from dominating weaker trees. In the USA, the occurrence of sprouts on the trunks of young, non-bearing trees can be significantly reduced by using a commercial sprout inhibitor containing naphthalene acetic acid (NAA) (Mudzunga, 2000). Although protective wraps and PVC pipes around the trunk can help reduce sprouting (Figure 11.3), they require careful monitoring to avoid insect and disease issues beneath the wraps.

Figure 11.3: PVC Protective Wrap for Young Plants.

Nagpur Mandarin

For Nagpur mandarin trees, it is ideal to have a low head and a dome-shaped crown. This involves pruning juvenile trees during the initial years of establishment. Training helps shape and size the young trees effectively. Initially, trees are guided to form a single trunk with 4-6 well-spaced branches, creating the basic framework. The lowest branches should not extend below 50 cm from the soil surface.

Sweet Orange

Trees are trained using a single-stem system, and shoots emerging from below the bud union should be regularly removed. In the first year, growth beyond a height of 0.7–1 m should be pruned to encourage side limb development. Only 4–6 well-spaced branches, with wide angles from the main trunk and distributed evenly should be allowed to grow up to 3–4 m. After this initial training period, no further training is necessary. Plant training should be completed within the first three years to ensure a mechanically strong canopy. Pruning of pre-bearing trees can be carried out at any time, avoiding peak winter.

Acid Lime

For young acid lime plants, a modified central leader system can be employed, with a smooth trunk growing to a height of 75-100 cm from the soil, with 4-5 well-distributed branches serving as scaffolding. All shoots on the trunk up to the specified height should be removed. Water sprouts on the main trunk and scaffolding branches of mature trees should also be removed promptly. Once a young plant is shaped as desired, minimal pruning is necessary. Light pruning in subsequent years promotes extensive root and shoot development, leading to earlier fruit production compared to heavily pruned trees.

Lemon

Young lemon trees tend to develop long, sprawling branches that produce fruits primarily at the tips, leading to drooping. Only a few of these branches are necessary for the tree's framework, while the rest, particularly those in the center, should be removed.

Canopy Management for Bearing Citrus Trees

Nagpur Mandarin

Bearing Nagpur mandarin trees require minimal pruning. The primary goal is to maintain the tree's framework and ensure higher yields with superior quality fruits. Pruning involves removing dead, diseased, damaged, intersecting, and weak branches, as well as water sprouts and rootstock suckers (Figure 11.4). Non-bearing trees can be pruned year-round, but for bearing trees, the optimal period is after harvesting, in late winter or early spring, when they are semi-dormant. In some regions of central and southern India, root pruning via deep ploughing during summer induces Mrig bahar flowering (Jhade *et al.,* 2021), but this can facilitate *Phytophthora* entry. Mandarins are prone to overbearing and alternate bearing. Pruning one-year-old shoots by half or completely can balance fruiting and vegetative growth, ensuring proper yield and high-quality fruits (Fonte *et al.,* 2022).

Figure 11.4: Dead Wood Pruning in Nagpur Mandarin.

Sweet Orange

Pruning of bearing sweet orange trees involves removing dead, diseased, criss-crossed, and weak branches, as well as water sprouts and suckers below the bud union. Thinning shoots to enhance sunlight penetration and aeration is essential. Treating cut ends with Bordeaux paste helps prevent fungal infections. The best time for pruning is after harvesting, during late winter or early spring. Spraying Bavistin (1 ml/liter of water) after pruning helps prevent secondary infections.

Acid Lime

Pruning practices for fruit-bearing acid lime trees focus on removing dead, dried, diseased, broken, and intersecting branches, as well as water sprouts. Pruning should be done immediately after harvesting, and cut ends should be treated with Bordeaux paste or Blitox (Figure 11.5).

Figure 11.5: Removal of Intersecting Branches and Bordeaux Paste Application.

Lemon

Mature lemon trees require more extensive pruning. Long shoots that have borne fruits at the tips should be pruned back to lower secondary shoots to encourage fruit production closer to the ground. Annual light pruning of branches that have borne fruits for several years stimulates new shoot growth and maintains high-quality fruit production. In West Bengal, light pruning (25 cm from the terminal portion of the shoot) combined with the integrated use of fertilizers (75% RDF + Vermicompost + Azotobacter + Vesicular Arbuscular Mycorrhiza) has shown significant improvement in fruit yield for Assam lemon (Ghosh *et al.,* 2016). Martin-Gorriz *et al.* (2021) found that mechanical pruning rotated with manual pruning is economically viable for the 'Fino 95' lemon cultivar in Spain.

Alternate Bearing Management through Pruning

Alternate bearing tendency present in few citrus cultivars can be alleviated by pruning post-harvest, with mild-intensity pruning in low flowering years and severe-intensity pruning in high flowering years (Mesejo *et al.,* 2020). Study conducted in Iran advocated that pruning alone or in combination with girdling can lower the impact of alternate bearing in kinnow mandarin. As it diminish competition for nutrients and induce accretion of carbohydrates in limbs, which enhanced yield of the next season (Ahmadpoor *et al.,* 2022). Another study performed in Spain advocated that annual mechanical pruning of flowering shoots in half-length diminishes alternate bearing tendency in mandarin cv. Nadorcott. In this experiment, collective yield augmented by 25 per cent as against unpruned trees for the period of the four years of the experiment (Mesejo *et al.,* 2020).

Flowering and Flower Regulation in Citrus

Citrus flowering is complex phenomenon and is induced either by low temperature stress or water deficit stress (Davenport, 2011). Cessation of root growth, confined roots, weak rootstocks and water stress are necessary for floral induction (Monselise, 2019). Citrus flowering is temperature dependant; flowering is one month earlier in high temperature zones like Florida when compared to cooler regions like California (Webber and Batchelor, 1979). In temperate climate, low temperature less than 15 °C for about 1.5 months, ceased the plant growth temporarily and when temperature increases during spring, the growth resumes and results in profuse flowering. Flowering is not temperature dependant in tropical climate. To induce flowering in these regions, water stress is important. Cassin *et al.* (1969) studied how orchards in eight tropical nations responded to prolonged periods of water stress affects flowering. They noticed that 20 to 28 days after the first rainfall or irrigation, a combination of vegetative and reproductive growth occurs. In sub-tropical areas, dry summers can occasionally cause flowering. In sub-tropical climates, summer production of lemons following water stress is locally called as Vardelli. Flowering was induced by withholding of irrigation in Italy, Israel (Oppenheimer, 1940) and California (Maranto and Hake, 1985).

In India, the citrus growing regions never experience freezing temperature to induce flowering. However, the temperature in North-East and North-West India experience low temperature (<10 °C) for 10-15 days which is sufficient to induce flowering. The low temperature prevailing in North-East India induce flowering in mandarins once a year and mandarin generally blooms during February to April (Deshmukh *et al.,* 2016). In hilly regions of South India, Coorg mandarin blooms twice a year during March to April and October to November (Raghupathi *et al.,* 2010). Cold stress during winter and water stress during August to September induces flowering in Coorg mandarin. Soil water deficit stress is required to induce flowering in Central India and South India where sufficient low temperature stress is lacking (Huchche and Ladaniya, 2014). In Central and South India, citrus bloom multiple times in a year, with profuse flowering in spring known as *Ambia bahar* (January-February flowering), while rainy season flowering is called *Mrig bahar* (June-July) and September-October flowering is known as *Hasta* bahar. *Ambia* flowering is preferred where assured irrigation facilities available, whereas *mrig* flowering is taken where water is scarce in Maharashtra and Madhya Pradesh for Nagpur mandarin and Mosambi sweet orange cultivars. A water stress period of 30 – 40 days is required to induce sufficient flowering in Nagpur mandarin. However, it may range from 30 to 60 days depends on the soil type. Water stress of 20-30 days in light soils (<30% clay content), 30 - 40 days in medium soils (< 40% clay content) and 40-50 days in heavy soils (>50% clay content) is required for inducing uniform flowering in Nagpur mandarin in *Ambia* season. To induce *Mrig* flowering the *ambia* flowers should be deblossomed and sufficient stress should be given during April-May. If any intermittent rainfall is received, application of one or two sprays of chlormequat chloride (1000-1500 ppm) is necessary to induce flowering. In Andhra Pradesh and Telangana, *Ambia* flowering and *Hasta* flowering is common for sweet orange cv. Sathgudi and in some areas *mrig* flowering is also preferred. The conditions for inducing flowering in mandarin and sweet orange varies with location, soil type and season (Table 11.2)

Table 11.2: Conditions Required for Induction of Flowering in Citrus in India

Parameters	*Nagpur mandarin and Sweet orange cv. Mosambi in Central India*		*Sweet orange cv. Sathgudi in South India*		
	Ambia	*Mrig*	*Ambia*	*Mrig*	*Hasta*
Water stress period	November-December-January	April - May	November-December	April - May	August-September
Water stress quantum	-3.5 Mpa	-3.5 Mpa	-	-	-
Flowering period	January – February - March	June - July	December-January	June - July	September-October
Temperature at the time of flower initiation	Min: >12 °C –Max: 27 – 32 °C	Min: >20 °C –Max -27–32 °C	Min: >17 °CMax: 24 29 °C	Min: >25-27 °CMax: 37 37.5 °C	Min: 22.5 °C Max: 32.7 – 33.5 °C -
Relative humidity	30– 80 per cent	>70 per cent	84-87 per cent	66-73 per cent	79-83 per cent
Fruit maturity	October - December	February – March - April	August-September	February – March	March-April - May

Irrigation Management in Citrus

Citrus, an irrigated crop requires assured irrigation during non-rainy seasons for optimum growth, yield and fruit quality. High moisture stress or water deficit stress during critical crop phenological stages adversely affects the crop growth, nutrient uptake, production and productivity in citrus (Shirgure, 2013). Irrigation in sub-tropical regions is an important cultural operation, whereas, it is costly operation in arid and semi-arid regions to obtain good quality fruits (Levy and Boman, 2004). Adequate soil moisture during flowering, fruit growth and maturation stages ensure good fruit set, fruit growth and high quality fruits. Any water stress during these critical stages results in inferior quality fruits (Ginestar and Castel, 1996). Therefore, water management is crucial for improving the citrus productivity. The water requirement of citrus varies with 1. Edaphic factors, 2. Climatic factors and 3. Crop factors (Table 11.3).

Table 11.3: Factors Regulating Water Requirement in Citrus

Factors	*Characters*	*Reference*
Edaphic factors	Rainfall, temperature, relative humidity, wind speed, solar radiation, sun shine hours	Shirgure, 2001
Soil factors	Texture, structure, organic matter content, water holding capacity, depth	Shirgure, 2001; Tennakoon and Milroy, 2003
Crop factors	Root architecture, species, age, stage of crop, canopy volume	Shirgure, 2001 Tennakoon and Milroy, 2003

Irrigation Methods

Surface irrigation and micro-irrigation are the common irrigation methods advocated in citrus orchards globally (Kusakabe *et al.,* 2006). Irrigation was thought to be least important cultural operation in citrus until renowned R.C.J. Koo proved irrigation in citrus improved the yield by 15–45 per cent in sandy loam soil in Florida. Thereafter researchers across the world initiated the research on irrigation on various citrus species. Surface irrigation or flood irrigation like basin, furrow, and border strip is traditional method of irrigation in citrus growing regions of the world. It supplies large quantities of water to the orchard. Though, surface irrigation is low cost, it leads to water loss, surface runoff, transfer soil borne pathogens like Phytophthora, soil loss, and leaches essential nutrients from the top soil (Phocaides, 2000). Due to negative effects of surface irrigation and scarcity of water, micro irrigation methods are becoming popular. Drip irrigation has the advantages and produced high yield with least water requirement in citrus (Panigrahi *et al.,* 2012) (Figure 11.6).

Drip and sprinkler irrigation systems generally have higher water-use-efficiency than surface irrigation methods (Sanchez and Peralta, 2003). Azzena *et al.* (1988) observed increased growth and yield of Valencia oranges irrigated through drip with emitters placed at 1 m distance from trunk. Likewise, 12 per cent yield increase and 9.3 per cent reduced water consumption was noticed in micro-sprinklers when compared to traditional irrigation method (Grieve, 1989).

Ring Method | Double Lateral Method | Single Lateral with micro-Laterals

Figure 11.6: Drip Irrigation for Citrus.

Drip irrigation increased the fruit yield and quality and saved water in sweet oranges (Kumar and Bojappa, 1994) and it increased the yield in Nagpur mandarin with benefit cost ratio of 2.07 (Gangwar and Singh, 1998). Drip irrigation @ 8 litres per hour resulted in high yield in acid lime with increased water use efficiency (Shirgure *et al.,* 2000). Drip irrigation at 80–100 per cent cumulative evapotranspiration (E_{cp}) showed higher yield, increased water use efficiency, net photosynthetic rate and stomatal conductance in Nagpur mandarin (Panigrahi *et al.,* 2012). Panigrahi (2023) reported that Nagpur mandarin plants at Central India irrigated through drip irrigation at crop water stress index of 0.4 with 60 per cent ET_c (crop evapotranspiration) resulted in high yield and fruit quality.

Critical Stages of Irrigation

Flowering, fruit development and maturation are the critical phenological stages in citrus. Water stress during these stages will adversely affect the citrus growth and development, production and productivity. Water deficit at flowering stage (near 50 per cent ET_C) lead up to 20 per cent yield loss; whereas same level of stress at fruit growth and ripening stage lead to 10 and 6 per cent yield loss, respectively (Garcia-Tejero *et al.,* 2012). In India, research works were carried to identify the stage wise water requirement in various commercial citrus species. Irrigation at different stages *viz.,* Stage I (January–February), Stage II (March-April), Stage III (May-June), Stage IV (July-August), Stage V (September-October), and Stage VI (November-December) based on evaporation replenishment (ER) with drip irrigation resulted in higher growth, improved water use efficiency, higher yield and quality fruits in mandarin, sweet orange and acid lime (Table 11.4). Based on the findings, the critical stages of water requirement for mandarin and sweet oranges are stage I to Stage III (January to June). Water stress during this period will hamper the yield and quality.

Table 11.4: Stage-wise Water Requirement in Citrus

Crop	*Results*	*Reference*
Kinnow mandarin	Drip irrigation scheduled at 60-80-60-80-60-80 per cent ER in stages I to VI resulted in higher yield (13 t/ha)	Singh *et al.*, 2020
Nagpur mandarin	Drip irrigation scheduled at 80-80-80-80-60-30 per cent ER in stages I to VI resulted in higher yield (21.48 t/ha)	Shirgure *et al.*, 2016
Sathgudi sweet orange	Drip irrigation scheduled at 60-80-60-80-60-80 per cent ER in stages I to VI resulted in higher yield (13 t/ha) and saved 13 per cent irrigation water	Lakshmi *et al.*, 2019
Acid lime cv. Phule Sharbati	Drip irrigation scheduled at 80-80-80-80-80-80 per cent ER in stages I to VI resulted in higher yield (10.2 t/ha) and saved 13 per cent irrigation water	Pawar *et al.*, 2020
Acid lime cv. Balaji	Drip irrigation scheduled at 60-80-60-80-60-80 per cent ER in stages I to VI resulted in higher yield (19.74 t/ha) and saved 12 per cent irrigation water	Lakshmi *et al.*, 2023

Regulated Deficit Irrigation (RDI)

These strategies are important in water scarcity areas and it has become common practices, particularly in arid zones. Recently, RDI is widely adopted in citrus. RDI is the application of less irrigation water than required to achieve the optimum yield, quality and enhanced water use efficiency (Garcia-Tejero *et al.*, 2010). RDI saved 6-22 per cent water without affecting the yield in Clementina de Nules budded on Carrizo citrange in Valencia, Spain (Gonzalez Altozano and Castel, 1999). Similarly, 10-12 per cent yield reduction was observed under RDI in sweet orange cv. Navalina. However, the water productivity was increased by 24 per cent (Garcia-Tejero *et al.*, 2010). Higher yield can be obtained in Nagpur mandarin when plants are irrigated with 80 per cent evapotranspiration using deficit irrigation method (Shirgure *et al.*, 2014). Panigrahi *et al.* (2014) reported that Kinnow mandarin plants irrigated with 50 per cent ETc (RDI_{50}) during early fruit growth period improved water-use-efficiency and attained the yield (59.6 -62.4 t/ha) equal to fully irrigated plants (61.9 -63.2 t/ha). Regulated deficit irrigation studies proved that the optimum yield and quality can be obtained with better water use efficiency using RDI method in citrus in water scarcity areas. By adopting this method, we can bring more area under cultivation.

Nutrient Management

Citrus, an evergreen perennial plant having long fruit growing period requires all 17 essential nutrients. For its optimum growth and production, 14 essential nutrients containing macro (N, P, K, Ca, Mg, and S) and micro (Fe, Zn, Mn, B, Cu, Mo, Ni and Cl) nutrients are critical. Deficiency in any one of the essential nutrients, tree growth is affected and produces characteristic deficiency symptoms (Vashisth and Kadyampakeni, 2020). Indiscriminate use of fertilizers not only wastage, but also cause toxicity to the plants, affects the absorption of other nutrients and also cause environmental hazards. Analysis of the nutrients is essential for optimum use of chemical fertilizers for sustainable production.

Soil and Leaf Analysis

Soil Analysis

Soil analysis is the most common method in determining the nutrient availability. It assesses organic matter content, pH, and extractable nutrients, which are crucial for developing and refining fertilization programs. Conducting soil analysis over several consecutive years is particularly beneficial for identifying trends. The accuracy of soil-test interpretations hinges on how well the soil sample represents the orchard or management unit being analysed. Each sample should consist of soil cores taken about 8 inches deep at the dripline of 15 to 20 trees, within the area wetted by the irrigation system and in the zone of maximum root activity. The optimum soil nutrients required for Nagpur mandarin (94.8-154.8 mg/kg N, 6.6-15.9 mg kg P, and 146.8-311.9 mg/kg K), Khasi mandarin (161.0-418.7 mg/kg N, 4.5-8.7 mg kg P, and 82.3-287.5mg/kg K), Kinnow mandarin (114.3-121.2 mg/kg N, 7.8-12.3 mg kg P, and 96.4-131.3 mg/kg K), Mosambi sweet orange (107.4-197.2 mg/kg N, 8.6-18.8 mg kg P, and 186.4-389.2) mg/kg K and acid lime (106.3-118.2) mg/kg N, 9.2-14.6 mg kg P, and 102.4-146.6 mg/kg K) for ideal growth, yield and quality (Srivastava *et al.,* 2022).

Leaf Analysis

Leaf analysis is crucial for detecting problems and optimizing fertilizer programs for citrus trees, as it accurately reflects the nutritional status of the plants (Smith, 1966; Embleton *et al.,* 1973). By examining leaf nutrients, growers can identify deficiencies or excesses that affect tree growth and fruit quality. Worldwide studies suggested that samples from 4-10 month old leaves should be used for analysis. 25-50 leaves from single tree or 100 leaves both from fruiting and non-fruiting shoots gave better leaf nutrient estimates (Reuther *et al.,* 1954; Jorgensen and Price, 1978; Srivastava and Singh, 2006). The optimum leaf nutrient contents for various citrus species are listed in the Table 11.5.

Optimum Fertilizer Requirements

The nutrient requirements of citrus largely depend on soil and stage of the crop. An optimum fertilization should benefit the crop both short and long term. Fertilization should focus on developing the canopy in young trees and in bearing trees; the focus should be on soil health, crop growth, production and quality. Several studies were carried out in different agro-climatic zones and on different

Table 11.5: Optimum Leaf Nutrient Content for Citrus

Nutrient	*Mandarins*				*Sweet Orange*	*Acid Lime*
	Nagpur Mandarin	*Coorg Mandarin*	*Kinnow Mandarin*	*Khasi Mandarin*	*Mosambi*	
Nitrogen (%)	2.24-2.40	2.27-2.69	2.22-2.32	2.26-2.83	2.36-2.48	1.53-2.10
Phosphorous (%)	0.07-0.10	0.46-0.74	0.11-0.15	0.11-0.16	0.13-0.15	0.10-0.15
Potassium (%)	1.18-1.56	1.30-1.80	1.10-1.41	1.86-2.07	1.62-2.82	0.96-1.66
Calcium (%)	1.32-1.55	1.65-2.15	2.32-2.79	1.85-2.12	1.72-3.17	3.05-3.43
Magnesium (%)	0.48-0.67	0.25-0.28	0.38-0.61	0.33-0.48	0.42-0.58	0.40-0.60
Iron (ppm)	110.24-132.12	119.9-211.4	22.4-58.3	170.1-225.1	132.1-148.2	117-194
Manganese (ppm)	29.36-43.41	28.0-30.4	26.3-56.2	74.03-83.43	52.4-62.0	21-63
Copper (ppm)	8.26-14.09	-	4.2-7.2	1.24-2.45	6.8-10.2	8.68-14.8
Zinc (ppm)	18.60-29.80	23.1-25.5	21.3-26.9	19.84-21.28	24.9-31.8	25-50

Source: Srivasta and Singh, 2004; Rymbai *et al.*, 2024; Srivastava and Singh, 2016; Raghupathi *et al.*, 2013; Varalakshmi and Bhargava, 1998.

species and varieties for their nutrient removal pattern using soil and leaf analysis and accordingly fertilizer doses have been recommended for growers. There are various fertilizer sources and formulations available for commercial citrus farming, as well as different application methods. Ensuring the correct type, rate, timing, and placement (within the root zone) of fertilizer is vital for enhancing nutrient uptake efficiency. Further, fertilizer requirement is highly dependent on soil type, species, and age and growth stages of the plant. "Feed the plants, not the soil." This modern view emphasizes plant mineral nutrition through both soil fertilization and foliar application.

Soil Application

Soil and foliar application are the two basic method of fertilizer application in citrus plants. Soil application depends on age of the plant, soil type, and growth stages. In order to increase the nutrient use efficiency, the fertilizers should be applied in split doses coinciding with flowering and fruit set, fruit growth and fruit maturation time. Most of the feeding roots are in top 15 cm and care should be taken during application. Fertilizers should be applied 1.0 to 1.35 m away from trunk and placed in 30 cm wide circular band with 10 cm deep trench. In hilly zones, fertilizers should be placed in top portion of half-moon terrace. The various recommendations from different parts of the country are summarized in the Table 11.6.

Table 11.6: Optimum Doses of Macro- and Micro-Nutrients for Citrus

Crop	*Dose*	*Source*
	Sweet Oranges	
Sweet orange cv. Mosambi	500 g N, 100 g P_2O_5, 400 g K_2O/tree	Ghosh, 1990
Sweet orange cv. Mosambi	800 g N, 170 g P_2O_5, 391 g K_2O/tree	Desai *et al.*, 1986
Sweet orange cv. Sathgudi	2.72 kg N, 1.81 kg P, 0.60 kg K/tree	Reddy and Swamy, 1986
	Mandarins	
Coorg mandarin	250 g N, 250 g P_2O_5, 500 g K_2O/tree	Kannan *et al.*, 1989
Kinnow Mandarin	1310 g N, 436 g P_2O_5, 436 g K_2O/tree	Monga *et al.*, 2004
Nagpur Mandarin	1200 g N, 600 g P_2O_5, 600 g K_2O/tree for shallow soil	Srivastava and Singh 2016
	600 g N, 400 g P_2O_5, 300 g K_2O/tree for deep soil	

Foliar Application

Foliar nutrition was controversial, contrasting with the established belief that plants primarily absorb minerals through their roots. Dating back to 1844 foliar nutrition gained traction in the mid-20th century, leading to agricultural recommendations for various crops (Srivastava and Singh, 2003), Although it cannot replace root nutrient uptake (Michalojc and Szewczuk, 2003), foliar application can meet significant portions of a plant's micronutrient needs (Elmer *et al.,* 2007). It addresses nutrient deficiencies, enhances plant nutrition, improves crop yield and quality, and mitigates stress from environmental factors like drought and frost. Besides minerals, foliar sprays deliver amino acids, sugars, organic acids, and growth regulators, contributing to plant bio-stimulation.

Fertigation in Citrus

Fertigation enhances fertilizer use efficiency, reduces nutrient leaching from rainfall or over-irrigation, and can lower fertilizer requirements (Steenwerth *et al.,* 2014). Nitrogen and potassium are the most commonly applied nutrients through fertigation in citrus production. Effective nutrient distribution requires water to travel from the injection point to the farthest emitter in 20 to 30 minutes, necessitating a minimum injection time of 30 to 45 minutes, followed by at least 30 minutes of additional watering to prevent clogging. Properly managed applications of plant nutrients through irrigation systems significantly enhance plant growth, fruit production, quality, and nutrient use efficiency (Srivastava and Singh, 2006). Conversely, poorly managed fertigation can lead to substantial yield losses. Fertigation studies in Nagpur mandarin showed high fruit yield efficiency when 500 g N + 140 g P + 70 g K per plant applied yearly through drip irrigation at Central India (Shirgure *et al.,* 2001a,b,c). Enhanced yield and higher quality fruits were obtained in sweet orange cv. Sathgudi (525:245:280 g NPK per plant with drip irrigation at 90 per cent ER) in South India (Lakshmi *et al.,* 2019) and cv. Phule Mosambi (640:240:480 g NPK per plant with drip irrigation at 80 per cent ER) in Central India (Hendre *et al.,* 2020) through fertigation.

Micro-nutrient Management

Micro-nutrients such as boron, zinc, manganese, iron, copper, and molybdenum are required in minute quantities and crucial for improving the movement and uptake of macronutrients. Deficiency of these micro-nutrients results in various symptoms (Zn-irregular green band along with midrib, thin twigs, narrow and pointed leaves, die-back; Fe-chlorosis in young leaves, reduction in leaf size; Mn - mottled leaves; B – necrotic leaves and premature shedding of leaves, hard fruits; Mo- yellow spots on the leaves), affects the plant function severely, thus affect the yield and quality of fruits. Soil test and leaf test are necessary to identify the exact level of micro-nutrients present in the plant and accordingly right quantities of micro-nutrients should be applied to the plants. Studies suggested that spraying of 0.5 per cent of $FeSO_4$, $ZnSO_4$, $MnSO_4$ or soil application of 200 g each of $FeSO_4$, $ZnSO_4$, $MnSO_4$ annually rectify the micro-nutrients problem in mandarin and acid lime in India. In Sweet orange, 300 g each of $FeSO_4$, $ZnSO_4$, $MnSO_4$ should be applied (Srivastava *et al.,* 2022).

Intercropping in Citrus

Intercropping in citrus orchard is not only remunerative but also conserves soil and moisture. Climate, soil type, water availability, species, spacing and age of plant are the important criteria for selecting the suitable intercrops. The exhaustive intercrops deprive all the nutrients and moisture from the soil and adversely affect the citrus orchards. They also alter the soil physico-chemical properties and cause injury to the citrus roots. Therefore, selection of intercrop is utmost important. The following are the characteristics of good intercrops.

1. It should be annual,
2. It should be tolerant to partial shade,
3. It should provide good soil cover,
4. It should improve the soil composition,
5. It should not compete for water and nutrient from main crop,
6. It should not be carrier for vectors and pathogens.

Citrus plants are planted at wide spaces (5 x 5 m or 6 x 6 m) which provide ample space for intercropping. Use of intercrops provides additional income during off-season without competing more for moisture and nutrient. Several intercrops such as forest trees, legumes, cover crops, grains, vegetables, *etc.*, were studied in citrus orchards in different countries (Singh *et al.*, 2018; Mulinge *et al.*, 2018; Selim *et al.*, 2020). Farmers in India grow several intercrops like wheat, maize, sugarcane, pearl millet, cotton, *etc.* without following proper spacing which depletes the soil fertility and soil moisture and interfere the citrus growth. In pre-bearing citrus orchards, intercropping with cotton and strip cropping with pulse is highly remunerative which helps the farmers economically during pre-bearing stages (Shivankar *et al.*, 2004). Further it is also observed that higher fruit yield was observed in citrus orchards intercropped with legumes when compared to monocropped citrus orchards (Srivastava *et al.*, 2007). Some of the suitable intercropping system identified for citrus is summarized below (Table 11.7).

Table 11.7: Suitable Intercropping Systems in Citrus Orchards in India

Citrus	*Intercropping system*	*Reference*
Nagpur mandarin	Soyabean -Chickpea	Srivastava *et al.*, 2007
	Soyabean – Gram	Shirgure, 2012
Kinnow	Guar – Wheat	Gill *et al.*, 2018
	Onion – Indian Squash	Soni *et al.*, 2021
Baramasi lemon	Fodder maize – Musk melon - Cauliflower	Dubey *et al.*, 2016
Mosambi	Groundnut	Ghosh and Pal, 2010

Weed Management

Several weeds infest citrus orchards and compete for nutrients, sun light, and water (Sullivan, 2003). It also harbours pest and diseases and nematodes.

Weeds create favourable environment of pest and pathogen and increase the risk of damages to the trees. Weed growth is slightly less in mature orchards due to shading effect. However, once established, weeds can hamper all the cultural operations. Hassan and Abd El-Naby (1998) reported 12.6 to 17.7 t/ha of weeds in one hectare orange orchards. Weeds control during critical phenological stages are important to increase the efficiency of field operations, reduce the insect pests and disease incidence and improve the production and quality of produce. Cultural, biological, mechanical and chemical methods are advocated to control the weeds in citrus orchards (Boman *et al.,* 2002). Cultural control includes live and plastic mulching to suppress the weed growth. Organic mulches (straw, grasses, wood chips, leaves, newspaper, *etc.*) are adopted by farmers in different countries and it is proved to be effective in controlling the weeds at early stages (Schonbeck *et al.,* 1991). Mostly grasses and straw are used for organic mulching and on decay; these organic mulches add nutrients to the soil. Plastic mulches, particularly black polythene sheet are used as inorganic mulch to control the weed growth. However, these mulches are effective against annual weeds and have little impact on established perennial weeds. Mechanical control involves disking which is effective between the rows (Figure 11.7). Care should be taken in citrus orchards while using disking, if not used carefully; it could damage the roots, tree branches and bury plant debris (Wright *et al.,* 2000). Chemical method is effective way of controlling the weeds in citrus orchards. Some of the commonly used herbicides are glyphosate, diuron, oxyfluorfen, simazine, *etc.* However, these herbicides possess serious environmental hazards. Leaching of herbicides pollutes water bodies and repeated use of single herbicide lead to resistance in weeds.

Figure 11.7: Weeding in Acid Lime.

Integrated Pest Management

The pests like citrus psylla, aphids, thrips, leaf miner, lemon butterfly, bark eating caterpillar, fruit sucking moth, mites and nematodes regularly attack the citrus starting from nursery to old orchards and cause economic losses to citrus growers (Figure 11.8). Timely management these pests are crucial for citrus cultivation (Ladaniya *et al.,* 2019).

Figure 11.8: Pests of Citrus.

Integrated pest management is a holistic way of controlling citrus pests. Identification of pest and its monitoring is the first step in IPM. Clean cultivation and pruning of intermingling branches allow good aeration and sunlight and reduces the buildup of insects. Use of pheromone traps and sticky traps can capture the insects, thereby reduces the usage of chemicals. Removal of fallen fruits and alternate host plants like *Tinospora cardifolia, Cocculus villosus, Cirampelos pareira* and *Convolvulus arvensis* from the orchard prevents the multiplication of fruit sucking moth larvae. Use of neem oil and mineral oils at initial stage helps to reduce the insect pest population. Several pesticides are used to control the citrus pests. Chemical pesticides have harmful impact on environment. Indiscriminate use lead to pesticide tolerance and residue in fruits adversely affect the human health. Use of proper pesticide, right insect/pests and right time of application for controlling the pests is important. The pesticides used in India for controlling the citrus pest are listed in the Table 11.8.

Table 11.8: Chemical Management of Citrus Insects and Pests

Category	*Pests*	*Chemicals for Control*
Sucking pests	Thrips, aphids, psylla, black fly, white fly, mealy bugs	Imidacloprid 17.8 per cent SL, Thiamethoxam 25 per cent WG, Spirotetrament 15.31 per cent OD
Foliage feeders	Leaf miner, citrus butterfly and leaf folder	Chloropyriphos 20 per cent EC, Imidacloprid 17.8 per cent SL
Bark/stem feeders	Bark eating caterpillar, stem borer	Chloropyriphos 20 per cent EC, Dimethoate 30 per cent EC, Quinolphos 25 per cent EC
Fruit feeders	Fruit sucking moth, fruit fly	Neem oil, Horticultural mineral oil
Mites	Flat mite, green mite, brown mite	Diafenthiuron 50 per cent WP, Spiromesifen 22.9 per cent SC, Propergite 57 per cent EC, Fenazaquin 10 per cent EC
Nematodes	Citrus nematode, Lesion nematode, Root knot nematode, Lance nematode, Spiral nematode, Ring nematode	Carbofuran 3G

Integrated Disease Management

Diversity of fungal, bacterial, and viral diseases affects citrus worldwide. These diseases are the main constraint in citrus production. Of the several diseases, Phytophthora, citrus greening or Huanglongbing (HLB), canker, citrus tristeza virus, ringspot virus, exocortis are causing economical losses to the citrus growers. Identification of disease at early stage and its appropriate management are crucial for control. The management of some of the important citrus diseases in India is discussed here.

Management of Citrus Canker

Citrus canker is an important bacterial disease affecting majority of the citrus species specifically acid lime. The severity of the disease varies with species and growing environmental conditions (Das, 2003). The pathogen *Xanthomonas axonopodis* pv *citri, X. axonopodis* pv. *aurantifolii* are widespread throughout the world and have wide range of host plants. Water soaked appearance and corky raised necrotic lesions will appear on leaves, stems and fruits (Goto, 1992) (Fig.9). Clean cultivation, application of recommended dose of nutrients, proper irrigation management, dead wood removal after harvest, and periodical removal of infected shoots and pre-monsoon spray of 1 per cent Bordeaux mixture are essential for controlling the canker. Once infected, spraying of 100 ppm streptocycline along with 1 per cent copper oxy chloride is recommended (Ladaniya *et al.*, 2019).

Canker on Leaves

Canker on Fruits

Canker on Twigs and Fruits

Figure 11.9: Citrus Canker on Acid Lime.

Management of *Phytophthora* spp.

Phytophthora (*P. nicotianae, P. palmivora, and P. citrophthora*), a soil borne fungal disease, is one of the most important disease in citrus which could cause mortality up to 25-62.5 per cent in mandarin and 15-20 per cent in mandarin (Das *et al.*, 2016). *Phythophthora* is active in soil for several years, and generally spreads through infected nursery plants and through irrigation water. *P. nicotianae, P. palmivora* are widespread in tropical areas, whereas *P. citrophthora* is common in cooler regions (Caccilo and Lio, 2008). Gummosis, collar rot, crown rot, root rot, and decline are the common symptoms of Phytophthora (Figure 11.10) Use of susceptible rootstocks, high clay soils, poor drainage, and prolonged contact of citrus trunk with water increases the risk of inoculum build-up. Integrated management is required to control this disease. Use of resistant rootstocks is one of the key control measures. *Poncirus trifoliata* and its hybrids offer tolerance/ resistance to *Phytophthora* root and foot rot. Wherever, trifoliate is used as rootostock for citrus, root rot is a minor problem. In India, the farmers use Rough lemon and Rangpur lime as predominant rootstocks and these rootstocks are susceptible to this pathogen.

In nursery condition sterilization of potting media, cultural and chemical control is required to control this disease. Due to this disease, about 20 per cent mortality is reported in nursery stage (Naqvi, 2000). Soil solarization is the best method to keep the inoculum under check (Figure 11.11). The solar radiation could be tapped in tropical areas and this non-chemical method offer good control over the pathogen. Once, the pathogen infected the plants, it is very difficult to eradicate the

Gummosis Decline

Figure 11.10: Symptoms of *Phytophthora.*

pathogen. Soil drenching and spraying of with metalaxyl and aluminum fosetyl will effectively control this disease (Naqvi, 2004; Ladaniya *et al.,* 2019) (Figure 11.12). In orchards, regular monitoring is crucial. Removal of dead woods, application of Bordeaux paste on the trunk up to 3 feet height (Figure 11.13) pre and post monsoon and spray of copper oxychloride are recommended for control.

Figure 11.11: Soil Solarisation for Potting Media.

Figure 11.12: Application of Fungicides at Infection Site.

Management of Citrus Greening

Citrus greening or Huanglongbing (HLB), caused by phloem limiting bacteria *Candidatus* Liberibacter *asiaticus* (CLas), transmitted through pysllid vector (*Diaphorina citri*) affects citrus production worldwide. Blotchy mottle, chlorosis

Figure 11.13: Bordeaux Paste Application on Stem.

with green veins, root decay, severe fruit drop, and half yellow half green colour elongated fruit are characterized symptoms of this disease (Hu *et al.*, 2021) (Figure 11.14). The leaves symptoms are always mimicking with nutrient deficiency symptoms, identification through laboratory techniques are important.

Blotchy Mottle Elongated Fruits

Figure 1.14: Citrus Greening Symptoms on Nagpur Mandarin.

Frequent monitoring of citrus pyslla and controlling them is vital for HLB control (Ladaniya *et al.,* 2019; Thakuria *et al.*, 2023). The effective way of controlling CLas is the use of antibiotics. The tetracycline is widely used for controlling the bacteria. Multiple stem injection or sprays are required and sometimes it leads to phytotoxicity. Alternate control measures are necessary. The combination of two anti-microbial agents *viz.*, 2S albumin and Nano- ZnO stem injection gave good control of CLas in sweet orange plants at controlled conditions (Ghosh *et al.,* 2018). Along with vector control and use of antibiotics, improved fertilization is necessary to control the disease and improve the overall health of the orchards.

Conclusion

Citrus cultivation requires a comprehensive understanding of advanced scientific orchard management practices. The establishment of citrus orchard needs well planned selection of location, soil preparation and selection of appropriate citrus varieties. The planting system plays a crucial role and it depends on soil, climate and variety. High density planting demands meticulous management and have the advantage of increased productivity per unit area. In modern scientific system, canopy management practices, including training and annual pruning are essential for optimization of light penetration, air flow and control of pest and diseases, thereby enhance the yield and fruit quality. Nutrient management and water management play a vital role in maintaining the tree health. Application of required doses of macro and micro nutrients and irrigation during critical period are essential to achieve the maximum production. Integrated pest and disease management strategies are critical to protect the citrus from various pest and diseases and minimize the use of harmful chemicals. In essence, this chapter underscores the complexity of managing citrus orchards and emphasizes the need for holistic approaches that balance productivity goals with environmental stewardship.

REFERENCES

Abobatta WF, 2021. Managing citrus orchards under climate change. *MOJ Eco Environ Sci.* **6(2)**: 43-44.

Ahmadpoor A, Salari M and Miri SM, 2022. Pruning and Girdling Influence Alternative Bearing of 'Kinnow' Mandarin (*Citrus reticulata* Blanco). *J. Hortic. Postharvest Res.* **5**: 13–20.

Azzena M, Deidda P and Dettori S, 1988. Drip and microsprinkler irrigation for young 'Valencia'orange trees. *Proc. Sixth Intern. Citrus Cong.*, Tel Aviv, Israel, **2**: 747-751

Bettini BA, Cavichioli TM, Cristofani-Yaly M, Azevedo FA, Martins ALM and Schinor EH, 2019. Performance and reaction to huanglongbing of 'Tahiti' acid lime grafted on citrandarins. *Acta Hortic.* **1230:** 99-105.

Bevington KB and Bacon PE, 1977. Hedging Valencia orange trees. *Aust. J. Exp. Agric.* **18(93):** 591-596.

Boman B, Wilson C, Vandiver Jr. V, Hebb J, 2002. Aquatic Weed Management in Citrus Canals and Ditches. UF/IFAS document CH181, the University of Florida Cooperative Extension Service. Available online: http: //edis. ifas.ufl.edu/ CH181

Bowman KD and Joubert J, 2020. Citrus rootstocks. **In:** Talon M, Caruso M, Gmitter FG Jr, editors. The Genus Citrus. pp. 105-128

Cacciola SO and Lio GMDS, 2008. Management of citrus diseases caused by Phytophthora spp. **In:** Integrated management of diseases caused by fungi, phytoplasma and bacteria Dordrecht: Springer Netherlands. pp. 61-84.

Cantuarias-Aviles T, Mourão Filho Fd AA, Stuchi ES, Da Silva SR and Espinoza-Núñez E, 2010. Tree performance and fruit yield and quality of 'Okitsu' Satsuma mandarin grafted on 12 rootstocks. *Sci. Hortic.* **123**: 318-322.

Cassin J, Bourdeaut J, Fougue A, Furon V, Gaillard JP, LeBourdelles J, Montagut G and Moreuil C, 1969. The influence of climate upon blooming of citrus in tropical areas. *Proc. 1st Intl. Citrus Symp.* **1**: 315–323.

Castle WS, Baldwin JC and Muraro RP, 2007. 'Hamlin' orange trees on Flying Dragon trifoliate orange, Changsha Mandarin, or Koethen Sweet Orange × Rubidoux Trifoliate orange citrange rootstock at three in-row spacings in a flatwoods site. *Proceedings of the Florida State Horticultural Society.* **120**: 92-96

Costa DP, Stuchi ES, Girardi EA, Moreira AS, Silva Gesteira AD, Coelho Filho MA, da Silva Ledo CA, da Silva AL, de Leão HC, Sampaio Passos O and dos Santos Soares Filho W, 2021. Less is more: A hard way to get potential dwarfing hybrid rootstocks for valencia sweet orange. *Agriculture* (*Switzerland*). **11**: 1-21.

Dalal RPS, Sangwan AK, Beniwal BS and Sharma S, 2013. Effect of planting density on canopy parameter, yield and water use efficiency of Kinnow mandarin. *Indian J. Hort.* **70(4):** 587-590.

Das AK, Nerkar S, Kumar A and Bawage S, 2016. Detection, identification and characterization of Phytophthora spp. infecting citrus in India. *Journal of Plant Pathology*. 55-69.

Das AK, 2003. Citrus canker-A review. *J. Appl. Hortic.* **5(1)**: 52-60.

Davenport TL, 2011. Citrus flowering. *Horticultural Reviews.* **12**: 349-408.

Desai UT, Choudhari KG and Choudhari SM, 1986. Studies on the nutritional requirements of sweet orange. *J. Maharashtra Agric. Univ.* **11(2)**: 145–147.

Deshmukh NA, Patel RK, Rymbai H, Jha AK and Deka BC, 2016. Fruit maturity and associated changes in Khasi mandarin (*Citrus reticulata*) at different altitudes in humid tropical climate. *Indian J. Agric. Sci.* **86(7)**: 854-859.

Dhillon WS and Thakur A, 2014. Canopy management and effects of pruning on flowering tendencies in fruit trees. In *National Seminar-cum-Workshop on physiology of flowering in perennial fruit crops. Lucknow, India.* pp. 182-201.

Dubey SK, Sharma N, Sharma JP, Sharma A and Kishore N, 2016. Assessing citrus (lemon) based intercropping in the irrigated areas of northern plains of Haryana. *Indian J. Hortic.* **73**(3): 441-444.

Elmer PAG, Spiers TM and Wood PN, 2007. Effects of pre-harvest foliar calcium sprays on fruit calcium levels and brown rot of peaches. *Crop protection.* **26(1)**, 11-18.

Embleton TW, Reitz NJ and Jones WW, 1973. Citrus fertilization. In: *The citrus industry, vol 3. University of California, Berkeley,* p 145.

FAOSTAT, 2023. *Data.* Retrieved from https: //www.fao.org/faostat/en/#data. Accessed June 10, 2023.

Ferrarezi RS, Jani AD, James III HD, Gil C, Ritenour MA and Wright AL, 2020. Sweet orange orchard architecture design, fertilizer, and irrigation management strategies under huanglongbing-endemic conditions in the Indian River citrus district. *Hort Sci.* **55:** 2028–2036.

Fonte A, Torregrosa A, Garcerá C, Mateu G and Chueca P, 2022. Mechanical pruning of 'Clemenules' mandarins in Spain: yield effects and economic analysis. *Agronomy.* **12(4)**, 761.

Forner-Giner MA, Rodriguez-Gamir J, Martinez-Alcantara B, Quiñones A, Iglesias DJ, Primo-Millo E and Forner J, 2014. Performance of Navel orange trees grafted onto two new dwarfing rootstocks (Forner-Alcaide 517 and Forner-Alcaide 418). *Sci. Hortic.* **179**: 376-387.

Gangwar LS and Singh S, 1998. Economic evaluation of Nagpur mandarin cultivation in Vidarbha region of Maharashtra. *Indian J Agric. Econ.* **53(4)**: 648-653.

García-Tejero I, Durán-Zuazo VH, Arriaga-Sevilla J and Muriel-Fernández JL, 2012. Impact of water stress on citrus yield. *Agron. Sustain. Dev.* **32***:* 651-659.

García-Tejero I, Jiménez-Bocanegra JA, Martínez G, Romero R, Durán-Zuazo VH and Muriel-Fernández JL, 2010. Positive impact of regulated deficit irrigation on yield and fruit quality in a commercial citrus orchard [*Citrus sinensis* (L.) Osbeck, cv. salustiano]. *Agric. Water Manag.* **97(5)**: 614-622.

Ghosh A, Dey K, Bhowmick N, Medda PS and Ghosh SK, 2016. Impact of Different Pruning Severity and Nutrient Management on Growth and Yield of Lemon cv. Assam Lemon (*Citrus limon* Burm.). *Vegetos.* **29**: 1.

Ghosh DK, Kokane SB, Kokane AD, Warghane AJ, Motghare MR, Bhose S, Sharma AK and Reddy MK, 2018. Development of a recombinase polymerase based isothermal amplification combined with lateral flow assay (HLB-RPA-LFA) for rapid detection of" *Candidatus Liberibacter asiaticus*". *PloS one.* **13(12)**: 0208530.

Ghosh SN, 1990. Nutritional requirement of sweet orange (*Citrus sinensis* Osbeck) cv. Mosambi. *Haryana J Hortic. Sci.* **19(1–2)**: 39–44.

Ghosh SN and Pal PP, 2010. Effect of inter-cropping on plant and soil of Mosambi sweet orange orchard under rainfed conditions. *Indian Journal Horticulture.* **67(2)**: 185-190.

Gill MS, Khehra S and Gupta N, 2018. Impact of intercropping on yield, fruit quality and economics of young Kinnow mandarin plants. *J. Appl. Nat. Sci.* **10(3)**: 954-957.

Ginestar C and Castel JR, 1996. Responses of young clementine citrus trees to water stress during different phenological periods. *J.Hortic. Sci.* **71(4)**: 551-559.

Girardi EA, Sola JGP, Scapin MDS, Moreira AS, Bassanezi RB, Ayres AJ and Peña L, 2021a. The perfect match: Adjusting high tree density to rootstock vigor for improving cropping and land use efficiency of sweet orange. *Agronomy.* **11(12)**: 2569.

Girardi EA, Ayres AJ, Girotto LF and Peña L, 2021b. Tree growth and production of rainfed Valencia sweet orange grafted onto trifoliate orange hybrid rootstocks under Aw climate. *Agronomy.* **11**: 2533.

Golomb A, 1988. High density planting of intensive citrus groves: A challenge and realization. **In**: *Proceedings of the Sixth International Citrus Congress.* pp. 921-930

Gonzalez-Altozano P and Castel JL, 1999. Effects of regulated deficit irrigation on â¿ clementina de nulesâ citrus trees growth, yield and fruit quality. In *III International Symposium on Irrigation of Horticultural Crops.* **537**: 749-758).

Gonzatto MP, Griebeler SR and Schwarz SF, 2022. Dwarfing rootstocks for high-density citrus orchards. **In:** Fruit Industry. *IntechOpen.*

Goto M, 1992. Citrus canker. **In:** Plant diseases of international importance. Vol. III (J. Kumar, H.S. Chaube, U.S. Singh and A.N. Mukhopadhyay, Eds.) Prentice- Hall, Englewood Cliff, NJ. pp. 170-208.

Grieve AM, 1989. Water use efficiency, nutrient uptake and productivity of micro-irragated citrus. *Aust. J. Exp. Agric.* **29(1)**: 111-118.

Hassan AAA and Abd El-Naby SKM, 1998. Menofiya J. *Agric. Res.* **23**: 1327-1343.

Hayat F, Li J, Iqbal S, Peng Y, Hong L, Balal RM and Chen J, 2022. A mini review of citrus rootstocks and their role in high-density orchards. *Plants.* **11(21)**: 2876.

Hendre P, Ranpise S and Durgude A, 2020. Effect of different irrigation and fertigation levels on yield and nutrient use efficiency in sweet orange (Citrus sinensis L. Osbeck) cv. *Phule Mosambi. Int. J. Chem. Stud.* **8**: 1577-1581.

Hu B, Rao MJ, Deng X, Pandey SS, Hendrich C, Ding F, Wang N and Xu Q, 2021. Molecular signatures between citrus and *Candidatus Liberibacter asiaticus. PLoS pathogens.* **17(12)**: 1010071.

Huchche AD and Ladaniya MS, 2014. Citrus Flowering and Fruiting—Recent Research Advances. In: *Proceedings of the Souvenir, National Seminar-Cum-Workshop on Physiology of Flowering in Perennial Fruit Crops* (pp. 74-88). Lucknow, India: Central Institute for Sub-tropical Horticulture (ICAR).

Jhade RK, Huchche AD, Pandey SK and Verma AK, 2021. Effect of growth retardants on intermittent irrigation during water deficit stress period in Mrig (Monsoon flowering) crop of Nagpur mandarin (*Citrus reticulata* Blanco). *J. Agric. Ecol.* **12**: 80-87.

Jorgensen KR and GH Price, 1978. The citrus leaf and soil analysis system in Queensland. In: P.R. Cary (ed.) *Proc. Int. Soc. Citriculture.* Aug. 15-23, Sydney, Australia. **1:** 297-299.

Kadam DM, Dubey AK, Sharma RM, Morade A, Sharma N and Bhardwaj C, 2022. Response of citrus (*Citrus spp.*) rootstock hybrids to PEG induced drought under hydroponic system. *Indian J. Agric Sci.* **92(10)**: 1230-1236.

Kamatyanatti M and Singh SK, 2019. Recent advances in citrus propagation and planting. *Advances in Horticultural Crop Management and Value Addition, 1.*

Kannan M, Thamburaj S and Seemanthini R, 1989. Studies on the manurial requirement of mandarins in Shevroy hills. *South Indian Hortic.* **37(4)**: 203–208.

Kumar APA and Bojappa KM, 1994. Studies on the effect of drip irrigation on yield and quality of fruit in sweet oranges and economy in water use. *Mysore J. Agric. Sci.* **28**: 338–44.

Kusakabe A, White SA, Walworth J, Wright CG and Thompson T, 2006. Response of microsprinkler-irrigated navel oranges to fertigated nitrogen rate and frequency. *Soil Sci. Soc. of Am. J.* **70**: 1623-1628

Ladaniya MS, Marathe RA, Das AK, Rao CN, Huchche AD, Shirgure PS and Murkute AA, 2020. High density planting studies in acid lime (*Citrus aurantifolia* Swingle). *Sci. Hortic.* **261**: 108935.

Ladaniya MS, Srivastava AK, Vijayakumari N, Huchche AD, Singh IP, Ghosh DK, Sonkar RK, Das AK, Dinesh Kumar, Rao CN, Shirgure PS, George Anjitha and Kommu KK, 2019. Good Agricultural Practices for Nagpur mandarin. *Extension bulletin No.* **63**, *ICAR Central Citrus Research Institute, Nagpur.*

Lakshmi LM, Ramana KT, Shigure PS, Srivastava AK and Patil PP, 2023. Stage wise water requirement for drip irrigated acid lime orchards in Rayalaseema region of Andhra Pradesh. *Int. J. Innov. Hortic.* **12(1)**: 87-91.

Lakshmi LM, Ramana KT, Shirgure PS and Patil P, 2019. Evaluating stage-wise irrigation water requirement of sweet orange (*Citrus sinensis*) cv Sathgudi grown on Alfisols. *Current Hortic.* **7(1)**: 22-27.

Levy Y and Boman BJ, 2004. Water management in citrus, p. 268–304. In: S. Singh, V.J. Shivanker, A.K. Srivastava, and I.P. Singh (eds.). Advances in citriculture. *Jagmander Book Agency, New Delhi, India.*

Lliso I, Forner JB and Talón M, 2004. The dwarfing mechanism of citrus rootstocks F and A 418 and 23 is related to competition between vegetative and reproductive growth. *Tree Physiol.* **24(2)**: 225-232.

Maranto J and Hake KD, 1985. Verdelli summer lemons: a new option for California growers. *Calif Agric.* **39** : 4.

Martin-Gorriz B, Martinez-Barba C and Torregrosa A, 2021. Lemon trees response to different long-term mechanical and manual pruning practices. *Sci. Hortic.* **275**: 109700.

Mesejo C, Martínez-Fuentes A, Reig C, Balasch S, Primo-Millo E, Agustí M, 2020. Mechanical Pruning Attenuates Alternate Bearing in 'Nadorcott' Mandarin. *Sci. Hortic.* **261**: 108-993.

Michalojc Z and Szewczuk C, 2003. Theoretical aspects of foliar nutrition. *Acta Agrophysica.* **85**: 9-17.

Monga PK, Vij VK and Sharma JN, 2004. Effect of N, P and K on the yield and fruit quality of Kinnow mandarin. *Indian J. Hortic.* **61(4):** 302-304.

Monselise SP, 2019. Citrus and related genera. In: *Handbook of flowering. CRC Press.* pp: 275-294.

Mudzunga MJ, 2000. Enhancement of vegetative growth in young citrus plantings (Doctoral dissertation, Stellenbosch: Stellenbosch University).

Mulinge JM, Saha HM, Mounde LÇ and Wasilwa LA, 2018. Effects of legume cover crops on orange (*Citrus sinensis*) fruit weight and brix. *International Journal of Plant and Soil Science, Hooghly,* v.21, n.4, p.1-9,

Murkute AA, Thirugnanavel A, Tejkumar J Prasanth, Sonkar RK, Roy Subhra Saikat and Dilip Ghosh, 2023. Citrus diversity in India. In: Ghosh D (Ed.), Indian Citrus Series (Vol. I). 151 pages. Today and Tomorrow's printers and publishers, New Delhi, India.

Naqvi SAMH, 2000. Managing *Phytophthora* diseases of Citrus. *Indian Hortic.* **44(4)**: 5-9.

Naqvi SAMH, 2004. Diagnosis and management of certain important fungal diseases of citrus. In: *Diseases of Fruits and Vegetables Volume I: Diagnosis and Management* (pp. 247-290). Dordrecht: Springer Netherlands.

Nawaz MA, Ahmed W, Iqbal Z and Khan MM, 2007. Evaluation of high density plantation on vigor and yield in Kinnow mandarin (*Citrus reticulata* Blanco). **In**: *Proceedings of the International Symposium on Prospects of Horticultural Industry in Pakistan, Faisalabad,* pp. 87–92.

Oppenhemier HR, 1940. How to produce summer lemons in Palestine. *Hadar,* **13** : 169- 170.

Panigrahi, P. 2023. Scheduling irrigation for improving water productivity and fruit quality in citrus (cv. Nagpur mandarin) under a dry tropical climate. *Irrig. Sci.* **41(4)**: 511-520.

Panigrahi P, Sharma RK, Hasan M and Parihar SS, 2014. Deficit irrigation scheduling and yield prediction of 'Kinnow'mandarin (*Citrus reticulata* Blanco) in a semiarid region. *Agric. Water Manag.* **140:** 48-60.

Panigrahi P, Srivastava AK and Huchche AD, 2012. Effects of drip irrigation regimes and basin irrigation on Nagpur mandarin agronomical and physiological performance. *Agric. Water Manag.* **104**: 79-88.

Pawar PS, Shirgure PS, Jagtap DD and Patil P, 2020. Acid lime (Citrus aurantifolia) production as influenced by different fertigation levels with irrigation schedules. *Indian J. Agric. Sci.* **90(4)**: 800-3.

Phocaides A, 2000. Technical handbook on pressurized irrigation techniques. *FAO, Rome, 372.*

Rabe E, 1996. Challenges of modern citriculture: Canopy management. *Proceedings International Society of Citriculture.*70-77

Raghupathi HB, Sakthivel T and Ravishankar H, 2010. Multivariate nutrient diagnostic norms for Coorg mandarin (*Citrus reticulata*). *Indian J. Agric. Sci.* **80(9):** 77-79

Raghupathi HB, Sakthivel T and Ravishankar H, 2013. Preliminary DRIS ratio norms for diagnosis of nutrient imbalance in Coorg mandarin. *Indian J. Hortic.* **70(1)**: 33-36.

Reddy PS and Swamy GS, 1986. Studies on nutritional requirement of sweet orange (*Citrus sinensis* Linn.) variety Sathgudi. *South Indian Hortic.* **34(5)**: 288–292.

Reuther W and Smith PF, 1954. Effect of method of timing nitrogen fertilization on yield and quality of oranges. *Proc. Fl. State Hort. Soc.* **67**: 20-26

Rymbai H, Deshmukh NA, Ramesh T, Verma VK, Baiswar P, Mawlein J and Mishra VK, 2024. DRIS standards for identifying yield-limiting nutrients in Khasi Mandarin and their relationship with fruit yield on acidic soil of the eastern Himalayas, India. *J. Plant Nutr.* **47(8)**: 1331-1346.

Sanchez CA and Peralta M, 2003. Response of Micro-Sprinkler Irrigated 'Lisbon'lemons to N Rate and Source on Superstition Sand. *Citrus Research Report.* http: //hdl.handle.net/10150/198115

Schonbeck M, Browne J, Deziel G and Degregorio R, 1991. Comparison of weed biomass and flora in four cover crops and a subsequent lettuce crop on three New England organic farms. *Biol. Agric. Hortic.* **8**: 123-143

Selim M, Hefny Y, Abdel-Wahab E and Mohamed M, 2020. Interplanting some soybean cultivars with mandarin trees in sandy soil. *Agricultural Sciences, Wuhan,* **11(1):** 88-110,

Shirgure PS, Srivastava AK and Huchche AD, 2016. Effect of drip irrigation scheduling on yield and quality of Nagpur mandarin (*Citrus reticulata* Blanco) fruits. *Indian J. Hortic.* **73(1)**, pp.30-35.

Shirgure PS, Srivastava AK and Huchche AD, 2014. Water requirements in growth stages and effects of deficit irrigation on fruit productivity of drip irrigated Nagpur mandarin (*Citrus reticulata* Blanco). *Indian J. Agric. Sci.* **84(3)**: 317-22.

Shirgure PS, 2013. Yield and fruit quality of Nagpur mandarin (Citrus reticulata Blanco) as influenced by evaporation based drip irrigation schedules. *Sci. J. Crop Sci.* **2(2)**: 28-35.

Shirgure PS, 2012. Effect of different intercropping systems on soil moisture conservation, fruit yield and quality of Nagpur mandarin (*Citrus reticulata*) in central India. *Scientific Journal of Agricultural.* **1(7):** 168-176

Shirgure PS, Srivastava AK and Singh S, 2001a. Effect of micro-jet irrigation system on growth, yield and fruit quality in Nagpur mandarin. *South Indian Hort.,* **49**(Special): 357-359.

Shirgure PS, Srivastava AK, Singh S, 2001b. Fertigation and Drip Irrigation in Nagpur mandarin (*Citrus reticulata* Blanco). *South Indian Hort.,* **49**(Special), 95-97.

Shirgure PS, Srivastava AK and Singh S, 2001c. Growth, yield and quality of Nagpur mandarin (*Citrus reticulata* Blanco) in relation to irrigation and fertigation. *Indian J. Agric. Sci.* **71(8)**: 547-550.

Shirgure PS, 2001. Water management and fertigation in citrus. In: Citrus. Eds. Shyam Singh and Naqvi, S.A.M.H. International Book Distributing Company, Lucknow, U.P. pp: 221-248.

Shirgure PS, Srivastava AK and Singh S, 2000. Water management in citrus–A review. *Agric. Rev.* **21(4)**: 223-230.

Shivankar VJ, Singh S, Huchche AD, Shirgure PS, Das AK, Mane PN and Patil SP, 2004. Effect of intercropping systems on insect pests complex of Nagpur mandarin (*Citrus reticulata* Blanco). Abstract International Citrus Congress held at Agadir, *Morocco.*

Singh G, 2010. Canopy management–an essential and integral element for healthy orchard establishment and sustainable production. *Prog. Hort.* **42(1)**: 15-17.

Singh G, Rattanpal HS, Gupta M and Sidhu GS, 2020. Standardization of stage wise water requirement in drip irrigated Kinnow mandarin orchards under sub-tropical conditions. *J. Agrometeorol.* **22(3):** 305-312.

Singh SK, Singh SK and Singh S, 2018. Vegetable crops as most efficient and economical intercrops: a brief review. *Plant Archives, Punjab,* 18(1): 923-929.

Smith PF, 1966. Leaf analysis of citrus. In: Childers NF (ed) Temperate to tropical fruit nutrition. Horticulture Publication, Rutgers University, *New Brunswick, pp* 208–228.

Soni ML, Saxena A, Nangia V and Yadava ND, 2021. Intercropping with vegetables on productivity and economic returns of Kinnow in arid region. *Indian J. Hortic.* **78(02)**: 211-215.

Srivastava AK, Hota Debashish, Dahat Shradhha and Sharma DE, 2022. Citrus nutrition: an Indian perspective. *Ann. Plant Soil Res.* **24***:* 1-15.

Srivastava AK and Singh S, 2016. Site-Specific nutrient management in Nagpur mandarin (*Citrus reticulata* Blanco) raised on contrasting soil types. *Commun. Soil Sci. Plant Anal.* **47(4)**: 447-456.

Srivastava AK, 2015. Soil Fertility Dynamics vis-à-vis Climate Change in Citrus. *Climate Change: The Principals and Applications in Horticulture Science, CRS Press (Apple Academic Press), USA.*

Srivastava AK, Huchche AD, Ram L and Singh S, 2007. Yield prediction in intercropped versus monocropped citrus orchards. *Sci. Hortic.* **114(1)**: 67-70.

Srivastava AK and Singh S, 2006. Biochemical markers and nutrient constraints diagnosis in citrus: a perspective. *J. plant Nutr.* **29(5)**: 827-855.

Srivastava AK and Singh S, 2004. Leaf and soil nutrient guide in citrus–A review. Agric. Rev, **25(4):** 235-251.

Srivastava AK and Singh S, 2003. Diagnosis and management of nutrient constraints in citrus. Manual no. 2. National Research Centre for Citrus, Nagpur, pp 1–70.

Srivastava, A. K. and Singh, S. 2002. Soil analysis based diagnostic norms for Indian citrus cultivar. *Commun. Soil Sci. Plant Anal.* **33(11-12)**: 1689-1706.

Steenwerth KL, Hodson AK, Bloom AJ, Carter MR, Cattaneo A, Chartres CJ and Jackson LE, 2014. Climate-smart agriculture global research agenda: scientific basis for action. *Agriculture and Food Security*. **3**: 1-39.

Sullivan P, 2003. Principles of Sustainable Weed Management for Croplands. http: //attra.ncat.org/attrapub/PDF/weed.pdf

Tennakoon SB, Milroy SP, 2003. Crop water use and water use efficiency on irrigated cotton farms in Australia. *Agric. Water Manag.* **61** : 179- 194

Thakuria D, Chaliha C, Dutta P, Sinha S, Uzir P, Singh SB and Singh D, 2023. Citrus Huanglongbing (HLB): Diagnostic and management options. *Physiol. Mol. Plant P.* **125***:* 102-016.

Varalakshmi LR and Bhargava BS 1998. Plant nutrient status and leaf diagnostic norms for acid lime (*Citrus aurantifolia*). *Indian J. Agric. Sci.* **68(7):** 375-376.

Vashisth T and Kadyampakeni D, 2020. Diagnosis and management of nutrient constraints in citrus. In *Fruit crops* (pp. 723-737). Elsevier.

Wani RA, Din S, Khan M, Hakeem SA, Jahan N, Lone RA and Khan J, 2021. Canopy management in fruit crops for maximizing productivity. *IJCS*. **9**(3): 160-165.

Webber HJ and Batchelor LD, 1979. *The citrus industry* (Vol. 1). *Univ of California Press.*

Wright GC, Matheron ME, McClure M, McCloskey B, Kerns D and Agnew K, 2000. Crop Profile for Citrus in Arizona. Pesticide Information and Training Office University of Arizona Tucson, AZ 85719. Available online: http: //www.ipmcenters.org/CropProfiles/docs/AZcitrus.html

Chapter 12

Advanced Techniques for Propagation of Citrus Plants

Tanjeet Singh Chahal and Vikramjit Singh

Punjab Agricultural University–Fruit Research Station, Jallowal, Jalandhar–144303, Punjab

Citrus Plants have a significant impact in various spheres of our life such as social, economic as well as cultural. The unique blends of nutrients in citrus fruits have made them un-detachable part of every corner of the world. Fortunately, India is one of the leading producers of citrus fruits in the world. The country is blessed with optimum climatic as well as soil conditions which helps in the cultivation of various citrus species, including mandarins, sweet oranges, grapefruits, lemons and limes (Anonymous, 2023). On the contrary their quality comes under influence of various unwanted stresses, such as diseases, pests, and harsh climatic conditions. To ensure the optimum quality citrus fruit production, the recent propagation techniques and suitable rootstocks are becoming increasingly important (Hayat *et al.*, 2022).

Large lots of orchards are established in India by production of high-quality genetically uniform planting materials in various citrus nurseries. High quality citrus plants are in huge demand due to the increased inclination towards citrus cultivation and ample need to replace the old orchards (Albrecht *et al.*, 2019). However, the large-scale cultivation of citrus fruits becomes cumbersome with this traditional method. As a result, quality and yield of the fruits gets compromised. (Dorji *et al.*, 2015).

Ensuring the long-term productivity, sustainability, and profitability of citrus orchards, availability of healthy saplings become crucial to eliminate various diseases. The ultimate tree health and fruit yield are the results of healthy citrus saplings. Any issues or diseases present in the nursery stage can have serious consequences. There are various diseases such as Citrus Tristeza Virus (CTV), Huanglongbing (HLB or citrus greening), and Phytophthora root rot which can infest the healthy plants and can even lead to complete breakdown of orchard (Albrecht *et al.,* 2017).

In addition, healthy seedlings with proper root system and vigorous growth can withstand under unfavourable environment conditions like stresses and pests encountered during the transplanting phase in the field. A strong start in the nursery establishment can enhance earlier fruit production and improved longevity of the orchard. (Kamatyanatti and Singh, 2019).To combat this issue, citrus propagation in India has taken the route of methods such as budding and grafting which ensure the genetic uniformity of the scion cultivar. But the success rate of these methods depends on the accessibility of suitable rootstocks which are adaptable to specific climatic as well as soil conditions (Bowman and Joubert, 2020).

Citrus seedlings are mainly produced from polyembryonic seeds, which contain multiple embryos derived from the maternal tissue (nucellus) in addition to the zygotic embryo resulting from sexual reproduction (Kepiro and Roose, 2007). Because they are genetically identical to their mother plant, nucellar embryos are preferred for producing rootstocks because they preserve the advantageous characteristics of the parent and provide uniformity (Grosser and Gmitter, 2011). Based on their performance and adaptability, a variety of introduced and native rootstocks have been suggested and evaluated for India's various citrus-growing regions (Ghosh, 2007). Dorji *et al.* (2015) states that the production and distribution of planting material free of diseases has been made possible by the citrus nurseries' construction in India. To guarantee the planting material's quality and health, these nurseries adhere to strict standards in the selection of rootstock and scion, collecting of rootstock seeds, health of mother plants and propagation methods. High-quality citrus nursery production is greatly aided by the integrated pest and disease management techniques and development of insect-proof screens.

Over the past few years, advancement in molecular and biotechnology techniques has unfolded various new paths for the betterment of citrus propagation in India. Techniques like micro-propagation, marker-assisted selection and somatic embryogenesis have been investigate for the accelerated multiplication of superior genotypes and the evolution of virus and disease resistant scions and rootstocks stated by Nageswara *et al.* (2008); Mustafa *et al.* (2021). These works connected with the traditional expertise of citrus multiplication, can be fruitful for the future of the citrus nurseries.

In India, establishment of well-managed citrus nurseries and utilization of suitable rootstocks are critical for the citrus production prosperity. The citrus industry's viability and profitability have been impacted by the use of techniques of contemporary propagation and the improvement in continuous perspectives of scion and rootstock cultivars. This chapter focuses on in-depth examination and provision of the citrus propagation's development, the emerging technologies, management of nursery, and importance of rootstocks to guarantee the high-quality citrus planting material.

Citrus Cultivation - Obstacles

Producing healthy citrus saplings is a complex process that involves various challenges, including:

Control of Diseases

Grahams (2020) states that the citrus nurseries must take precautions to keep their facilities free from different infections *i.e.*, parasites, viruses, and bacteria. Moreover, the diseases can spread through contaminated equipment and tools, root rot of Phytophthora, rootstocks, and budwood from contaminated mother plants.

Insect Vector Control

According to Bove (2006), some insect vectors, notably the Asian citrus psyllid (*Diaphorina citri*), can spread fatal illnesses like HLB. Nurseries must implement strict insect controlling procedures to prevent the spread of illnesses.

Preserving Genetic Integrity

To preserve the intended fruit quality and orchard performance, it is crucial to make sure that the propagated cultivars are true-to-type and free from genetic mutations or off-types (Lee *et al.*, 1993).

Environmental Control

It is essential to provide ideal growing conditions, including temperature, humidity, and irrigation management for production of robust and healthy seedlings (Handreck and Black, 2002).

Management Techniques for Nurseries

There are various factors which decide the successful nursery production, which includes putting in place appropriate sanitation procedures, employing sterilized growing media, and sticking to best practices for transplanting, budding, seed germination, and pest management (Hayat *et al.*, 2022).

The objective of this chapter is to provide a detailed guidance for producing healthy citrus rootstock seedlings and disease free grafted plants under modern nursery conditions.

It will address the following important points:

1. The infrastructure required for disease-free citrus nursery production, such as concrete platforms, greenhouses, and screen houses.

2. Establishing and maintaining seed gardens, mother blocks, and foundation blocks in order to collect disease-free rootstock seeds and budwood.
3. Growing media preparation, seed germination methods, and transplanting approaches for best seedling growth.
4. Propagation techniques such as budding and bud forcing, employed for superior growth of citrus nursery plants.
5. Various cultural methods for rearing nurseries, such as fertilization, irrigation, and integrated disease management techniques.

Nursery owners can contribute to the citrus growers and other industry stakeholders by providing them with healthy planting material and thereby increasing quality yield and support the citrus industry's long-term feasibility by following the principles and practices described in this chapter.

Structures for Disease Free Citrus Nursery Production

Using the right structures at each stage of propagation and focusing on nursery hygiene are essential for producing healthy citrus plants. In current citrus nurseries, a variety of protected structure types are used to control disease and create the ideal growing conditions for plants.

Greenhouse for Production of Primary Nurseries

In the early stages of citrus nursery production, rootstock seeds are germinated and immature saplings are grown in greenhouses. Primary nursery plants can be produced in a greenhouse free of diseases like tristeza virus and citrus greening, which can spread through insect vectors. The greenhouse regulated environment incorporates filtered air, the ideal temperature and humidity, and netting for insect exclusion. According to Shyam *et al.* (2004), when seedlings of rootstock reach height of 10 to 12cm, the transformation occurs in the form of larger pots after being grown in the greenhouse in tiny tubes or cells.

Figure 12.1: Greenhouse Structure for Production of Primary Nursery.

Screen Houses for Foundation and Mother Blocks

For the pathogen-tested budwood source trees' foundation blocks and premium scion varieties' mother blocks to be preserved, the requirement is of net or screen homes. Kamatyanatti and Singh (2019) states that to keep away the insects, the screen house keeps its surrounding air pressure higher as compared to the internal air pressure. By utilizing the screens' double door entryways and walls, these structures are intended to keep out insects and vectors of disease-carrying aspects. Khan *et al.* (2005) presents a study in which he states that the mother block trees provide material of scion for rootstocks grafting, while further reproduction takes place in the original trees' buds.

Shade Net Houses for Seedling Growth and Budding

The seedlings of rootstock are transferred into shade net homes after development under greenhouse conditions, where they can grow until they attain right size for budding operations. Chahal (2019) states that the fifty percent shade net provide intense heat and sunlight protection, extending the frame of time for budding operations. The budding process involves grafting bud from the scion of suitable variety onto rootstock seedling. According to Khan *et al.* (2005), the net house's airflow and shade provide the graft union to heal under perfect environment.

Figure 12.2: Shade House for Budding and Bud Forcing Operations in Citrus Rootstocks.

Concrete Platforms for Media Sterilization

For containerized plants, nurseries typically make their own media for potting, incorporating a mixture of soil, sand, manure of farmyard, and other materials. Before using this potting mixture, it must be exposed to either steam sterilization or solarization for eradication of media borne diseases. Kamatyanatti and Singh (2019) states that concrete platform or floor is built in an exposed portion of nursery. After media's arrangement in a thin layer, plastic sheet is placed over it. The increased temperature under the plastic sheet and injection of steam successfully eradicates the hazardous bacteria. Khan *et al.* (2005) states that to destroy any microbes, they should be exposed to sunlight until it reaches an enough high temperature.

The net houses, greenhouses, screen houses, concrete platform and disease free elite planting material are required for a modern nursery production system for citrus. By keeping out disease-carrying organisms and dangerous germs, these structures shield the nursery plants against infection and maintain congenial environment for producing healthy planting material.

Plantation Blocks

Mother Blocks Establishment

In setting up a hi-tech nursery infrastructure, a crucial step is the establishment of mother blocks which ensure true-to-type and disease-free bud stick production for quality nursery plants. According to Chahal and Kamran (2017), these blocks serve as the main bud wood source for commercial nurseries. To check the entry and spread of diseases and to stop contamination, certain protocols are put in place to protect mother blocks.

Monitor the mother plants regularly and keep a close eye for infections and as soon as possible remove any afflicted plants.

Site Selection and Preparation

1. Select a location with soil that has adequate drainage, high fertility, and low incidence of diseases.
2. Ensure that the property is physically separated from potential sources of contamination, such as surrounding citrus orchards or nurseries.
3. Prior to construction, clear the site of any existing vegetation, flatten the terrain, and establish an irrigation system, preferably utilizing drip or micro-sprinklers.
4. Construct structures, such as a screen houses or net houses that provide proof against insects, in order to safeguard the mother plants from disease-carrying insect vectors.
5 Regularly examine and uphold the insect-proof construction in order to ensure its structural integrity and efficiency in preventing insect carriers.

Bud Wood Collection and Handling

1. At the optimal growth stage, harvest bud wood from the mother plants, taking care to ensure that the bud wood is fully grown and free of diseases.
2. In bud wood harvesting operations, to check the spread of diseases, sterilize the instruments and equipment regularly.

3. Ensure that the gathered bud wood is handled and stored in cool and humid place to preserve its viability and prevent it from drying out.
4. Keep up to data records on the origin of the bud wood, date of collection, and any appropriate treatments to ensure traceability and maintain quality control (Albrecht *et al.*, 2019)

Figure 12.3: Mother Block of Scion Varieties under Protected Conditions.

Rootstock Block for Seed Production

It is very important to establish a seed garden or a rootstock block for the production of superior, reliable and truthful rootstock seeds. It is foundation for the propagation of citrus nursery plants. A well establish and properly maintained seed garden regularly provides authentic seeds which amplifying the consistency and the overall quality of the nursery plants (Chahal, 2019).

Acquire certified rootstock mother plants from prestigious organizations, such as government approved certification programs or recognized research institutions.

Plant the rootstock mother trees in the allocated seed garden, adhering to the appropriate spacing and planting instructions (Singh, 2015).

Maintenance and Management

1. Enforce rigorous sanitary measures, disinfect tools and equipment, and prohibit unauthorized individuals from accessing the seed garden area.
2. Adhere to a consistent irrigation and fertilization regimen that aligns with the distinct needs of the rootstock type and the prevailing local conditions.

3. Trim and cultivate the rootstock mother trees to encourage ideal development, fruit production, and seed generation. Eradicate any undesired suckers or water shoots.
4. Employ integrated disease and pest management approach to monitor and regulate pests and illnesses, prioritizing biological and cultural control methods over chemical pesticides.

Figure 12.4: Seed Garden for Regular Supply of Rootstock Seeds.

Seed Collection and Extraction

1. Collect the fruits from the rootstock trees after they have reached the right level of maturity, making sure that the seeds are completely formed and capable of germination.
2. Employ suitable techniques, such as hand extraction, to separate the seeds from pulp of the collected fruits.
3. Thoroughly cleanse and sterilize the extracted seeds to eliminate any attached particles or possible disease-causing microorganisms.
4. Ensure that the extracted seeds are completely dried and stored in ideal conditions to preserve their potential to grow until they are required for nursery propagation.
5 Keep complete records of the origin of the seeds, the date of collection, and any treatments administered to ensure accountability and maintain high standards of quality control (Khan *et al.,* 2005).

Indicator Plants for Disease Detection

Indicator plants play an important role in citrus nursery production for detecting and monitoring the presence of several diseases that might affect the quality and productivity of nursery stock. Indicator plants are highly valuable for identifying viruses, viroids, and other systemic diseases that might be transmitted by bud wood or propagation materials.

Selection of Indicator Plants

1. Select indicator plants that are proven to be susceptible to particular pathogens of interest, such as viruses, viroids, or bacterial infections.
2. Common indicator plants utilized in citrus nurseries encompass Mexican lime (*Citrus aurantifolia*) for Citrus Tristeza virus (CTV), sweet orange (*Citrus sinensis*) for Citrus Greening disease, and etrog citron (*Citrus medica*) for Exocortis.
3. To prevent the introduction of potential pathogens into the nursery environment, make sure you get indicator plants from approved sources that are free from diseases (Chahal and Kamran, 2017).

Planting and Maintenance

1. Place the indicator plants strategically in the nursery or near the mother blocks and seed gardens to provide convenient monitoring and observation.
2. Ensure that the indicator plants are kept in ideal growing conditions by adhering to approved cultural methods for irrigation, fertilization, and pest control.
3. Conduct routine examinations of the indicator plants to detect any indications of illness or irregularities, such as changes in leaf color, stunted growth, or tissue death.
4. Record any noted symptoms and send samples for laboratory testing and verification, if needed.

Disease Management and Response

1. In the case that a pathogen is identified in the indicator plants, immediately execute suitable control actions to hinder its transmission to other nursery stock.
2. Employ rigorous cleaning measures to decrease the danger of contamination by isolating and eliminating any sick plants.
3. Perform comprehensive examinations of the nursery stock and mother blocks, and apply requisite treatments or preventative measures to guarantee the production of plants that are free from diseases.

4. Evaluate and modify nursery management procedures, such as sanitation protocols, bud wood sources, and propagation strategies, in order to minimize the potential for future disease introduction and dissemination (Kamatyanatti and Singh, 2019).

Citrus nurseries can ensure the long-term sustainability and productivity of commercial citrus orchards by creating and maintaining plantation blocks, which include mother blocks, seed gardens, and indicator plants. This practice ensures the production of disease-free nursery plants of superior quality.

Media Preparation

The choice of a potting mixture for propagating citrus nursery plants is crucial in ensuring the propagated plants' optimal health and superior quality. A highly effective potting mixture should consist of well-balanced components that offer enough aeration, water retention, porosity, and water drainage while also provide vital nutrients for optimal plant growth.

Composition of Potting Media

An optimally formulated potting mix generally comprises a combination of organic and inorganic substances, meticulously chosen according to their accessibility, expense, and desired characteristics.

Mainly organic components utilized in citrus nurseries include farm yard manure, coco peat, peat moss, and composted sawdust derived from untreated softwood or hardwood.

In addition, the mixture used for potting often includes inorganic elements like coarse sand or silt for better aeration and drainage, vermiculite for water retention and aeration, and perlite, a lightweight material that enhances aeration and drainage.

The specific ratios of these elements may fluctuate depending on various circumstances, including the type of rootstock, the grower's preferences, and the prevailing meteorological conditions. It is crucial to verify that the potting media is devoid of weeds, soil-borne diseases, and other impurities.

However, an equal ratio of virgin soil and cocopeat mixture is employed for the primary nursery. Furthermore, the production of media for secondary nursery in north-western Indian nurseries involves using a constant percentage of farmyard manure and cocopeat, together with double the amount of soil (1:1:2), as stated by Chahal (2019).

Methods of Sterilization

In order to eliminate disease-causing microbes, weed seeds, and other potential pollutants from the potting mix, a range of sterilizing processes can be utilized:

Figure 12.5: Media Preparation for Primary and Secondary Nursery.

Solarization

This process involves exposing the potting mix ingredients or the completed mix to prolonged exposure to direct sunlight, usually for duration of one to two weeks in the summer months. To facilitate the inactivation of pathogens and weed seeds, the mixture should be distributed in thin layers and then covered with transparent polyethylene sheets in order to trap heat and moisture (Hayat *et al.*, 2022).

Steam Sterilization

This method involves the use of electric furnaces that may reach temperatures sufficiently high to eliminate microorganisms by heating the media. Electric furnaces are commonly built on a concrete platform in an exposed section of the nursery. The potting mix is placed into the furnace and subjected to a temperature of 60 °C (140°F) for thirty minutes in order to eliminate germs (Durner, 2020). This technique efficiently eradicates soil-borne pathogens, while simultaneously conserving beneficial microbes that are crucial for nitrogen cycling and decomposition.

Chemical Fumigation

Specific fumigants, such as chloropicrin or dazomet, can be employed to treat either the potting mix components or the prepared mix. These compounds function as fungicides, herbicides, insecticides, and nematicides, successfully eradicating a wide variety of soil-borne infections and pests (Albrecht *et al.*, 2017).

It is crucial to exercise caution when handling and using chemical fumigants, adhering to the safety regulations and guidelines provided by the manufacturers. Following sterilization, the potting mixture must be stored in a completely sanitary and protected location to avoid any possibility of recontamination. It is advisable to regularly check and analyse the potting mix for possible pathogens and impurities in order to make sure the cultivation of robust and disease-free citrus nursery plants.

Production of Rootstock Seedlings

It is essential that healthy rootstock seedlings should be used for the propagation of citrus plants, as they supply the root system and impact different characteristics of the grafted scion cultivar. The production of rootstock seedlings goes through multiple important stages, such as seed selection, seed extraction, sowing techniques, and transplantation. Each step in the process must be executed with care and attention so as to ensure the generation of healthy rootstock seedlings. As mentioned above, the seed selection and extraction from the rootstock fruits must be done carefully. Now the next step is rootstock sapling raising. There are multiple techniques available for propagating rootstock seedlings, which include:

Open Field Sowing

Seeds can be directly sown in properly prepared nursery beds or fields. This strategy is appropriate for regions with advantageous environmental conditions throughout the sowing time.

Greenhouse or Screenhouse Sowing

Seeds are sown in portrays or containers in controlled conditions, such as greenhouses or screenhouses. This technique enables more effective disease protection and control over temperature, humidity, and light parameters, leading to increased rates of germination and the development of healthier seedlings (Hayat *et al.,* 2022).

Trays and Protrays

Trays and protrays, which are multi-cell containers, are a modern and efficient method for cultivating rootstock saplings. Protrays with 20 cells, each measuring around 18.5 cm in length, 14.7 cm in breadth, and 12 cm in depth, are widely used for propagating rootstock seeds (Chahal, 2019). These protrays provide the effective handling and precise monitoring of individual seedlings, so ensuring consistent growth and development.

Transplanting of seedlings: Once the seedlings have grown to an appropriate height, ideally between 10-12 cm, then they are transplanted into larger pots, polybags, containers, plastic citri pots. The polybags must have enough height and width, usually one foot in height and four to six inches at in width, to accommodate the growth of the saplings till grafting or budding operations. On the other hand, citri pots are made of light weight plastic and generally the size is 34 cm in height and 10 cm in width. These pots are ideal for propagating saplings. It is important to select saplings with a single stem for transplantation. Trimming the tap root to promote the growth of a fibrous root system can be done. After transplanting, the soil around the sapling should be firmly pressed to obliterate air pockets and promote root and plant growth (Khan *et al.,* 2005).

Figure 12.6: Sowing of Rootstock Seeds in Protray Cells.

Figure 12.7: Carrizo and Rough Lemon Seedlings in Individual Protray Cells.

Propagation Techniques

Propagation methods have an important role in the production of citrus nursery plants. Choosing virus and disease-free bud wood and properly performing grafting or budding processes are crucial for obtaining genetically identical seedlings to the parent plant. T-budding is the predominant method used in citrus propagation. This technique involves inserting a bud from the appropriate scion cultivar onto a rootstock seedling.

In addition, bud forcing techniques are used to stimulate early bud break and promote faster scion growth, hence reducing the entire duration of nursery production.

Selection of Bud Wood

The careful selection of suitable bud wood is a crucial stage in the process of propagation. Optimal bud wood, which is free from diseases and viruses, should be carefully chosen from certified mother plants in the designated mother block. When choosing branches for bud wood, it is important to avoid suckers and water sprouts. These branches are highly robust, thorny, and often come from the lower trunk of the plant.

The ideal branches for collecting bud wood are those that are between six months to one year old or current growth. These branches exhibit maturity signs, such as gray-colored streaks on the bark surface, and have swollen but dormant buds, indicating active cell sap flow. Straight branches with completely developed, round and healthy buds should be selected and prepared by removing leaves, leaving about 8-10 inches of the branch.

The cultivar name, date, source, and any other relevant information should be labelled on the bud wood to keep traceability and secure the production of true-to-type nursery plants. According to Rajput *et al.* (1985), fresh bud wood can be used immediately or stored properly at 4°C in sealed plastic bags or alkathene sheets or moist cloth containing damp saw dust to avoid drying for longer periods, up to several months. In addition to this, bud wood can be stored for two weeks by keeping them carefully wrapped in moist sphagnum moss or saw dust (Aiyappa and Srivastava, 1968). Before storage, the bud wood should be sanitized by dipping it in a bleach solution and then drying it to remove excess moisture, preventing condensation and mold growth within the storage bags. Moreover, under the northwestern Indian conditions budding is generally carried out in the month of July-September and February- March in this time the bud wood becomes scarce due to initiation of new growth (Chahal, 2019).

T-Budding Method

T-budding is the most used technique for citrus propagation due to its higher success rate. Firstly, remove all the sprouts, leaves and thorns on the stem of the rootstock up to just below the point of operation site. Further step is making a vertical T-shaped cut on the bark of the rootstock seedling, approximately 1.5-2.0 cm long vertical cut is made on pencil thickness rootstock 15- 20 cm from ground level. Next move is making another cut above the vertical cut. Then the flap of bark is lifted carefully without injuring the tissue below, by blunt end of the budding knife. Then a shield-shaped bud of suitable size about 2.5 cm with or without a slice of wood from the scion cultivar is inserted beneath the bark of the rootstock with the bud facing upwards. Then the bud is tightly wrapped with budding tape or suitable wrapping material like alkathene sheet above and below the bud, exposing the bud and petiole stub. Make sure the wrapping should be tight, but at the same time it should not press or girdle the bud (Chahal, 2019).

After 3-4 weeks, the bud union should be checked for successful grafting. If the bud appears green, it indicates a successful union, and the budding tape can be removed. Once the scion growth has reached approximately 6 cm above the union, the top portion of the rootstock seedling can be pruned (Khan *et al.,* 2005).

Figure 12.8: T-budding Method for Citrus Propagation.

Bud Forcing Techniques

Bud forcing techniques are employed to induce early bud break and promote faster scion growth, thereby reducing the overall nursery production time. Two commonly used methods for bud forcing are cutting and bending (Williamson, 1997). Different methods which can be used for forcing usually consist of complete removal of the rootstock top or modification of the attached rootstock shoot in a manner which reduces apical dominance. Forcing methods which leave the rootstock shoots attached to the plants are expected to result in more scion growth. The leaves of the stock continue to furnish the roots with nutrients, while the flow of sap would force the bud into growth. The three commonly used methods in commercial nurseries are discussed as below:

Cutting

In this method, the rootstock shoot above the bud union is partially or completely removed, stimulating the bud to break and grow. Cutting of the rootstock top is the most common method used in Indian nurseries because it is easy to perform and less labour intensive. This method completely destroys the apical dominance of the rootstock. It includes complete removal of stock 1.5 inches above the bud union. However, this technique takes away a good part of the assimilating apparatus of the plant. This method of forcing induces faster stimulation of the scion bud, but this shoot finds it difficult to maintain growth as it soon runs out of nutrients. Further, the roots of the plants under cutting treatment starve for carbohydrates and stop producing Cytokinin which is necessary for sustained growth of the shoot (Chahal and Kamran, 2017).

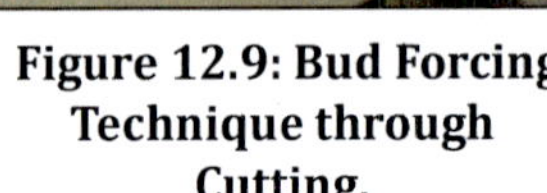

Figure 12.9: Bud Forcing Technique through Cutting.

Figure 12.10: Bud Forcing technique through Lopping.

Figure 12.11: Bud Forcing Technique through Bending.

Lopping

Lopping is a crucial step in the propagation of citrus plants through budding. It is the process of partially severing upper portion of the rootstock above the bud union to force the scion bud to sprout and grow. The timing and method of lopping play a vital role in the successful establishment of the scion and the overall health of the nursery plant.

Time of Lopping

The appropriate time for lopping depends on the season in which the budding is performed. Generally, buds inserted in the fall or cooler months take longer (6-8 weeks) for the bud union to form compared to those inserted in the spring (3 weeks) (Srivastava, 2012). It is recommended to wait until the bud union is established before lopping, which can be determined by the green color of the bud (Singh *et al.*, 2018).

Methods of Lopping

1. Lopping after bud union: This is the preferred method, where lopping is performed after the scion bud has formed a union with the rootstock tissues. The rootstock is partially severed 5-8 cm above the bud joint and bent, allowing the top to continue nourishing the roots while forcing the scion bud to grow (Srivastava, 2012).
2. Lopping at the time of budding: Some nurseries practice lopping simultaneously with bud insertion. Although, this may cause the bud to sprout earlier, it reduces the 'take' or success rate of the bud union (Ollitrault and Navarro, 2012).

3. Defoliation: Instead of lopping, an alternative method suggested by Jacobs (1960) involves defoliating the rootstock, retaining only one leaf above the bud. This leaf can provide sufficient nourishment to the roots while allowing the bud to grow.

After Lopping

Once the scion bud has grown to a length of 25-30 cm, the lopped portion of the rootstock can be completely removed by cutting it back to the bud union. The cut surface is then sealed with wax or a similar material to prevent desiccation and entry of pathogens (Randhawa *et al.,* 1961; Srivastava, 2012).

Bending

This technique involves bending the rootstock shoot above the bud union, leaving it partially attached to the plant. The bending of the shoot disrupts the apical dominance, triggering the bud to break and grow (Williamson, 1997).

Figure 12.12: Bud Forcing Through Bending: A Technique for Faster Scion Growth.

In this method, rootstock photosynthates are translocated to scion and roots of nursery plants. The leaves of the stock continue to furnish the roots with nutrition, while the sap flow forces the bud into growth. The movement of photosynthates to the root system register production of cytokinins which are brought up in the wood vessels and break apical dominance. This is preferred method of bud forcing in kinnow mandarin and kagzi lime budded on rough lemon rootstock under Punjab conditions as it produces higher scion growth thereby early production of nursery plants in comparison to cutting (Chahal, 2019).

By employing proper bud wood selection, budding techniques, and bud forcing methods, nursery growers can produce high-quality citrus nursery plants in a timely and efficient manner, contributing to the overall success of the citrus industry.

Irrigation and Fertigation

The production of high-quality citrus nursery trees depends on effective fertigation and irrigation control. While fertigation enables the precise delivery of nutrients to satisfy the unique requirements of the growing trees, proper irrigation guarantees that the plants receive the ideal amount of water. Many irrigation and fertigation methods used in citrus nurseries along with their pros and cons, and the significance of appropriate management for optimum tree growth and development will all be discussed in this part.

Irrigation Methods for Rootstock Liners in Seedbeds

Traditionally, hose pipes, rain wands, and perforated polyvinyl chloride (PVC) pipes have been used to irrigate citrus rootstock liners manually in seedbeds (Salvador *et al.,* 2016). These techniques are utilised to saturate the plant canopy and guarantee that water reaches the substrate, together with spray bars and overhead microsprinklers. (Ferrarezi *et al.,* 2019).

Nevertheless, there are several disadvantages to these techniques, such as the risk of leaf diseases, ineffective application, significant loss of fertilizer solution, and high labor costs (Carvalho *et al.,* 2019).

Subirrigation is one of the creative irrigation techniques that have been created to address these issues. Subirrigation is a technique that enables the fertilizer solution to be circulated while adding water to the bottom of containers. The ebb-and-flow bench is the most popular type of subirrigation system, which has been shown to have an irrigation efficiency of 76 as compared to 9 per cent for sprinkler irrigation and 53 per cent for drip irrigation.

Irrigation of Transplanted Rootstock and Grafted Trees in Citrus Pots

For the production of nursery trees in citrus pots, drip irrigation using drippers with multiple outlets connected to microtubes is the most common method (Carvalho *et al.,* 2019). This system is simple to operate and maintain, and it can have one emitter with one or multi-outlet small tubes to route water from the emitter to the trees (Ferrarezi *et al.,* 2017).

On one side, most trees are watered daily to maintain the substrate at field capacity and minimize leaching, Girardi *et al.* (2018) showed that regulated deficit irrigation can lead to significant water savings and decrease the risk of certain diseases. On the contrary, the potential negative effects on tree quality require further investigation.

Moreover, drip irrigation is becoming popular in water scarce areas. This method of irrigation improves uniformity and quality of plants.

Excessive irrigation is detrimental to the citrus nursery as it may lead to citrus foot and root rot, a very serious fungal disease caused by *Phytophthora spp.* Waterlogged conditions around the sapling should be avoided.

Fertigation in Seedbeds and Nurseries

With the introduction of containers, soilless substrates, and protected environments in citrus rootstock and tree production, several studies have been conducted to determine the optimal fertilization and mineral nutrition requirements (Rajput *et al.*, 1985).

Nurseries typically use granular, controlled release, and/or water-soluble fertilizers for fertigation. A balanced fertilizer solution is applied once or twice per week in soluble forms (8N-P-8K, or 20N-20P-20K, or 9N-3P-6K) (Shyam *et al.*, 2004). However, it is important to consider the chemical composition of the irrigation water and substrate when adjusting the nutrient solution and determining the concentrations of micro-nutrients to be incorporated.

Irrigation and Fertigation of Mother Blocks

In mother blocks, drip irrigation is the most common method, followed by manual irrigation using hose nozzles, rain wands, and overhead micro-sprinklers (Carvalho *et al.*, 2019). Trees in these blocks are cultivated either in the soil or in large pots for prolonged periods. Although nurseries often use rigid schedules based on timers to control irrigation, this approach can be problematic as it does not consider water demand, water availability, environmental conditions, and plant growth stage.

Maintaining an adequate balance of nutrients is crucial for the nutrition of budwood blocks to avoid deficiencies. The same types of fertilizers used in nurseries can be applied, but at larger rates. It is recommended to maintain adequate concentrations of N, P, Mg, and micro-nutrients to stimulate the growth of sprouts that will produce the buds for grafting.

To encapsulate, proper irrigation and fertigation management in seedbeds, nurseries, and mother blocks are essential for accelerating tree production and ensuring that growers receive trees that will result in profitable crop yields. Further research is needed to refine water requirements, irrigation methods and frequency, fertilizer types, sources and rates, and updated guidelines for different varieties and rootstocks (Albrecht *et al.*, 2019).

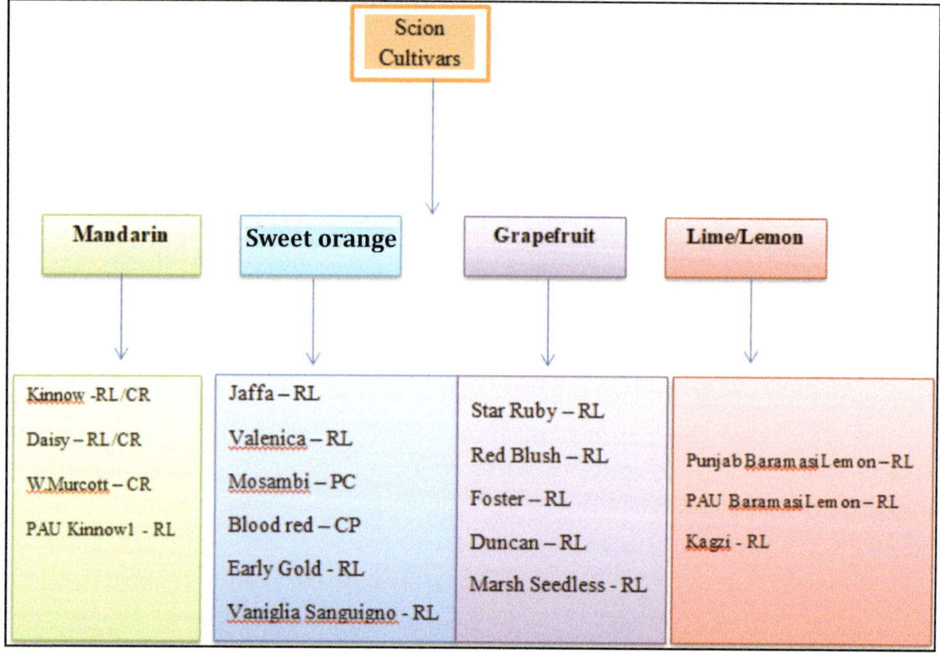

Rootstocks – RL Rough Lemon, CR Carrizo Citrange, PC Pectinifera, CP Cleopatra

Figure 12.13: Citrus Rootstocks Recommended for different Scion Cultivars.

Disease Management in Citrus Nursery

Several diseases may adversely affect the quality and general health of the nursery stock in citrus nurseries. The production of healthy and diseases and virus free citrus nursery plants is dependent upon the use of effective disease management strategies. In this section, we will discuss some of the considerable diseases impacting citrus nurseries along with some management strategies.

Citrus Foot Rot/Gummosis (*Phytophthora parasitica* var. nicotianae)

It is also known as Phytophthora rot, collar rot, and brown rot. Gummosis refers to the rotting of the bark anywhere on the tree. This disease infects the roots and trunks of citrus trees when there is high soil moisture. Symptoms of foot rot are easily identifiable as the bark is destroyed near the ground level of the trunk. An important symptom is the profuse exudation of gum from the trunk bark. In some instances, the bark dries, shrinks, cracks, and sheds into vertical lengthwise strips. The color of the bark and wood below the bark becomes dark brown. Eventually, the tree dies due to girdling caused by the destruction of bark at the root. Girdled plants exhibit typical pale green leaves with heavy defoliation. Dark stains on the bark that spread into the wood are the initial signs of this disease (Shyam *et al.*, 2004)

Management

- ☆ When propagating citrus plants, make sure to use disease-free bud-wood.
- ☆ Take preventive measures in mother blocks to avoid the spread of diseases.
- ☆ Obtain rootstock seeds from healthy plants and discard any dropped fruit or fruit near the soil.
- ☆ Ensure that the bud union is positioned 6 to 9 inches above the ground level.
- ☆ To eliminate diseased parts, remove the infected portion of the bark along with some surrounding healthy tissue using a sharp tool (Pathak, 2001).

Citrus Canker (*Xanthomonas axonopodis pv. Citri*)

Affects leaves and Twigs in nursery plants. It shows up as small, circular, water-soaked yellowish spots that get bigger over time, leading to a rough, corky, and brownish appearance surrounded by a chlorotic yellow halo. These lesions appear on both sides of the leaves. When the lesions affect twigs, they girdle them, causing the twigs to die (Pathak, 2001).

Management

- ☆ Please make sure to spray the nursery plants with a mixture of 50g of Streptocycline and 25g of Copper Sulphate in 500 litres of water. This should be done specifically during the months of October, December, and February. Additionally, you can use Bordeaux mixture (2:2:250) or copper oxychloride 50 WP (3g/litre of water) to reduce canker. It's important to spray the nursery plants, especially during the rainy season in July and August.

Citrus Greening (*Liberobacter asiaticus*) (Fastidious bacterium)

Symptoms of this disease affects almost all citrus varieties irrespective of root stock. It causes stunted leaves, sparse foliage and twig dieback. Sometimes, only a portion of the tree is affected. There is a diverse range of foliar chlorosis, with a mottling pattern that resembles zinc deficiency being common. Young leaves initially appear normal but soon become leathery, develop prominent veins, and a dull olive green color. Green circular dots can be seen on the leaves, and many twigs become upright and produce smaller leaves (Rajput and Hari babu, 1985). This disease is transmitted by grafting. In the nursery and orchard, further spread of the greening is through insect-vector, citrus psylla (*Diaphorina citri*).

Management

- ☆ Use bud-wood from greening free plants.
- ☆ Prune and burn dead wood and branches showing greening symptoms.
- ☆ To control citrus psylla spray 200 ml Confidor 17.8 SL (Imidacloprid)/160 g Actara/Dotara 25 WG.

Citrus Tristeza (*Citrus tristeza virus*)

Lime can be susceptible as a seedling or buddling on any root stock. However, mandarin and sweet orange seedlings, when buddled on rough lemon and trifoliate orange root stocks, are tolerant. Moreover, rangpur lime root stocks are also tolerant, while grapefruit and sour orange are susceptible. There is fine pitting of the inner face of bark of sour orange stock. Grapefruit and acid lime are susceptible, regardless of the root stock used. Acid lime leaves exhibit a large number of vein flecks, which are elongated translucent areas. Trees become stunted and may die (Pathak, 2001). The virus is transmitted through infected bud- wood in nurseries. Further spread in nursery and orchard is by aphid vector.

Management

- ✰ For sweet orange and mandarin, avoid using susceptible rootstocks.
- ✰ For acid lime, use seedlings that have been pre-immunized with a mild strain of tristeza.
- ✰ Ensure that the budwood used for citrus propagation is free of viruses.
- ✰ Utilize certified elite planting material that is free of tristeza and consider using tolerant rootstocks such as Rough Lemon (*Citrus jambhiri*).
- ✰ To prevent further spread, promptly eradicate and destroy infected trees.
- ✰ For aphid control, spray 200 ml Confidor 17.8 SL (Imidacloprid)/160 g Actara/Dotara 25 WG. (Anonymous, 2023).

Citrus Exocortis (*Citrus exocortis viroid*)

It only affects rangpur lime and trifoliate orange rootstocks. Symptoms include vertical cracking and scaling of the bark in the entire root stock portion, yellow blotching on twigs, and extreme stunting of the plant (Anonymous, 2023). The disease can spread through infected bud wood, dodder (amber bel), and sap. Exocortis is also readily transmitted to healthy plants by contaminated tools used in nursery for budding (knife) or pruning of trees (secateur).

Management

- ✰ Plant viroid-free elite mother blocks on tolerant rootstocks such as Rough lemon, Cleopatra, and Pectinifera.
- ✰ Regularly disinfect horticultural tools (knife and secateurs) with a solution of sodium hypochlorite (Anonymous, 2023).

REFERENCES

Aiyappa KM, Srivastava KC and Sulladmath UV, 1968. Studies on citrus root system iii. lateral spread and depth of penetration, relationship between top and root positions, dry-matter content of various components of roots, *etc.* in case of healthy, non-chlorotic and variously chlorotic citrus reticulata blanco (Coorg Mandarin-Seedling Trees). *Indian J. Hortic.*, **25**(3and4): 126-39.

Albrecht U, Bordas M, Lamb B, Meyering B and Bowman KD, 2017. Influence of propagation method on root architecture and other traits of young citrus rootstock plants. *Hortic. Sci.*, **52**(11): 1569-76.

Albrecht U, Rossi L and Zekri M, 2019. Citrus Rootstock Propagation: Traditional Techniques and Recent Advances: *Hortic. Sci.*, **1329**, 8/2019. *EDIS, 2019*(4), pp. 1-4.

Anonymous, 2023. Jaswinder Singh Brar and Gurmeet Singh Buttar (ed) Package of Practices for Cultivation of Fruits, Punjab Agricultural University Printing Press, Ludhiana, India.

Bove JM, 2006. Huanglongbing: a destructive, newly-emerging, century-old disease of citrus. *J. Plant Pathol.*, pp. 7-37.

Bowman KD and Joubert J, 2020. Citrus rootstocks. In *The genus citrus*. Woodhead Publishing. (pp. 105-27).

Carvalho SAD, Girardi EA, Mourão Filho FDAA, Ferrarezi RS and Coletta Filho HD, 2019. Advances in citrus propagation in Brazil. *Rev. Bras. Frutic.*, pp *41*, e-422.

Chahal TS, 2019. Advances in commercial citrus cultivation under North Western Indian conditions. In Compendium of training program on "Advance Technologies for Improving Fruit Production in Jammu Region" Sher-e-Kashmir University of Agricultural Science and Technology – Jammu (pp. 18-25).

Chahal TS and Kamran A, 2017. Disease free citrus nursery production. In ICAR–Winter School on "Hi-tech intervention in fruit production towards hastening productivity, nutritional quality and value addition" [E-book].

Dorji K and Lakey L, 2015. Citrus nursery management – A technical guide. Department of Agriculture, Thimphu, Bhutan. pp vi+32.

Durner EF, 2013. Principles of horticultural physiology. CABI. pp 216-65.

Ferrarezi RS and Testezlaf R, 2017. Automated ebb-and-flow subirrigation for citrus liners production. I. Plant growth. *Agric. Water Manag.*, **192**: 45-57.

Ferrarezi RS, Meadows T, Stephens J, James HT, Eckman MA, Macan NP and Gast T, 2019. July. Accelerated Production of Citrus Nursery Liners Using Automated Ebb-and-Flow Subirrigation. In *2019 ASHS Annual Conference*. ASHS.

Ghosh SP, 2007. Citrus fruits. ICAR Publications. pp. 23.

Girardi EA, Brandão AD, Coelho RD, do Couto HT, Buckeridge MS and Mourão Filho FDAA, 2018. Regulated deficit irrigation benefits the production of container-grown citrus nursery trees. *Trees*, **32**: 1751-66.

Graham J, Gottwald T and Setamou M, 2020. Status of huanglongbing (HLB) outbreaks in Florida, California and Texas. *Trop. Plant Pathol.*, **45**: 265-78.

Grosser JW and Gmitter FG, 2011. Protoplast fusion for production of tetraploids and triploids: applications for scion and rootstock breeding in citrus. *Plant Cell, Tiss. Org.*, **104**: 343-57.

Handreck KA and Black ND, 2002. *Growing media for ornamental plants and turf*. UNSW press.

Hayat A, Haque EU, Nasir M, Afzaal S, Sabir S and Ahmad S, 2022. Citrus Nursery Management and Modern Trends. In *Citrus Production* (pp. 95-108). CRC Press.

Jacobs WW, 1960. Skin-budding offers new possibilities for citrus propagation. *Farming in South Africa*, **35**(11): 12-13.

Kamatyanatti M and Singh SK, 2019. Recent advances in citrus propagation and planting. *Advances in Horticultural Crop Management and Value Addition*, pp. 1-12

Kepiro JL and Roose ML, 2007. Nucellar embryony. *Citrus genetics, breeding and biotechnology*, pp. 141-149.

Khan MM, Usman M and Abbas M, 2005. Citrus Nursery Raising: Principles and Practices.

Lee AT and Roxburgh K, 1993. *Guidelines for the Production of Container-grown Citrus Nursery Trees in Southern Africa*. Citrus Improvement Programme.

Mustafa G, Usman M, Joyia FA and Khan MS, 2021. Citrus Biotechnology: Current Innovations and Future Prospects. In *Citrus-Research, Development and Biotechnology*. IntechOpen.

Nageswara Rao M, Soneji JR, Chen C, Huang S and Gmitter FG, (2008). Characterization of zygotic and nucellar seedlings from sour orange-like citrus rootstock candidates using RAPD and EST-SSR markers. *Tree Genet. Genomes*, **4**: 113-24.

Ollitrault P and Navarro L, 2012. Citrus. *Fruit breeding*, 623-62.

Pathak PN, 2001. *Diseases of Fruit Crops*. Oxford and IBH Publications.

Rajput CBS and Hari Babu R, 1985. Citriculture. Kalyani Publishers, pp. 98.

Randhawa SS, Bajwa BS and Bakshi JC, 1961. Studies on the Behaviour of Seedlings of Some Common Citrus Rootstocks in the Nursery. *Indian J. Hortic.*, **18**(1): 70-80.

Salvador CA, Ferrarezi RS, Barreto CV and Testezlaf R, 2016. Method to evaluate the efficiency of manual overhead irrigation in citrus rootstock liner production. *Eng. Agric.*, **36**: 724-35.

Shyam S, Shivanker VJ, Shrivastava AK and Singh IP, 2004. Advances in citriculture. *Citrus Propagation; Jagmander Book Agency: Delhi, India*, pp. 178-205.

Singh KK, Krishan C and Singh K, 2018. Propagation of citrus species through cutting: A review. *J. Med. Plants Stud.*, **6**(1): 167-72.

Srivastava AK, 2012. Integrated nutrient management in citrus. *Advances in citrus nutrition*, pp. 369-89.

Williamson JG, 1997. Forcing method and time of rootstock shoot removal affect growth of containerized citrus nursery trees. *Hort. Technology*, **7**(3): 244-46.

Chapter 13

Nursery Production, Orchard and Canopy Management in Sub-temperate Fruits

Harpreet Kaur, Mandeep Singh Gill and Nav Prem Singh

Department of Fruit Science, Punjab Agricultural University, Ludhiana–141004, Punjab

India dominates the world's production of both fruits (73.9 MT) and vegetables (139.8 MT) (Saxena *et al.*, 2021). However, the country's productivity of most fruits and vegetables is less than that of leading nations. The lack of good planting material and the prevalence of aged and deteriorating orchards inhibit the productivity of fruits (Sharma and Jindal, 2018). Sub-temperate regions are characterized by mild winters and warm summers, which causes many challenges for fruit growers. The cultivation of sub-temperate fruit crops, such as apples, pears, cherries, plums and apricots, requires a proper understanding of various horticultural practices. Among these practices, the production of nurseries, orchard establishment, and canopy management are critical for successful fruit production and long-term orchard sustainability. These interrelated aspects impact the tree vigour, precocity, yield, fruit quality and overall orchard profitability (Bhat *et al.*, 2021). Nursery production is one of the significant steps in establishing orchards, where suitable rootstock and scion cultivars are selected and propagated. This step should be planned correctly by adopting scientific techniques. The plant material for various fruit crops has to be raised efficiently, which needs skilled knowledge. The availability of quality planting material is one of the most essential requirements for increasing the production and productivity of any fruit crop. Rootstock selection is an important decision, as it influences the architecture and vigour of the tree, precocity, fruit size, and resistance to biotic and abiotic stresses (Fazio *et al.,* 2015;Musacchi *et al.,* 2021). Various propagation techniques, such as budding, grafting, and layering (Hartmann *et al.,* 2011), allow the combination of desirable scion cultivars with compatible rootstocks, which enables the transfer of desired traits to the nursery stock. Nursery management practices like irrigation, fertilization, pest and disease management, and pruning are crucial in producing high-quality nursery plants in containerized or field-grown systems. These practices ensure the plants are well-equipped for successful orchard establishment and subsequent growth and development.

Once high-quality nursery stock is obtained, the focus of the fruit grower shifts to the establishment of the orchard. Many factors, such as soil type, drainage, the slope of the land, and exposure to light and wind, should be kept in mind while deciding the site for orchard establishment, as they directly affect the growth and fruit quality (Musacchi, 2021). Sub-temperate fruits generally thrive in well-drained, fertile soils and moderate climates. Proper planting techniques, including spacing, depth, and orientation, are critical for successful establishment and future growth (Ferree and Warrington, 2003). Newly planted orchards require adequate irrigation and fertilization to promote root establishment and early vegetative growth, with drip irrigation systems being a common choice for efficient water delivery (Mushtaq *et al.,* 2022).

Canopy management becomes the focal point for optimizing productivity and fruit quality. Various training systems, such as central leader, modified central leader, or vertical axe, shape the canopy, ensuring optimal light interception, air circulation, and fruit development (Musacchi, 2021; Fazio, 2015). Regular pruning practices, including dormant and summer pruning, are essential for maintaining the desired canopy structure, removing unwanted growth, and promoting fruit bud formation. Through chemical or manual methods, proper fruit thinning is necessary to reduce crop load and improve fruit size, quality, and return bloom (Byers and Costa, 2002). This chapter aims to comprehensively understand nursery production, orchard establishment, and canopy management practices for sub-temperate fruit crops.

tIn a nursery, seedlings, saplings, or other planting materials are raised and sold in gardens and orchards. The main component of an ideal nursery may be a primary nursery, which is the place where seeds are sown and maintained for the production of rootstocks (Nieuwenhuis *et al.,* 2000). A secondary nursery is a section where scion cultivars are propagated onto rootstocks through techniques like budding, grafting, or layering. A containerized nursery is a type where containerized planting material is raised, facilitating easy handling and transportation (Nieuwenhuis *et al.,* 2000). An insect-proof and disease-free nursery is designed to produce planting material free from insect pests and diseases, ensuring healthy and vigorous plants

(Panth*et al* 2020). To deliver quality planting material for sub-temperate fruit crops, the following nursery production practices should be implemented:

1. Selecting disease-free, true-to-type and high-yielding mother plants is crucial for obtaining quality scion wood and rootstocks.
2. Appropriate propagation techniques, such as budding, grafting, or layering, should follow recommended procedures and use disinfected tools and propagation.
3. The use of well-drained, fertile and disease-free soil or soilless growing media is essential for raising healthy nursery plants, and proper sterilization or fumigation of soil and growing media should be carried out to eliminate soil-borne pathogens and nematodes (Nieuwenhuis *et al.*, 2000).
4. Adequate irrigation and balanced fertilization are necessary for the optimal growth and development of nursery plants.
5. Fertigation techniques involving the application of water-soluble fertilizers through irrigation systems can ensure precise nutrient delivery (Srivastava and Singh, 2020)
6. Implementing an integrated pest and disease management (IPDM) strategy is crucial for maintaining the health and quality of nursery plants, which includes regular monitoring, cultural practices, biological control methods, and judicious use of eco-friendly pesticides (Panth*et al* 2020)
7. Maintaining optimal environmental conditions, such as temperature, humidity, and light levels, is essential for nursery plant growth, and the use of shade nets, greenhouses, or protected structures may be required to regulate these conditions (Nieuwenhuis *et al.*, 2000)
8. Proper transportation and handling techniques should be employed to minimize stress and ensure the viability of nursery plants during transportation from the nursery to the planting site (Nieuwenhuis *et al.*, 2000). By implementing these nursery production practices and adhering to the nursery registration and certification guidelines, growers can ensure the availability of high-quality, disease-free and true-to-type planting material for establishing productive and sustainable sub-temperate fruit orchards.

Layout of Nursery Model

Layout is the arrangement of different critical elements of a nursery, including the paths, buildings, beds, irrigation channels, *etc.* It is prepared to use the resources efficiently. By the specifications, a location-specific model for the nursery setup should be created. While designing and organizing nurseries, there are a few points that should be kept in mind, which are as follows:

Fencing

Before starting a nursery, the area surrounding it has to be well-fenced with barbed wire to prevent animals from trespassing and from being stolen. Adding a living hedge with prickly fruit trees may strengthen the barrier even further.

Roads and Paths

Besides adding beauty, well-planned roadways and trails inside the nursery will facilitate simple, cost-effective operations. This could be achieved by creating distinct parts and blocks within the nursery. However, there shouldn't be any needless road and walkway construction that wastes land. Every trail or road should take customers to a location of interest inside the nursery.

Progeny Block/Mother Plant Block

The nursery should have a mother plant block or scion bank of commercial and recommended varieties and well-kept progeny orchards. When starting progeny orchards, the mother plants should ideally come from a reputable nursery or the original research facility from where they are released. It is commonly known that the initial selection of mother or progeny plants for additional multiplication is crucial to the success of any nursery. Any errors committed in this area will damage the reputation of the nursery. Fresh scion material is available throughout the year within the nursery, and there is no waiting period between scion separation and grafting/budding. A well-managed progeny block or mother plants block will inspire customer confidence, lower production costs, and increase grafting, budding, and layering success rates.

Irrigation System

An ample quantity of water is necessary for the watering of nursery plants. Therefore, to achieve more significant development and success in the production of planting material, preparation for guaranteed irrigation facilities must be arranged well in advance. In areas with low water tables and frequent power outages, water storage or rainwater collecting tanks should be built to supply nursery plants with life-saving irrigation. Since water scarcity is a limiting factor in most of the areas in the state, a well-laid out PVC pipeline system will solve the problem to a greater extent. A qualified agricultural engineer might be contacted to arrange the pipeline in this case. This makes it easier to distribute irrigation water to the various blocks of the nursery economically and effectively and also lowers the seepage losses.

Office-cum-Store

An office-cum-store is essential for effective management of the nursery. The office building should be constructed where both adequate supervision and customer access are ensured. The office building may be decorated with attractive photographs and essential characteristics of fruit varieties propagated in the nursery. A spacious store room is required to accommodate polybags, tools, implements, packaging material, labels, pesticides, fertilizers, *etc.* There should also be a provision for a separate sales counter.

Seedbeds

This component is vital for raising the seedlings and rootstocks in a nursery. Since they require frequent irrigation, these must be laid out near the water source. Raise the beds 15 cm from the ground level to ensure adequate drainage. These have to be placed in an open area with enough sunshine. Beds of a 1-meter width of any desired length should be constructed. There must be a 60-cm working gap between each bed. This makes it easier to plant seeds, pull weeds, water, spray, and raise seedlings. The layout of irrigation canals should be practical. An alternative for irrigating the beds would be to install a sprinkler irrigation system, which promotes greater seedling development and consistent germination.

Nursery Beds

After being removed from the seed beds, seedlings are moved onto the nursery beds. Nursery beds are required for establishing rootstock seedlings for grafting/ budding, planting cuttings, and layered plants. A sprinkler system can also be used to irrigate nursery beds.

Propagation Structures

Adequate space should be set out for contemporary propagation structures, such as net homes, mist chambers, and polyhouses. These structures offer the ideal conditions for cuttings to root, germinating seeds, and plants to harden off before planting in the field.

Potting Mixture and Potting Yard

A high-quality potting mixture is essential for the successful growth of nursery plants. Different mixes of well-rotted farm yard manure, leaf mould, and fertile soil can be combined to create potting mixtures for various uses. Adding enough super phosphate to the potting mixture will decompose and solubilise more readily. The potting yard, where the potting is done, is a good place to store the potting mixture. Even on a wet day, grafting/budding procedures or seedling potting are made easier with constructing a potting yard of an appropriate size.

Packing Yard

Plant material is packed in a packing yard before sale or delivery to outlying locations. We can combine the working shed with the yard. There should be enough room for many employees to sift and bundle the plants carefully. Name, location, and plant variety description should all be appropriately marked on the packaged bundles. The selling counter should be close to the packing yard.

Compost Unit

The nursery needs a massive amount of organic manure to produce good planting material. Consequently, it is best to build vermi-compost pits so that garbage and weeds may be used at the nursery site itself to produce organic manure. This will lower the cost of purchasing manures. It ought to be built close to the potting shed.

Nursery Production: Practices for Quality Planting Material

Implementing best practices in nursery production ensures the availability of high-quality planting material for sub-temperate fruit crops. This includes selecting disease-free mother plants, employing appropriate propagation techniques, and maintaining optimal environmental conditions in nurseries (Srivastava and Singh, 2020).

Rootstock Preparation

Rootstocks are pivotal in determining the propagated plants' overall vigour and growth characteristics. Selection of suitable rootstocks involves factors like compatibility with the scion and adaptability to soil conditions (Graukee*et al* 2003). Rootstocks are prepared by selecting healthy and vigorous rootstock plants. Proper care is taken to ensure that rootstocks are free from pests and diseases, as these can adversely affect plant growth (Tsering*et al* 2018).

Hardening Off and Transplanting

Before transplanting to the field, propagated plants undergo a hardening-off period to acclimatize to outdoor conditions. Gradual exposure to natural light and temperature variations prepares the plants for successful transplanting (Nazir*et al* 2023). Once propagated plants have developed sufficient root and shoot growth, they can transplant to the field orchard. Proper planting techniques and spacing are crucial for optimal growth and productivity.

Maintenance and Care

After transplanting, continued monitoring and care are essential to promote healthy growth and establishment. Adequate irrigation, fertilization, and pest management practices are implemented to support plant growth and development.

Irrigation and Fertigation

Maintaining optimal soil moisture levels is vital for the healthy growth of nursery plants. Drip or micro-sprinkler irrigation systems efficiently deliver water directly to the root zone, minimizing water wastage and promoting uniform moisture distribution (Kabashima *et al.*, 1993). Fertigation, applying water-soluble fertilizers through the irrigation system, ensures precise nutrient delivery to the plants, promoting vigorous growth and development (Ahmed *et al.*, 2022).

Integrated Nutrient Management

An integrated approach to nutrient management is essential for maintaining soil fertility and plant health. Incorporating organic and inorganic nutrient sources optimizes nutrient availability and enhances soil structure. Organic amendments such as farmyard manure, vermicompost, and biofertilizers improve soil health by improving its physical, chemical, and biological properties.

Integrated Pest and Disease Management

Implementing an integrated pest and disease management (IPM) strategy is crucial for minimizing pest and disease damage while reducing reliance on chemical pesticides. This holistic approach involves regular monitoring of pest and disease populations, employing cultural practices, biological control agents, and selective pesticide applications only when necessary (Grauke *et al.*, 2003). Bio-control agents like Trichoderma and Bacillus species effectively control soil-borne diseases, reducing the need for synthetic chemicals.

Environmental Control

Creating optimal environmental conditions within the nursery is essential for maximizing plant growth and minimizing stress factors. Shade nets, greenhouses, or protected structures help regulate temperature, humidity, and light levels, creating a favourable microclimate for plant growth (Davis *et al.*, 2021). Adequate ventilation and sanitation measures are also essential to prevent the build-up of pests and diseases, ensuring the overall health of nursery plants.

Propagation Techniques and Nursery Production in Sub-temperate Fruit Crops

Pear

In sub-tropical regions, the key to successful propagation of Pear lies in the choice of rootstock, particularly Kainth (*Pyrus pashia*). This rootstock, known for its non-production of root suckers and reduction of precocity, provides a strong foundation for the plant's growth.

Collection of the Seeds of Kainth (*Pyrus pashia*)

Fully ripe fruits are harvested from the end of September to the first week of October. In December, the seeds from the softened fruits are layered to promote germination. They are kept in wooden boxes with alternating layers of moist sand for 30 days. During January, the stratified seeds are sown in a nursery bed. Within ten days, the seeds begin to germinate. When the seedlings become one month old, they should be placed in the nursery at a spacing of 10-60 cm. During the following January, these seedlings will become graftable.

Rootstock Raising for Kainth without Stratification

Seeds collected from ripe fruits can be sown in the last week of September or the first week of October. Freshly removed seeds should be placed in wooden boxes with damp sand for germination. The seeds will germinate within 10-12 days. When these seedlings reach the stage of two to four leaves, they are planted in the field in lines 10 cm apart, with 60 cm space after every four lines. These seedlings can be grafted in December or January, or they can become buddable in May or June.

The pears are tongue-grafted or T-budded onto Kainth seedlings. The best time for tongue grafting in Pear is December-January. After this, pears should be planted in winter, when the plants remain dormant. Planting should be finished by January.

Peach

Peach is commonly propagated by budding and grafting on Sharbati and Flordaguard rootstock. Flordaguard rootstock is resistant to root-knot nematodes, and fruits on this rootstock will ripen from the end of June to the first week of July.

Raising of Rootstock

Harvest ripe fruits of Flordaguard/Sharbati in June and July. Stones should be removed from the pulp, cleaned, and dried in the shade for four to five days. For each quintal of stones, apply 300g of Ziram, Thiram, or Captan on the stones. The stones should be packed in dry gunny sacks or plastic bags and kept in a cold, dry area until they are needed for stratification in November through January. Stratification is a low-temperature treatment in which stones are layered alternately in damp sand at 7.2°C for 100–120 days or until the seed dormancy is broken and germination begins. The seedlings can also be raised by using kernels or seeds in place of stones. Stratification can be done in wooden boxes or open ditches with good drainage. The dormancy of the seed is said to be broken when germination occurs in the top layer. The radicle or plumule of seeds that are sprouting is not harmed in the process of removing stones or seeds from each layer of sand. Keep the uncracked and cracked stones apart from the germinating stones, and plant each one in a nursery plot that has been carefully prepared. These are planted in rows at a distance of 15 cm, spaced 30 cm apart. Seeds are generally sown during the first two weeks of February. In the first week of May, around 40 per cent of the seedlings from the nursery plots are ready for T-budding. In December and January, the successful budded plants are prepared for transplantation, while the unsuccessful budded seedlings are used for grafting in December and January. The scions produce new growth in mid-February. In general, a scion may produce three to four shoots that can continue to grow until April. One of the vigorous shoots is kept in May, while the rest are chopped off. This method produces vigorous nursery plants. It is best to remove the polythene tape surrounding the graft union before it causes girdling. Young grafts must be staked to avoid breaking.

Peach can also be propagated by stenting technique, *i.e.* simultaneously grafting and rooting. The basal portion of cutting of Sharbati rootstock is dipped in 2000 ppm IBA solution for 2 minutes during the first week of January. This method shortened the propagation period by one year.

Figure 13.1: Cordon System in Pear.

Figure 13.2: Espalier System in Pear.

Figure 13.3: Espalier System in Plum.

Figure 13.4: Y System in Peach.

Plum

Kala Amritsari and Sutlej Purple are mainly cultivated plum varieties under sub-tropical regions. Kala Amritsari is propagated by hardwood cuttings, while Satluj Purple is by simultaneous grafting and rooting of rootstock cuttings. Cuttings and grafting are prepared between the end of December and the end of January. The ideal length for hardwood cuttings of Kala Amritsari should be 15-20 cm, with 4-5 buds per cutting and pencil thickness. In order to promote rooting success, the basal parts of the cutting, which are 4-5 cm long, are immersed in a 100 ppm IBA solution for 24 hours.

Satluj purple is propagated by the stenting technique. Rootstock is made from the cuttings of Kabul Green Gaze. During the first fortnight of January, scion wood is tongue-grafted onto rootstock cuttings that are around 20 cm long. The basal portion of the grafts is immersed in a 100 ppm (100 mg/litre) IBA solution for a whole day. The cuttings or grafts are planted in nursery plots that have been carefully prepared, with rows spaced 30 cm apart and a distance of 15-20 cm. After a year, in January, the cutting/graft plants become ready for planting.

Orchard and Canopy Management in Temperate Fruit Crops

Establishing a fruit orchard is a long-term venture, and any early-stage mistake or inattention can adversely affect the productivity and proceeds. Planning and site preparation are prerequisites for successfully establishing a fruit orchard. This is paramount, but researchers and experienced orchardists alike often ignore this aspect. Cultural practices encompass a range of activities aimed at optimizing growing conditions, enhancing fruit quality, and minimizing pest and disease pressure. This comprehensive guide delves into five vital cultural practices essential for successful orchard management.

Site Selection and Soil Preparation

Choosing a suitable site with well-drained soil and appropriate climatic conditions is the foundation for successful sub-temperate fruit production. Factors such as elevation, slope, aspect, and air drainage should be considered to minimize the risk of frost damage and optimize microclimate conditions. Soil analysis and appropriate soil amendments, including organic matter and pH adjustment, are necessary to create an ideal growing environment for fruit trees.

Variety Selection

Selecting fruit varieties that are well-adapted to the local sub-temperate climate is critical. Varieties should be chosen based on their chilling requirements, disease resistance, and market demand. The following varieties are recommended in sub-temperate regions:

Fruit	*Varieties*
Pear	**Hard Pear**- Punjab Nakh, Patharnakh **Semi Soft Pear**- Punjab Gold, Punjab Nectar, Punjab Beauty. Baggugosha **Hard Pear**- Punjab Nakh, Patharnakh
Peach	**Yellow Fleshed-** Earli Grande, Florida Prince, Partap, Shan-i-Punjab **White Fleshed**- Prabhat, Sharbati **Nectarine**- Punjab Nectarine
Plum	Satluj Purple and Kala Amritsari

Canopy Management

Proper canopy management in fruit trees is crucial for maximizing yield and improving quality through shaping and size control. The goal in managing the canopy of a perennial tree is to maximize sunlight exposure and utilize the land and environmental factors for increased production. Canopy management aims to minimize extreme shading. Rigid canopy management is necessary to increase light penetration and productivity since almost all fruit crops in temperate and tropical regions receive less than 70 per cent of the radiated light that reaches their canopies (Whiley *et al.*, 2013). Effectively managing the tree canopy, particularly in controlling its size, is now crucial for minimizing production costs and boosting fruit crop yield and quality. The canopy design and shape significantly influence the orchard's light interception and economic returns. The production and quality of fruits heavily depend on tree vigour, light, temperature and humidity. Effective canopy management enhances production and improves fruit quality by optimizing tree vigour and maximising available temperature and sunlight.

Canopy Management in Sub-Temperate Fruit Crops

Training: Structuring a Plant

Training techniques for temperate fruit crops play a crucial role in shaping tree architecture, optimizing light interception, and maximizing fruit production in sub-temperate regions. Various training systems are employed based on climate, soil conditions, orchard layout, and grower preferences. Here, we explore the training methods commonly used in sub-temperate areas and their implications for orchard management.

Modified Central Leader System

This method is commonly used for canopy management in sub-temperate fruit crops (Anonymous 2021). It combines elements of a central leader with an open system. It works similarly to a central leader by allowing the stem to grow for the first two years and then head back at a height of 75 cm. With this method, the tree's height also decreases as the main trunk's length decreases. The scaffold branches are encouraged to grow longer and bigger. The scaffold branches should expand and elongate. The trees have robust crotches and a sturdy structure. The system is ideal for commercial fruit trees due to their relatively short height, enabling efficient spraying, pruning, harvesting, and other operations.

Training and Pruning in Pear, Peach and Plum

Trees should be trained using the modified leader system to create a strong scaffold system. At the time of planting, the plants should be headed back to a height of 90 cm. The buds start sprouting in February or March. No branch should grow higher than 45 cm above the ground. The top shoot in the centre is retained as the leader for the second year, but it is trimmed down where the immature section begins. Choosing three to five laterals that are appropriately spaced 15 to 20 cm apart and oriented in various directions is recommended. It is beneficial to bend these laterals downward by tying them to the ground with thick thread or rope. The growth of secondaries will be promoted by tying these laterals. The laterals that remain at the conclusion of the second growing season should be headed back to encourage the emergence of buds that will eventually develop into tertiary branches.

High Density Planting in Sub-Temperate Fruit Crops

The pear cv. Patharnakh is planted at an 8 m x 4 m distance, and the trees are trained by bending the main and two secondary scaffolds. This also contributes to a two-year reduction in juvenility. Punjab Agricultural University, Ludhiana, recommends planting peach trees using a "Y" training method at a 6 m × 1.5 m distance. This technique ensures a higher yield and better-quality fruit than conventional planting. Satluj Purple plum trees can also be planted with summer pinching (10–15 cm) and a modified leader system at a distance of 6×1.5 m. This method ensures a large production of fruits with higher quality.

Pruning: Shaping Trees for Optimal Growth

Pruning is the process of removing any vegetative plant part in order to regulate growth and keep the balance between vegetative and reproductive development. Pruning is a science and an art: It takes art to prune properly and science to know when and how to prune for the best results. Proper pruning techniques aim to remove dead or diseased wood, improve light penetration and airflow within the canopy, and shape trees to facilitate harvesting and maintenance operations. To correctly prune fruit plants, it is essential to have a thorough understanding of their fruiting habits.

The pear trees produce spurs, and they last for around eight years. If the spurs are over-shaded, they will stop bearing fruit or perhaps die. A spur that is lost on any part of the tree cannot be replaced. Therefore, pear trees do not require annual pruning. In the case of old and bearing trees, January is the best time to prune the dead wood and old non-bearing spurs. Thinning out and heading back of laterals can also promote the formation of more fruiting spurs.

Peach and nectarines bear fruit laterally on a year-old shoot. It is thus necessary to promote the growth of these branches via pruning. The best month to prune peaches is January. The best way to prune is to remove some of the one-year-old branches expected to bear fruit. It is necessary to head back some old and improperly positioned branches and the sterile parts of flower bud-bearing shoots. It is recommended that around 40 per cent of one-year-old shoots be thinned down to promote healthy tree development and enhance the quantity and quality of the fruit. Lengthy and hanging branches should also be shortened by heading back.

Plum trees also bear fruit on one-year-old growth and short spurs, depending upon the variety. Light yearly pruning should usually be performed in January. It consists of removing slender, crowded branches and twigs from the tree. To improve the colour of the fruits by allowing sunlight to reach the tree's centre, branches that are crisscrossed, long, or out of position should be cut down. Additionally, water sprouts or suckers that emerge from the crown of the tree should also be pruned regularly. After every 4 or 5 years of fruiting, heavy pruning of the trees is done by heading back the lengthy branches about half of their length. This involves cutting back the long branches by around half of their length. Cover any incisions larger than 4–5 cm with Bordeaux paste.

Thinning: Enhancing Fruit Quality and Tree Health

Thinning is a crucial orchard management practice aimed at reducing fruit load to optimize fruit size, improve colouration, and prevent branch damage caused by overbearing (Costa *et al.,* 2018). By thinning excess fruit, growers ensure the remaining fruits receive adequate resources for proper development and maturation.Thinning can be achieved through hand thinning or chemical thinning, depending on orchard size, labour availability, and fruit type (Radivojevic*et al.,* 2011). Hand thinning involves manually removing excess fruit clusters, while chemical thinning utilizes plant growth regulators to induce fruit drop and thin the crop load. Proper thinning promotes even fruit distribution, reduces biennial bearing tendencies, and enhances overall fruit quality attributes such as sweetness, flavour, and marketability (Ahmed *et al.,* 2022).

Irrigation and Nutrient Management

Under sub-tropical regions, the fruit development period of peaches starts after the fruit set in March and continues from April to June till maturity, depending upon the variety. This is the crucial time for irrigation of the trees. Trees should not experience any moisture stress, especially in 25-30 days before fruit maturity, because the maximum weight gain occurs during these days. The irrigation scheduling depends upon the type of soil and the irrigation source. It should be given weekly for the first three to four weeks after the fruit set. After that, irrigation should be given at 3-4 days intervals from the second week of April to the start of harvesting. Being a deciduous tree, plums can be irrigated infrequently throughout the winter. Pear trees need frequent irrigation after transplanting. During the summer, the irrigation interval may be 5 to 7 days; in July and August, it may be 15 days. Irrigation should be restricted after the harvesting of fruit. No irrigation should be given in January. Bearing trees should be irrigated through the flood method during summer months at 5 to 7 days so that fruit size may be increased. Efficient irrigation systems, such as drip or micro-sprinklers, help optimize water use and reduce water waste. Soil moisture monitoring and scheduling irrigation based on crop water requirements are recommended for efficient water management.

Pest and Disease Management: Safeguarding Orchard Health

Pest and disease management is critical for protecting fruit trees from damage and ensuring sustainable orchard productivity. Integrated pest management (IPM) approaches combine chemical, biological and cultural control methods to minimize pest populations and environmental impacts (Grauke *et al.,* 2003). IPM strategies involve regularly monitoring pest populations, implementing cultural practices to reduce pest habitat and food sources, and deploying biological control agents such as predatory insects, parasitic wasps, and microbial pathogens (Tworkoski *et al.,* 2007). Based on economic thresholds and pest life cycles, selective pesticide applications are used as a last resort to control pest outbreaks and prevent crop losses. Disease management focuses on sanitation, proper pruning to improve air circulation, and timely application of fungicides to prevent fungal infections and reduce disease pressure (Ahmed *et al.,* 2022).

Weed Control: Maintaining Orchard Cleanliness

Effective weed control is essential for decreasing the competition for light, water, and nutrients and preventing weed-mediated pest and disease transmission in orchards (Graukee*t al.,* 2003). Cultural weed management practices aim to suppress weed growth while improving soil health and biodiversity. Cultural weed control methods include mulching, mowing, cover cropping, and manually removing weeds. Mulching with organic materials such as wood chips, straw, or compost helps suppress weed growth, conserve soil moisture, and enhance soil organic matter content (Costa *et al.,* 2018).

Conclusion

Nursery production, orchard establishment, and canopy management are inextricably linked components that collectively shape the success of sub-temperate fruit production systems. By integrating these practices through a holistic approach that embraces cutting-edge research, innovative technologies, and sustainable practices, growers can navigate the unique challenges of sub-temperate climates and meet the ever-increasing demand for high-quality fruits. This comprehensive strategy optimizes yields and fruit quality. It contributes to long-term orchard sustainability, resource efficiency, and environmental stewardship, ensuring a resilient and profitable sub-temperate fruit industry for generations.

REFERENCES

Ahmad S, Malik AR, Kirmani S, Shah MA, Ganie MA and Mushtaq R, 2022. Orchard Planning, Establishment and Soil Management. In Apples (pp. 9-24). CRC Press.

Ahmad S, Ashraf I and Anjum MA, 2017. Fruit and vegetable nurseries: Establishment and management. Book Chapter: *Horticulture: Science and Technology*. Publisher: University of Agriculture, Faisalabad, 133-59.

Anonymous, 2021. *Package of Practices for Fruit Crops.* Punjab Agricultural University, Ludhiana.

Bhat R, Bhat KM, Hussain S, Mir MM, Iqbal U and Bashir M, 2021. Canopy Management in Stone Fruits. Production Technology of Stone Fruits, 103-29.

Byers RE, Costa G and Vizzotto G, 2002. Flower and fruit thinning of peach and other Prunus. *Hortic Rev* **28**: 351-92.

Costa G, Botton A and Vizzotto G, 2018. Fruit thinning: Advances and trends. *Hortic Reviews,* **46**: 185-26.

Fazio G, Robinson TL and Aldwinckle HS, 2015. The Geneva apple rootstock breeding program. *Plant Breeding Reviews*, **39**: 379-24.

Ferree DC and Warrington IJ, (Eds.). 2003. Apples: botany, production, and uses. CABI Publishing.

Grauke LJ and Thompson TE, 2003. Rootstock development in temperate nut crops.

Hartmann HT, Kester DE, Davies FT and Geneve RL, 2011. Hartmann and Kester's plant propagation: principles and practices (8th ed.). Prentice Hall.

Kabashima JN, 1993. Innovative irrigation techniques in nursery production to reduce water usage. *Hort Science*, **28**: 291-93.

Lea-Cox JD, Ross DS and Teffeau KM, 2002. Developing water and nutrient management plans for container nursery and greenhouse production systems. In XXVI International Horticultural Congress: Protected Cultivation 2002: In Search of Structures, Systems and Plant Materials, **633**: 373-79.

Musacchi S, Iglesias I and Neri D, 2021. Training systems and sustainable orchard management for European pear (*Pyruscommunis* L.) in the Mediterranean area: *A review. Agronomy*, **11**: 1765.

Mushtaq R, Majid I, Din S, Malik AR, Sharma MK, Ahmad S and Been TJ, 2022. Orchard Cultural Practices: Irrigation, Fertigation and Weed Management. *In Apples*. **3:** 119-31.

Nazir N, Yaseen I, Been TJ, Khalil A, Iqbal U, Mir MM, Rehman MU, Banday SA, Malik AR and Bhat S, 2023. Improved Propagation Techniques in Temperate Nuts. In Temperate Nuts 113-48 *Singapore: Springer Nature Singapore.*

Nieuwenhuis M and O Connor N, 2000. Challenges and opportunities for small-scale tree nurseries in the East African highlands. UNASYLVA-FAO. 56-60.

Panth M, 2020. Sustainable Approaches to Managing Soil-borne Diseases in Nursery Production (Master's thesis, Tennessee State University).

Radivojeviæ D, Zabrkiæ G, Milivojeviæ J, Velièkoviæ M and Oparnica È, 2011. Effect of chemical and hand thinning young apple tree on yield and fruit quality. 6th International Symposium on Agriculture. Opatija. Croatia (1044-47)

Tworkoski T and Miller S, 2007. Rootstock effect on the growth of apple scions with different growth habits. *SciHorti* **111**: 335-43.

TseringDolkar MA, Banoo A, Slathia D, Stanzen L and Lakdan S, 2018. Mitigation of temperate fruit crop problems through use of rootstock. *Int. J. Communi. Sys*. **6**: 880-87.

VerCauteren KC, Lavelle MJ and Hygnstrom S, 2006. From the field: fences and deer–damage management: a review of designs and efficacy. Wildlife Society Bulletin, **34**: 191-200.

Chapter 14

Protected Nursery Production of Citrus Plants

Uday Raj Patial and Anirudh Thakur

Department of Fruit Science, Punjab Agricultural University, Ludhiana–141004, Punjab

Globally, India is the third largest producer (14.3 million tons) of citrus fruits after China and Brazil (FAO, 2021). Citrus is the second important fruit crop in India after mango with an area of 1.09 million ha (NHB 2022). The area under citrus has expanded 2.89 times in the last two decades. One of the most major constraints in citrus production is non-availability of quality planting material. There is a huge gap between the demand and supply of quality citrus planting material. In citrus, rough lemon (*Citrus jambhiri* L.) is the best rootstock having deep root system and resistance to citrus tristeza virus, citrus xyloporosis virus and citrus exocortis viroid, however, it is vulnerable to *Phytophthora* (Cao *et al.*, 2013). The production cycle of citrus plants is around two years which involves growing of rootstock plants for one year and subsequent maintenance of budded plants for another one year. There is an urgent need for the fast multiplication of citrus nursery plants on large scale to meet the huge demand of quality planting material. The citrus plants are nowadays produced with solarized media in containerized system to prevent the spread of *Phytophthora* root rot from nurseries.

In India, citrus is cultivated in 26 States, however, only 9 states cover more than 70 per cent of the area and accounts for more than 85 per cent of total production. There is a huge demand for quality planting material. One of the most important constraints of citrus production is non-availability of quality planting material. Pedigreed true to type disease free planting material is the basis of the fruit industry. However, still the planting material available to farmers is often of insufficient quality. In India, there are more than 4,409 fruit nurseries including 1,575 under government sector and 2, 834 under private sector. These nurseries are producing around 1, 387 million fruit plants annually. However, there are only 140 fruit nurseries having NHB rating 3 or above.

Citrus nursery can be is a good source of income for small farmers. However, keeping in view the presence of bud transmissible and soil borne diseases in citrus containerized nursery under protected condition is the most viable option. Citrus is infested with bud transmissible diseases such as trizteza, exocortis, and ring spot virus; and greening bacterium. The rough lemon rootstock is resistant to citrus tristeza and exocortis hence, greening and ring spot are the major bud transmissible diseases which concern nurserymen in India. Citrus greening is transmitted by an insect 'Asian citrus psylla' hence, propagation of citrus plants must be done on a healthy site and in structures specially designed to exclude citrus psylla. In Florida State of USA, due to the threat of bud transmissible diseases like citrus greening, propagation of citrus plants outside protective structures is prohibited since 1st January, 2007. Fruit crop propagation is also hampered by weather vagaries like high and low temperature; drought, frost and hot winds. Protected nursery production provides protection to the nursery plants from these extreme weather conditions and gives additional benefit for off-season propagation. Screen net houses protect the mother plants from insect vectors. The shade net houses (50-75% shade) offer advantages like possibility of raising the plants in containers, enhancing the grafting or budding duration, protecting the buddlings and grafts from extreme hot and dry weather. Smaller nursery area is required for containerized nursery production and there is no need to change nursery sites every year. Use of sterile growing medium in containers eliminates the spread of soil-borne diseases, insects and nematodes. The article discusses a complete package for the containerized nursery production of citrus plants under protected structures.

Mother Plants and Rootstock Block

The citrus nurseries should be away (>1.6 km for citrus) from commercial orchards. Mother blocks of different varieties of citrus should be maintained in a healthy condition and free from bud transmittable diseases like greening and ring spot. The mother plants should be planted at higher density to accommodate greater number of plants and fruiting should not be taken from them to get continuous supply of bud wood. To protect from vector transmitted and bud transmissible diseases the mother block should be grown inside insect proof screen net houses 40 mesh size. Mother plants should be pruned during the pruning season (winters) to keep them in vegetative phase and to enhance the bud supply capacity of the plants. The mother plants should be monitored regularly for the presence of major viruses and virus-like agents such as citrus tristeza virus, ring spot virus, citrus greening and exocortis. This can be is done by biological indexing using indicator plants like acid lime for tristeza virus, sweet orange for citrus greening bacterium and ring spot virus and Etrong citron for exocortis under insect proof-controlled conditions. Serological indexing is also possible by using ELISA methods. Nucleic acid-based assay using DNA extracted from the leaves and amplifying it in PCR with virus specific primers and detection of bands after the gel electrophoresis is the most reliable technique. However, in case of absence of facility for conducting these tests the mother plants showing the symptoms of these bud transmissible diseases should be removed from the mother block and buds should not be taken from such plants. The symptoms of citrus greening and ring spot are most prominent during the coldest months. Under the modern production systems for the fruits, a fruit crop variety is generally budded/grafted on a rootstock which provides the root system to the composite plant. The rootstock influences plant vigour, fruit yield and quality; and resistance to root diseases (Hayat *et al.,* 2022). In India, rough lemon (*Citrus jambhiri* lush) is the major citrus rootstock used for the commercial citrus varieties. Rootstock block should be maintained with plants of recommended rootstocks for continuous supply of seeds or cuttings for raising rootstock.

Preparation and Sterilization of Potting Mixture

In the containerized nursery production system (polybags, root trainers, plug trays *etc.*) potting mixture is used by following solarization and fumigation (with Basamid 50g per m^2) of potting mixture on a concreate floor to prevent the spread of soil borne diseases. The potting mixture, consisting two parts of soil, one part of farmyard manure and one part cocopeat should be prepared on a concrete floor and spread as a flat bed of 45 cm height. These beds should be drenched with water and covered with 100 µ UV stabilized transparent polythene sheet during the last week of April to May for 4-6 weeks. The solarized soil is further fumigated with Basamid Granular (Dazomet 98%) @ 50 g/m^2 and completely covered with 100 µ UV stabilized transparent polythene sheets during May-June. After three weeks, the polythene sheet should be removed and the mixture should be given atleast three turnings at an interval of one week to remove the fumigant.

Selection of Rootstock

There is a vast diversity of rootstocks used for raising citrus around the world. However, the choice of rootstock depends on certain factors such as the scion cultivar, soil type and environmental conditions, with each rootstock best suited to specific niches. Some of the most commonly used citrus rootstocks in India are:

Rough Lemon

It is the most widely used rootstock under Indian conditions. It offers a high yielding potential with a wider adaptability. It performs good even in alkaline soil conditions with pH higher than 8.0. Most of the Kinnow, Daisy, Jaffa, grapefruit, lime and lemon orchards in North India are raised on Rough lemon. Meanwhile, citrus orchards of Khasi mandarin, sweet oranges, lime, lemon in North-East Indian region are also based on Rough lemon. However, Rough lemon imparts good resistance to tristeza and exocortis virus; and can tolerate saline or drought conditions, but is susceptible to root rot.

Rangpur Lime

It is a major rootstock for mandarins and sweet oranges in central and South India. It offers a good compatibility with high yield potential to the scion cultivars. It imparts good resistant to tristeza virus, can tolerate saline and drought conditions, but is susceptible to exocortis and root rot.

Carrizo Citrange

It is a good option for the raising of Daisy, Kinnow, W. Murcott orchards, in those regions where pH of soil is below 8. Kinnow orchards raised on Carrrizo are observed to be tolerant to root rot and produce better quality fruits, but the yield potential is lower as compared to Rough lemon.

Cleopatra Mandarin

It is recommended for raising sweet orange cv. Blood Red orchards. Cleopatra plants are tolerant to root rot and can resist the exocortis and tristeza viruses. It can tolerate moderate salt levels but could not withstand drought conditions.

Raising of Rootstock and Budding

The rough lemon seeds (2 seeds per bag) should be sown directly in black polythene bags (size 12" x 7", 250 gauge) filled with sterilized potting mixture consisting of FYM + Soil + Cocopeat (2:1:1 v/v) in the second fortnight of August under screen house/polycarbonate roof screen/shade net house. The seeds start germinating in about 3 weeks and when the seedlings become 10 cm high retain only one nucellar seedling per bag by uprooting all other seedlings. Then the seedlings should be sprayed with 1.5 per cent urea at monthly interval. The polythene bags containing seedlings should be shifted in glasshouse/polyhouse during winter (mid-November till end of February) which should again be shifted to screen house/shade net house (50% shade) during first week of March. More than 75 per cent seedlings will become buddable in the first fortnight of May. The budding operation can be started from May onwards under shade net house which will become saleable during September-October and remaining plants will become saleable in next March. The propagation should be performed under strict phytosanitary conditions. Sterilize the pruning tools and budding knives by cleaning the tools with warm, soapy water and then with ethyl alcohol. Use a small bottle of the ethyl alcohol for periodic re-treatment of propagation tools during nursery operations.

TIME LINE FOR PRODUCTION OF CITRUS PLANTS

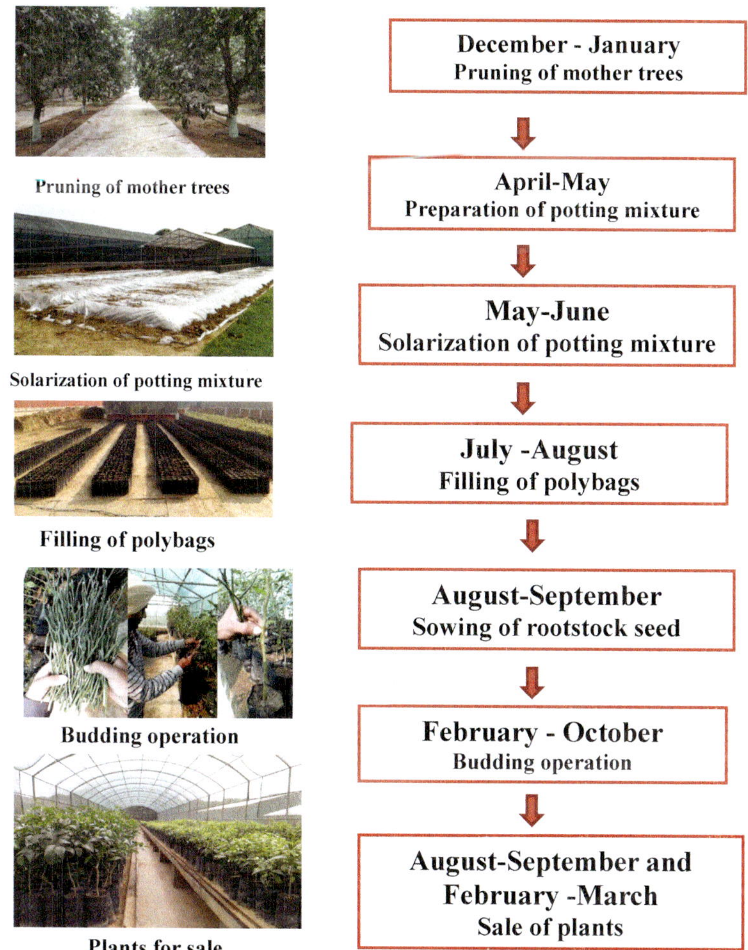

Pruning of mother trees

Solarization of potting mixture

Filling of polybags

Budding operation

Plants for sale

Budding should be performed when the seedlings attain the girth of 0.7 to 0.9 cm at more than 9" height above the soil surface or polybag as it provides ample height for the graft or bud union to be above the soil to prevent it from rotting after the planting of plant in orchard. Bud forcing by bending the rootstock after removal of polythene sheet helps in early sprouting and fast growth of the sapling.

Shoot Tip Grafting

Shoot tip grafting (STG) or micro-grafting is a widely practiced technique in different citrus producing areas around the world. The purpose behind STG is to produce virus-free citrus plants by establishing mother blocks of citrus grafts created through STG, which in turn supply disease-free buds. It offers an advantage over conventional propagation methods (T-budding) by producing large number of healthy, true to type plants within a short period of time. However, plants grown conventionally from the nucellar tissues often result in juvenile plants which are less desirable in terms of growth and are prone to large number of viruses and vectors. Though thermotherapy has been employed for the elimination of viruses, but it is not effective against all viral diseases. Furthermore, citrus meristems are not easily cultivated using standard tissue culture methods. Later, a Spanish scientist Navarro proposed a technique called 'shoot-tip grafting' in 1970's (Navarro, 1981), that could be employed in citrus improvement programs to produce disease-free, true-to-type planting material along with a reduced risk of diseases transfer across borders.

Table 14.1: Standards for Quality Nursery of Fruit Crops (Ministry of Agriculture and Cooperation, Government of India, New Delhi, 2012)

Fruit crop	Citrus, Mango, Guava, Ber, Amla, Pear, Peach and Plum
Method of propagation	Budding/grafting
Plant age	Not less than one year old
Plant height	> 60 cm
Stem diameter	> 0.5 cm
Grafting/budding height	Not below 8 inches
Disease pest incidence	free from pest, fungal

The procedure of micro-grafting begins with preparing rootstocks, by *in-vitro* germination of seeds in a sterile environment. Once the seedlings are ready, they are carefully removed from agar medium and are cut about 1 cm above the cotyledonary node. Aseptic scions are excised, 2 cm below the apex of the terminal shoots from the desired plant, followed by the trimming of leaves and sterilization. Both the rootstock and scion are kept in sterile water until they are grafted. A precise incision is made at the top of the decapitated rootstocks and the shoot tips are inserted, ensuring proper contact with the vascular ring. The plants are placed in a nutrient-rich medium post grafting, to encourage development and once few new leaves emerge, they are transferred to soil for further growth. The entire process is carried out in a sterile environment to prevent contamination. The success rate of STG ranges from 30-50 per cent or even higher sometimes (Juarez *et al.*, 2015).

Micro-budding

Micro-budding is another innovative propagation technique that offers a faster and more efficient alternative to the traditional methods of citrus propagation, such as shield or T-budding. In conventional approaches, 1 – 1.5 years old rootstock seedlings are used for the budding which takes approximate 1.5 – 2 years to become saleable after sowing. Moreover, these traditional approaches are season specific, making it more time-consuming and less flexible. Addressing these limitations, Skaria and Zhang (2000) developed a technique called micro-budding, which was later standardized by Vijayakumari and Singh (2003) for different *Citrus* species like Nagpur mandarin, sweet orange and grapefruit. This method allows year-round production of high-quality, budded seedlings at a lower cost, in a limited time, space and resources. Additionally, micro-budding is highly effective in producing virus-free plants, particularly in combating diseases like citrus tristeza virus.

Protected Nursery

Protected Mother Plants

Protected Nursery Plants

Saleable Plants

The seeds of suitable rootstocks (say, Rough lemon or Rangpur lime) are germinated in nursery and raised till plant reaches 20–25 cm height for around 45 days. Newly emerging buds of uniform size (3-5 mm) are carefully excised from the tender flush of desirable elite mother plants when they are 45-60 days old. The rootstock seedling is beheaded at 15-20 cm height, making a 2–2.5 cm wedge-shaped cut in a downward direction, and delicate scion bud is inserted into it. Now, a micropipette tip is placed around to keep the bud intact and make an optimum microclimate for bud union. However, micro-budding is often confused with micrografting or shoot tip grafting (STG). STG is conducted under *in vitro* conditions with the aim of virus elimination using a very small bud of 0.2– 0.5 mm. Whereas, micro-budding is conducted under *in vivo* conditions (preferably, shade net/screen houses) using a relatively larger size bud of 3 -5 mm.

Bud Wood Certification

The successful propagation of healthy citrus plants needs budwood from vigorous, highly productive, true to type and consistently yielding mother plants. Since, viruses can be transmitted through infected budwood, this poses a serious threat in multiplication of citrus plants. The spread of citrus decline in India is largely attributed to the lack of an effective budwood certification program, leading to widespread in viral infections. A proper budwood certification program involves selecting mother plants based on their genetic authenticity (true-to-type), vigour and productivity. These plants should be at least 10 years old and uniform in age and growth.

During the first flush, the parent trees undergo rigorous inspections for virus symptoms using field indexing methods. Trees which show no disease symptoms for a period of 2-4 years are then registered for further propagation, under the supervision of certified state agencies and growers. In citrus production, virus-free plants are used to establish virus-free mother or foundation blocks, typically housed in insect-proof screen houses. Budwood is collected from these certified blocks, and the plants are continuously monitored for major viruses and virus-like pathogens such as Citrus tristeza virus (CTV), citrus greening, ring spot virus and exocortis to ensure the highest standards of plant health and disease resistance. The budwood certification schedule suggested by CCRI, Nagpur is illustrated below:

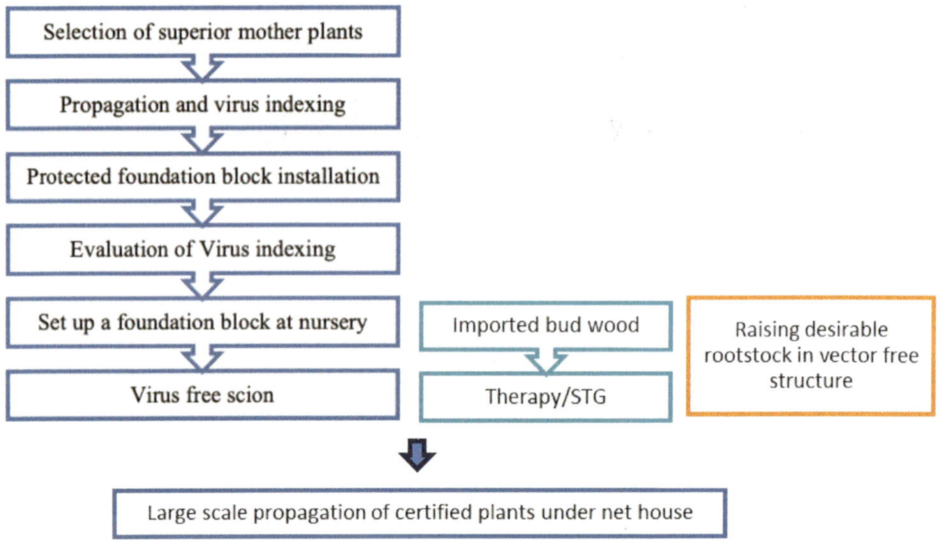

Indexing Virus and Virus like Diseases

Viral diseases were not a major concern in *Citrus* species, before 1940 (Iftikhar *et al.,* 2021). However, over the last 60 years, approximately thirty economically significant viruses and virus-like diseases have been identified as major contributors to citrus decline worldwide. Citrus orchards, which ideally have a lifespan of over 50 years, are now often facing premature decline within just 15 years. This early decline is largely attributed to the presence of graft-transmissible viruses and virus-like diseases, inadequate orchard management and improper nursery practices, making nurseries being the primary source of these problems.

To tackle this issue, citrus nurseries must adopt advanced technical and scientific approaches to provide certified, disease-free plants to growers. Indexing is an essential process practiced for diagnosing and producing disease-free citrus plants. Indexing consists of several techniques which are selected on the basis of the availability of resources. Some of the commonly used indexing methods include tissue grafting, budding, insect transmission for detecting pathogens, as well as more advanced laboratory techniques like ELISA and PCR. The indexing techniques are divided into two classes broadly:

Field/Biological Indexing

Biological indexing involves introduction of an infected plant material into healthy indicator plants to detect and identify the presence of pathogens. The infected scions are grafted/budded onto indicator plants. Alternatively, mechanical inoculation or transmission through insect vectors (psyllids for citrus greening, leafhoppers for prokaryotic diseases and aphids for citrus tristeza virus) is practiced. Though biological indexing is effective, but it is time-consuming, requiring greenhouse facilities and takes 6 to 12 months for results. A minimum of 3-4 indicator plants are needed per treatment to ensure accurate detection of graft-transmissible pathogens. The details of biological indexing for graft-transmissible pathogens, including the use of indicator plants and the exhibited symptoms are summarized as:

Sl.No.	*Disease*	*Indicator Plant*	*Symptoms*	*Reference*
1	Citrus tristeza virus	Mexican lime, Sour orange, Duncan grapefruit	Vein clearing on leaves, leaf cupping, stem pitting	Hancevic *et al.*, 2018
2	Greening	Kagzi lime, sweet orange	Leaf chlorosis, blotches	Katoch *et al.*, 2022
3	Psorosis	Sweet orange, citron	Leaf flecking, mottling	*Simeone et al., 2021*
4	Xyloporosis/ Cachexia	Parson's Special mandarin Sweet lime	Gumming at bark and bud union, wood/stem pitting	Krueger *et al.*, 2022
5	Exocortis	Etrog citron Rangpur lime	Leaf epinasty, tip browning	Alcantara-Mendoza *et al.*, 2017
6	Citrus bark cracking viroid/citrus viroid IV	Etrog citron	Leaf drooping, leaf midrib necrosis	Steyn *et al.*, 2016
7	Citrus stubborn disease	Sweet orange, grapefruit	Stunted growth, leaf mottling, little leaf (Zn deficiency symptoms)	Umar *et al.*, 2020

Sl.No.	*Disease*	*Indicator Plant*	*Symptoms*	*Reference*
8	Yellow vein clearing	Sour orange, lemon	Leaf yellowing, vein clearing	Liu *et al.*, 2020
9	Bud union crease	Sweet orange	Brown lines at union	Vives *et al.*, 2002
10	Cristacortis	Sweet orange, Grapefruit	Flecks	Lee *et al.*, 2021

The another technique is indexing using insect infection, where vector insects are collected from the field and reared in the lab conditions for a few days. Artificial diet is provided to these insects to observe the fasting period on the basis of transmission (persistent, semi-persistent or non-persistent), before allowing them to feed on virus-infected plants to acquire the virus. In non-persistent transmission, long fasting period (30–90 minute) increases the likelihood of virus acquisition from infected plant. The insects are transferred on the young leaves of virus infected plants and allowed to feed for a few minutes. Afterwards, these insects are transferred to healthy indicator plants, kept in controlled insect-free environment to avoid contamination. The insects remain on the indicator plants for at least 24 hours before being eradicated using insecticides. However, there is no effect of fasting in persistent and semi-persistent transmission, but a longer feeding period improves virus transmission. In semi-persistent transmission, insects are maintained for a week and then eradicated, while in persistent transmission, they are continuously transferred to new indicator plants until they die.

Laboratory Based Advanced Indexing Techniques

The mother plants recovered through nucellar embryony (*in vivo* or *in vitro*), micrografting, micro-budding, thermotherapy can be indexed using different lab techniques like ELISA, PCR, electron microscopy or serological assays for the presence of diseases. Many commercial disease specific kits are also available for the rapid indexing of the diseases. The major techniques used for the indexing of citrus diseases are summarised as

Sl.No.	*Technique*	*Indexed Diseases*	*Advantages*	*Limitations*
1	ELISA	Citrus tristeza, ringspot, mosaic, polyviruses, exocortis, greening	Rapid, economic for large sample units	Requires sensitive conditions
2	ISEM	Citrus tristeza virus	Rapid, specific	Requires specific antiserum, limited sampling units.
3	Molecular hybridization	Viral diseases	Routinely applicable	Highly sensitive, time consuming

Sl.No.	Technique	Indexed Diseases	Advantages	Limitations
4	RT-PCR	Greening, Tristeza, Gummosis and many other diseases	Easy, routinely applicable	Requires gene specific primers and other equipment's
5	PAGE	Exocortis, Xyloporosis	Best for detection and characterization of viroids	Requires proper conditions
6	Immuno-fluorescence	Greening, tristeza, yellow vein virus	Easy, economic	Limited sampling units, non-specific

Serological methods enable rapid detection of plant viruses based on antigen-antibody reactions. ELISA is a widely used serological technique for the detection of diseases (viz., Citrus tristeza, ringspot, mosaic, polyviruses, exocortis, greening bacterium) due to its cost-effectiveness and ability to process large sampling units. However, ELISA is not applicable to viruses and viroids that have lost their coat proteins, as these pathogens do not produce the antigens required for antibody recognition. Additionally, not all viruses react with antiserum developed against a primary virus. In such cases, cDNA or cRNA probes, as well as RT-PCR assays, are employed for accurate detection. The use of ELISA has been considered as a well proven technique for the mass detection of CTV. The use of another serological techniques like DTBIA and LFIA has been popular among the citrus nurseries around the world for the detection of tristeza virus. Around 500,000 samples plant can be tested at once using DTBIA for the CTV, charging 0.24 euros per sample (Shang *et al.,* 2024). Likewise, LFIA is another technique which can be performed using a strip, even by a non-skilled person, directly on the field for the evaluation of CTV within 10 minutes (Maheshwari *et al.,* 2017).

The molecular detection of citrus viruses has transformed diagnostics, enabling the identification of infections at early seedling stages to minimize economic losses. PCR-based methods based on high-throughput sequencing, allow for the identification of viral strains and their evolutionary relationships. RT-PCR or real time RT-PCR are commonly used for the rapid screening and identification of citrus viroids. However, in case of the presence of multiple viroids, use of multiplexing RT-PCR is preferred. Osman *et al.* (2015) developed a multiplex RT-qPCR assay that simplifies virus testing in citrus nurseries by combining three separate tests into one (for tristeza, psorosis and leaf blotch virus) under, "Section 3701: Citrus Nursery Stock Pest Cleanliness Program", saving time and labour while ensuring high sensitivity and specificity.

Several procedures have been developed for the detection of different pathogens *viz.*, CTV (tristeza), CLas (*Candidatus* Liberibacter asiaticus), HSVd (hop stunt viroid) using multiplexing-PCR; while, CPsV (Psorosis), CLBV (leaf blotch virus), CVV (variegation virus), CTLV (Tatter leaf virus) and CEVd (exocortis viroid) using RT-qPCR (Yao *et al.,* 2023). ddPCR is another modern technique which is 100-fold sensitive to RT-qPCR and exhibits a greater accuracy for the detection of CTV (Wang *et al.,* 2020). RT-LAMP (reverse transcription loop-mediated isothermal amplification) has emerged a s a new technique which offers specific pathogen detection without the need for thermal cycling. RT-LAMP protocol is 100 times more sensitive than conventional RT-PCR. Selvaraj *et al.* (2019) observed that IC-RT-LAMP allows rapid on-site detection by capturing CTV virions from crude leaf sap directly, bypassing nucleic acid extraction, and achieving amplification in just 6 minutes 45 seconds.

Conclusion

India is 3rd major producer of citrus fruits around the world and despite the expanding citrus cultivation area over the past two decades, availability of insufficient quality planting material is still a major problem. The lack of high-quality, disease-free planting material has been a critical constraint to citrus production. There are numerous citrus nurseries across the country, but they do not meet the necessary standards, with only a small fraction achieving high NHB ratings. To ensure the healthy nursery production, containerized nursery production under protected structures, microbudding and shoot tip grafting are pivotal for rapid multiplication and virus elimination in citrus propagation. Moreover, the adoption of budwood certification and rigorous virus indexing techniques ensures that mother plants and rootstocks are free from harmful viruses like greening, tristeza, and exocortis. By implementing these advanced nursery protocols, we can bridge the gap in planting material quality, safeguard citrus industry and boost the productivity of citrus fruits.

REFERENCES

Alcántara-Mendoza S, Vergara-Pineda S, García-Rubio O, Cambrón-Sandoval VH, Colmenares-Aragón D and Nava-Díaz C, 2017. Characterization of Citrus exocortis viroid in different conditions of indexing. *Revista Mexicana de Fitopatología*, **35**(2): 284–303.

Cao MS Atta, Su H, Wang X, Wu Q, Li Z and Zhou C, 2013. Molecular characterization and phylogenetic analysis of Citrus viroid V isolates from Pakistan. *European Journal of Plant Pathology*, **135**: 11–21. https: //doi.org/10.1007/s10658-012-0034-1.

FAO, 2021. Statistical database of Food and Agriculture Organization of the United Nations, Rome, Italy. Accessed on 6th May, 2023. https: //www.fao.org/faostat/en/#data/QCL.

Hancevic K, Radiæ T, Paskoviæ I and Urliæ B, 2018. Biochemical and physiological responses to long-term Citrus tristeza virus infection in Mexican lime plants. *Plant Pathology*, **67**(4): 987–994.

Hayat F, Li J., Liu W, Li C, Song W, Iqbal S, Khan U, Umer Javed H, Ahsan Altaf M, Tu P, Chen J and Liu J, 2022. Influence of citrus rootstocks on scion growth, hormone levels, and metabolites profile of 'Shatangju' mandarin (*Citrus reticulata* Blanco). *Horticulturae*, **8**: 608. https: //doi.org/10.3390/ horticulturae8070608.

Iftikhar Y, Majeed MZ, Vadamalai G and Sajid A, 2021. Indexing virus and virus-like diseases of citrus. In: *Citrus-Research, Development and Biotechnology*. IntechOpen.

Juarez J, Aleza P and Navarro L, 2015. Applications of citrus shoot-tip grafting in vitro. *Acta Horticulturae*, **1065**: 635–642.

Katoch S, Arora A, Kaur A, Singh H and Hunjan MS, 2022. Graft transmission and confirmation of *Candidatus Liberibacter asiaticus* using PCR with species-specific primers. *Plant Disease Research*, **37**(2): 187–191.

Krueger RR and Vidalakis G, 2022. Study and detection of citrus viroids in woody hosts. In: *Viroids: Methods and Protocols*, pp. 3–21.

Lee RF, Keremane M and Ramadugu C, 2021. Use of young plants for biological indexing of graft transmissible pathogens of citrus. *Crop Protection*, **143**: 105524

Liu C, Liu H, Hurst J, Timko MP and Zhou C, 2020. Recent advances on Citrus yellow vein clearing virus in citrus. *Horticultural Plant Journal*, **6**(4): 216–222.

Maheshwari Y, Selvaraj V, Hajeri S, Ramadugu C, Keremane ML and Yokomi RK, 2017. On-site detection of Citrus tristeza virus (CTV) by lateral flow immunoassay using polyclonal antisera derived from virions produced by a recombinant CTV. *Phytoparasitica*, **45**: 333–340.

Navarro L, 1981. Citrus shoot tip grafting in-vitro (STG) and its applications. In: *Proceedings of the International Society of Citriculture*, **1**: 452–456.

Osman F, Hodzic E, Kwon SJ, Wang J and Vidalakis G, 2015. Development and validation of a multiplex reverse transcription quantitative PCR (RT-qPCR) assay for the rapid detection of Citrus tristeza virus, Citrus psorosis virus, and Citrus leaf blotch virus. *Journal of Virological Methods*, **220**: 64–75.

Selvaraj V, Maheshwari Y, Hajeri S and Yokomi R, 2019. A rapid detection tool for VT isolates of Citrus tristeza virus by immunocapture reverse transcriptase loop-mediated isothermal amplification assay. *PLoS ONE*, **14**: e0222170.

Shang P, Xu L and Cheng T, 2024. Serological and molecular detection of Citrus tristeza virus: A review. *Microorganisms*, **12**: 1539.

Simeone M, Gómez C, Bertalmío A, Ruiz E, Hauteville C, Godoy Suarez L and García ML, 2021. Detection of citrus psorosis virus by RT-qPCR validated by diagnostic parameters. *Plant Pathology*, **70**(4): 980–986.

Skaria M and Zhang T, 2000. Field performance of micro-budded citrus trees in Texas. In: *ISC Congress-2000*, Orlando, Florida, pp. 130.

Steyn C, Cook G, Burger JT and Maree HJ, 2016. Construction and application of infectious citrus viroids for biological indexing. *Journal of Citrus Pathology*, **3**(1): 112-116.

Thakur A, Dhaliwal HS and Kashyap P, 2017. Protected citrus nursery production is a profitable enterprise. *Indian Horticulture*, **62**: 14–19.

Umar UDU, Naqvi SAH, Ahmed I, Rehman A, Zulfiqar MA, Ullah S and Rehman A, 2020. Prevalence and serological detection of *Spiroplasma citri*, a cause of citrus stubborn disease in southern Punjab, Pakistan. *Pakistan Journal of Agricultural Research*, **33**(4): 748–753.

Vijayakumari N and Singh S, 2003. Standardization of micro-budding technique in Citrus. *Indian Journal Horticulture*, **60**: 127–130.

Vives MC, Galipienso L, Navarro L, Moreno P and Guerri J, 2002. Citrus leaf blotch virus: A new citrus virus associated with bud union crease on trifoliate rootstocks. In: *International Organization of Citrus Virologists Conference Proceedings (1957-2010)*, **15**: 15.

Wang Y, Yang Z, Zhao J, Li R., Wang Q, Li J, Li Z and Zhou Y, 2020. Development of a sensitive and reliable reverse transcription droplet digital polymerase chain reaction (RT-ddPCR) assay for the detection of Citrus tristeza virus. *European Journal of Plant Pathology*, **156**: 1175–1180.

Yao SM, Wu ML and Hung TH, 2023. Development of multiplex RT-PCR assay for the simultaneous detection of four systemic diseases infecting citrus. *Agriculture*, **13**: 1227.

Chapter 15

Orchard and Canopy Management in Hill Horticulture

N.C. Sharma, Pramod Verma and Tanzin Ladon

Department of Fruit Science, Dr. Y.S. Parmar University of Horticulture and Forestry, Nauni, Solan, Himachal Pradesh

Hill horticulture involves cultivating crops in rocky landscapes, addressing challenges like steep slopes and harsh conditions. Despite fertile terrain and suitable climates, organized horticulture in hills remains sparse. Traditional terraced cultivation is crucial for the economy but faces limitations in expansion, leading to low economic returns. Historical evidence suggests fruit cultivation is favored over cereal farming (Atkinson, 1882).

In hilly regions, sparse population density limits per capita land availability due to unsuitability for traditional crops, necessitating diverse horticultural planning for economic development. Perennial vegetation cultivation aids soil conservation and ecology (Ghosh, 1985). Throughout the entire Himalayan Mountain range, spanning from the Jammu and Kashmir Himalaya to the Assam Himalaya, horticulture has been practiced for centuries, offering a diverse array of fruits (Papola, 1996). Hilly horticulture drives economic growth and environmental restoration. The Himalayan range supports centuries-old horticulture, primarily domestic, with varying climates ideal for diverse fruit cultivation. The climatic conditions in the Himalayas vary from sub-tropical in low-lying river valleys to Alpine and cold in highly elevated regions, providing suitable conditions for a wide range of fruit cultivation.

Various fruits like apple, pear, stone fruits (peach, apricot, plum, almond and cherry), nut crops (walnut and pecan nut), kiwifruit, persimmon, citrus (lemon, lime and kinnow), pomegranate and guava thrive at different elevations in hilly regions along with vegetable cultivation, and flower crops. Besides, commercial tea plantations exist in specific hilly areas, however, horticulture sector's growth in these regions is limited and hindered by factors like poor transportation and marketing infrastructure.

Orchard Management

It refers to management of orchard soil and plants by proper utilization of the resources for higher production of quality fruits on sustainable basis without adversely affecting the quality of the environment. An orchard, typically enclosed, is dedicated to growing fruit trees, encompassing resources such as land, water, trees and external inputs. Efficient utilization of these resources is essential for maximizing production per unit area sustainably while maintaining environment quality. Orchard site selection and preparation are the first management decisions that affect the productivity of a tree fruit orchard. Whether considering a new site, replanting a former tree fruit site, or updating to current varieties or production systems, there are many important decisions to make.

Selection of Orchard Site

Setting up an orchard is a significant long-term commitment, necessitating careful selection of site to optimize the production of high-quality crops. While establishing an orchard, the climatic conditions in the area should be taken into consideration. Current weather and historic environmental data should be assessed, if available, historic temperature extremes, frost risk, average bloom date, consistency of cropping may be instructive. Orchard sites should be free from spring frosts. Sites having gentle slops with good air drainage are ideal, as cold air can move down slope and not accumulate in a low spot or "frost pocket" causing problems especially at the time of bloom. South facing slopes warm up faster in the spring than north facing slopes, so planting early blooming cultivars at south facing locations would increase their risk for spring frosts. In southern aspects full day sunlight trigger the evapo-transpiration rates and planting of dwarfing fruit varieties should be avoided in orchard sites facing south.

Replant Site Considerations

Replant disease is a serious consideration in the hilly sites, as most of the temperate fruit crops especially apple and peach are prone to replant disease. Small and fragmented land holdings are major constraints in hilly areas necessitate the plantation on old sites. Prior to planting on old sites, roots and other residues of old plantation must be removed, as soil-borne disease are caused by replant pathogens like nematodes, bacteria and fungi *viz., Rhizoctonia, Pythium* and *Phytophthora* (Brown and Schimanski, 2002) causing reduction in survival, growth and yield of replanted trees (Yao *et al.,* 2006). Orchard soil should be tested to determine nematode levels before the trees are uprooted, especially if the previous plantations had signs of decline. The following steps are taken while planting on a replant site:

- Displace the site of the planting pit from the old site.
- Larger size planting pit (5 ft wide and 3ft deep) should be prepared well in advance and kept open before the actual planting time.
- Soil of the pit should be disinfected before filling the pits using formalin solution (1:9 v/v). Fumigation in a replant site should ideally be done in the fall, as it does not work well when soil temperature is below 5.5°C (Sallato, 2022).
- If possible, change the soil of the pit using forest soil. While filling the pit, soil after every 6-inch layer should be sprayed with formalin solution.
- Keep the fully prepared treated pit untouched for three weeks before planting.
- Intercropping with antagonistic plants like marigold, creeping red fescue (*Festuca rubra*) and red top (*Agrostis alba*) also decreased the population of soil borne pathogen at replanting sites (Edwards *et al.,* 1994).
- Use only healthy and disease free planting material (Sharma *et al.,* 2020). In case of apple, replant disease resistant root stocks such as M.793 (Soni *et al.,* 2011; Singh and Sharma, 2017), CG.30, CG.210 (Isutsa and Merwin, 2000; Rumberger *et al.,* 2007; Laurent *et al.,* 2010) and CG.935, CG.202 (Robinson *et al.,* 2004) can be used.

Planting Systems and Design

Systematic planting of fruit trees is very important to optimize tree density and streamline orchard operations effectively. The irrigation system, tree support system, alleyway grass or cover crop, and understory weed strip areas are all planting design considerations. The space and maneuverability logistics for workers, tractors, trailers, bins, trellises, overhead nets, wind machines, *etc.* need to be considered. In valley areas on flat lands, trees are planted following square or rectangular systems. High density plantations especially in apple, is gaining popularity using dwarfing rootstocks and following these planting systems. Ladon *et al.* (2024) reported a productivity of 42.13 t ha^{-1} in apple cv. Jeromine grafted on M.9 and planted at a density of 2.5 × 1.00 m and trained to the Tall Spindle training system.

Tree rows are commonly oriented in the North-South compass direction to maximize sunlight availability to the orchard trees. The sunlight penetration and interception into the tree canopy is very important for adequate growth and fruit quality. Contour system of planting is practiced on undulating lands with gentle slopes. The primary purpose of contour planting is to combat soil erosion and conserve moisture, making it suitable for establishing orchards on undulating terrain. Contour lines are strategically designed and graded to slow down the flow of water in irrigation channels, allowing it more time to penetrate the soil without causing erosion. However, under steep slopes, terrace system of plantation is recommended. Terracing is implemented across the slope of the hill, following the contour lines and width of these contour terraces is adjusted according to the steepness and extent of the slope. The slope of the terrace is kept inward which helps not in prevention of runoff and soil erosion but also in conservation of moisture in the orchard. The level of erosion prevention correlates directly with the width of the terrace.

Some fruit crops (apple, plum, cherry, almond and kiwifruit *etc.*) grown in hilly regions, require cross pollination and rely on pollinators/pollinizers for fruit set. Thus, placement of pollinizers is another crucial consideration in formulating the planting scheme. In such cases, ideally, every tree of main variety should be located close to the pollinizer variety and the preferred arrangement is in solid rows. One scheme is to alternate two rows of pollinizers between four rows of the major cultivar and an exception is Delicious apple where a pollinizer row should be set at every third row. Depending on the species, different percentage of pollinizing variety is planned in the orchard. Some of the common planting plans for different fruit crops are given in Figure 15.1.

Soil Management/Orchard Floor Management

Soil management aims to preserve or enhance soil quality, which involves shielding it from direct sunlight and erosion caused by rainfall and wind. Effective soil management methods are crucial for weed control, integration of organic and inorganic fertilizers, and enhancing water absorption in the soil. A particular management practice is followed depending on factors like crop, type of soil, climate, rainfall of the area, topography, tree spacing, rooting depth of the crop, planting system/orchard design and economic condition of the farmer *etc.* Orchard floor management should be combined with canopy, nutrient and water management, provision for pollination, fruit thinning and disease & pest management in order to obtain desirable tree growth and yield of better quality fruits.

In sloppy lands particularly when the gradient of the slope is greater than 10 per cent, sod culture is practiced allowing the grasses to grow in the interspaces of the orchard. Among the grasses, orchard grass (*Dactylis glomerata*) is suitable for northern aspects, tall fescues (*Festuca arundinaceae*) for southern aspects and Timothy (*Phleum pretense*) for cold temperate region. Mixing of red clover (*Trifolium pretense*) and white clover (*Trifolium repense*) with grasses is useful for improving soil fertility. The grasses are mowed periodically to reduce competition for water and nutrients in orchard soil. Sod culture prevents soil erosion to the greatest extent besides maintaining organic matter and soil structure. The practice of sod culture should be avoided in orchards having young or dwarf trees because such trees are shallow rooted and orchard grasses compete for nutrients and water with tree roots. The roots of large trees on the other hand penetrate the soil to a greater depth and are less likely to suffer from such competition.

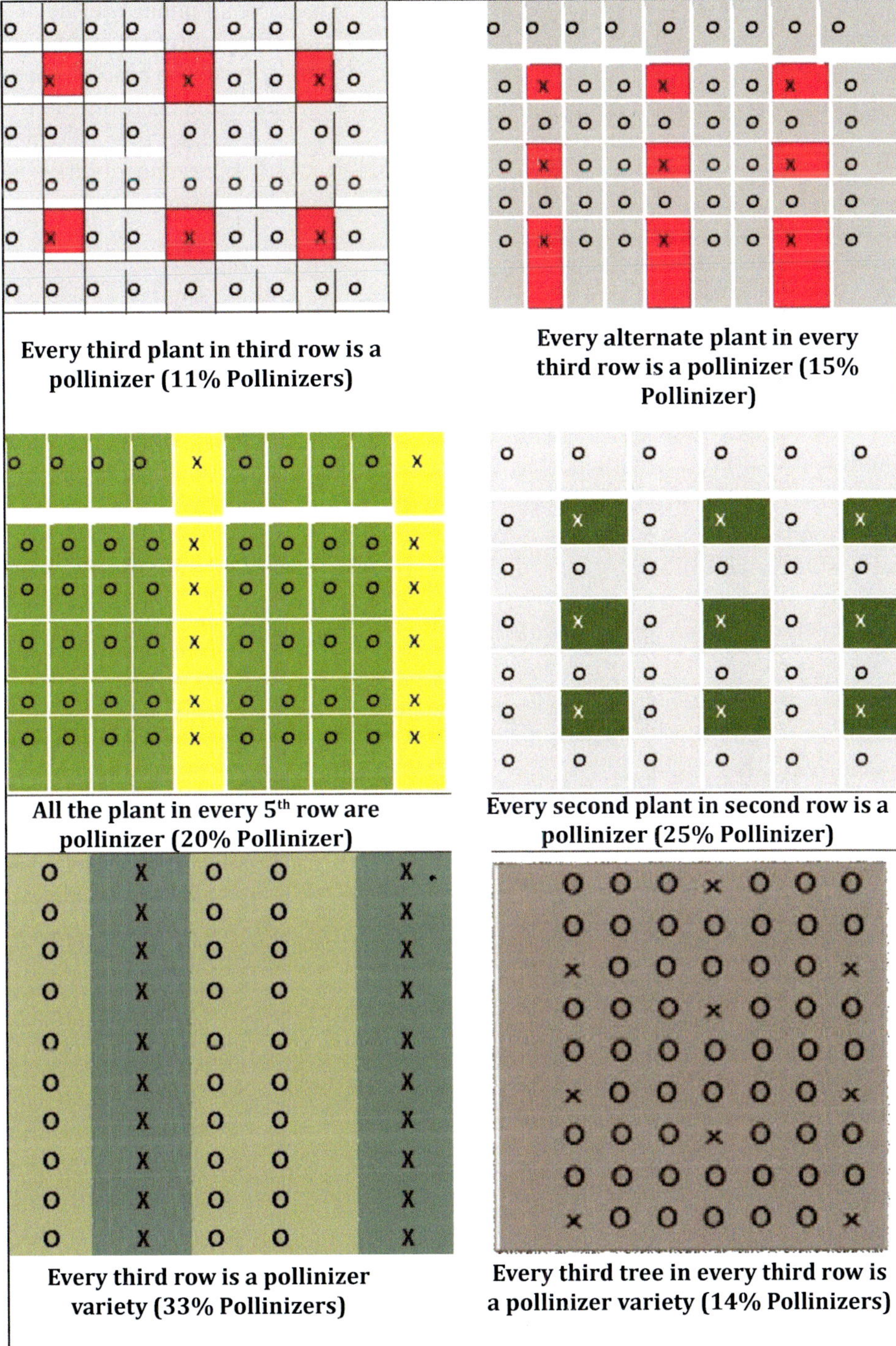

Every third plant in third row is a pollinizer (11% Pollinizers)

Every alternate plant in every third row is a pollinizer (15% Pollinizer)

All the plant in every 5th row are pollinizer (20% Pollinizer)

Every second plant in second row is a pollinizer (25% Pollinizer)

Every third row is a pollinizer variety (33% Pollinizers)

Every third tree in every third row is a pollinizer variety (14% Pollinizers)

Figure 15.1: Planting Plants for different Percentage of Pollinizers in the Orchard.

In valley areas with gentle slopes, sod culture + mulching is better orchard floor management practice. Using this management practice, tree basins are kept free of weeds by spreading the organic or inorganic mulches in the basin. Among the organic materials, straw, hay, crop residue, leaves, saw dust and composts *etc.* can be used as mulching material, however, among the inorganic mulches, black/colored polythene or nylon sheets are popular mulching material. Choice of organic or inorganic mulch depends on easily availability of the material, cost of mulch material and economic conditions of the farmer. Organic mulches are more useful for valley and low/mid-hills areas as these help in keeping the soil temperature cooler by 2- 3° C in summer. Inorganic mulches are usually preferred in fruit orchards located in the high hill areas having cooler climate and sufficient Organic matter. It helps in raising the soil temperature in summer, required for greater nutrient absorption and roots metabolic activities. Both kind of mulches suppress weed growth, conserves soil moisture, control soil erosion and regulate soil temperature.

In flat lands clean cultivation and intercropping orchard management practices are followed. Intercropping is practiced especially during non-bearing stage of the trees in the orchard. Preference should be given to short duration vegetable and pulse crops to be grown as intercrops in the orchards. At least 120 cm radius must be left from the base of the growing fruit trees for taking intercrops.

Water Management

The effectiveness of orchard management significantly hinges on the judicious use of water, especially during pivotal stages of plant growth and development. Fruit trees rely on irrigation water to sustain adequate growth, fruit quality and yield, especially during dry months. However, water availability remains a constraining factor in fruit cultivation, particularly in hilly regions. Even in regions with high rainfall (>1500 mm), sufficient moisture may not reach the root zone during dry periods, leading to potential drought stress. To ensure the provision of necessary water quantities during critical growth stages of fruit crops, effective water management is imperative. In hilly regions, most of the orchards are rainfed where three fundamental components of water management are in situ moisture conservation, water harvesting and efficient utilization of conserved/harvested water through improved cultural practices.

Two natural water resources that can effectively be harvested are rainfall and low water springs or rivulets. About 10 to 14 per cent of total rainfall, depending on soil and rainfall characteristics may be lost as surface runoff. Water harvesting can be done through *in situ* rainwater harvesting and water harvesting in farm ponds. In fruit plantations, rain water can be harvested in the tree basin areas by soil working techniques like crescent bunding with open catchpits, V-ditch and preparation of tree basins in special manner (Half-moon basin and trench basin) *etc.* These techniques provide satisfactory *in-situ* harvest of rain water and moisture conservation in sloppy lands. In half moon basin, a kind of bund is made on down side of the slop in half circle manner below plant and water is stored inside the bund (Fig. 15.2). In trench basin, a trench is made on down side of the slop below plant and water is stored in the trench. The stored water inside the bund or trench is available to the plant roots continuously and slowly. These techniques are found to be useful for conserving soil moisture in the root zone of trees for the dry months.

Figure 15.2: Half-Moon and Trench Basin

Figure 15.3: Trench Basin Farm Ponds.

The run-off water from orchards and water flowing from other sources like spring and small streams that though cannot serve as water source for irrigating crops due to their meagre flow rates but, can be effectively stored in suitable small or big reservoirs. Small ponds may be used as water reservoirs in which water is stored during lean period and used when required by crops. The provision of a small pond in one corner of the orchard to collect runoff water during high intensity rains and its utilization as lifesaving irrigation or during critical periods of the crop is an age old practice. Ponds vary in shape, size and mode of construction (Fig. 15.3). The water harvested from natural resources must be used very efficiently for fruit crops, possibly through drip irrigation. Drip irrigation system is ideal for fruit crops, because it enable to achieve higher water use efficiency and also help in maximizing crop production within limited water resource. This system of irrigation has been found 2.27, 2.87 and 2.62 times more efficient for apple, plum and apricot as compared to convention irrigation system (Anonymous, 2014).

Nutrition Management

Nutrients are compounds essential for plant growth and development. In hill horticulture, nutrition management involves not only supplying nutrients but method and timing of their application is also important for better nutrient uptake and maintenance of optimum nutritional status of the plant. The quality of the produce is directly related to the nutritional status of the tree and orchard soils. The key to mineral nutrition of orchard is judicious application of fertilizers and manures on the basis of soil and leaf analysis. Nutrients play an important role in the physiological activities of the plant and thereby affect the growth, yield and quality of the fruits. Thus, nutritional inputs have direct role in overall plant growth, yield and quality of the produce. However, deficiency of any one of these mineral nutrients results in decreased growth and yields with poor quality of the produce besides development of various nutritional/physiological disorders.

Most of the orchards in hilly regions are rainfed and established on sloppy lands. The orchard soils are shallow and have poor fertility status due to which root system of the plants fails to develop properly and leads to serious problems of water and nutrient losses. In these areas rainfall is not well distributed and frequent dry spells during April to June (summer) and September to November (autumn) make the nutrients unavailable to the trees. Rains during monsoon and winter season result in leaching of nutrients especially when rains are heavy and continuous. Due to steep slopes, the size of the basin is invariably small (radius is < 90 cm) and root system of plants cannot be fully exploited for the uptake of nutrients. Heavy rains during monsoon may cause leaching of nutrients and soil erosion, whereas, drought periods during May-June and September-October may check the availability of nutrients (Verma, 2002). Under such situations, nutrient inputs even if applied in optimum quantities will not be properly utilized by the plant and plants lead to the nutrient element deficiencies because of the continuous depletion of soil nutrients. Therefore, integrated nutrient management practices should be adapted to provide balanced nutrition and reduce the nutrient losses. Chemical fertilizers, organic manures and bio fertilizers are used in complimentary manner to maintain soil fertility and provide the nutrients to the plants for sustaining the production. This ensure the efficient use of applied fertilizers, harnessing the best possible positive and synergistic interactions amongst the various factors of production, least adverse effect on environment by minimizing nutrient losses, maintain productivity and sustaining higher yields of quality fruits under specific agro-ecological conditions. It is a dynamic approach which is aimed on providing a congenial ionic environment at the soil-root interface with respect to essential plant nutrients so that the nutrient mobility and uptake is optimized. It includes all the processes, factors and mechanism of nutrient availability such as quality, intensity, buffering power of soil, water relations in soil, diffusion parameters, root properties and root system of the crop. When the soil does not furnish adequate quantities of nutrient elements necessary for normal growth and development, it is essential that required amount may be supplied from the external sources (Verma, 2002). The doses of organic manure and nutrients recommended for some fruit crops grown in hilly regions are given below in Table 15.1.

Table 15.1: Nutritional Input Recommendations for some Fruit Crops Grown in Hill Regions

<table>
<tr><th>Fruit Crop</th><th>Age of Tree</th><th>FYM (kg)</th><th>N (g)</th><th>P_2O_5 (g)</th><th>K_2O (g)</th></tr>
<tr><td rowspan="4">Pome fruits (apple and pear</td><td>1 year</td><td>10</td><td>70</td><td>35</td><td>70</td></tr>
<tr><td colspan="5">The dose of FYM, N, P and K increased by equal proportion every year up to an age of 10 years and doses are stabilized after that.</td></tr>
<tr><td>10 years or above</td><td>100</td><td>700</td><td>350</td><td>700</td></tr>
<tr><td colspan="5">In pome fruits, there may be 'on' and 'off' year fruiting pattern, therefore, nutrient doses should be reduced by 50 per cent in bearing trees during off year.</td></tr>
<tr><td rowspan="3">Stone fruits (peach, plum, apricot and almond)</td><td>1year</td><td>10</td><td>70</td><td>35</td><td>100</td></tr>
<tr><td colspan="5">The dose of N, P and K is increased by equal proportion and FYM @ 5 kg every year up to an age of 7 years and doses are stabilized after that.</td></tr>
<tr><td>7 years or above</td><td>40</td><td>500</td><td>250</td><td>700</td></tr>
<tr><td rowspan="3">Cherry</td><td>1year</td><td>10</td><td>50</td><td>25</td><td>60</td></tr>
<tr><td colspan="5">The dose of N, P and K is increased by equal proportion and FYM @ 5 kg every year up to an age of 10 years and doses are stabilized after that.</td></tr>
<tr><td>10 years or above</td><td>60</td><td>500</td><td>250</td><td>600</td></tr>
<tr><td rowspan="2">Kiwifruit</td><td>Fully grown</td><td>40-50</td><td>750-800</td><td>400-500</td><td>700</td></tr>
<tr><td colspan="5">This dose is for vines planted at 6 × 6 m distance; however, in vines planted at 6 x 3 m distance, doses should be reduced to half of the above dose.</td></tr>
</table>

Fruit Crop	*Age of Tree*	*FYM (kg)*	*N (g)*	*P_2O_5 (g)*	*K_2O (g)*
Walnut	1year	10	20	30	18
	The dose of N, P and K is increased in varying proportions and FYM increased @ 5 kg every year to an age of 20 years and doses are stabilized after that.				
	20 years or above	100	2000	500	1500
Pecannut	1 year	10	75	75	75
	The dose of N, P and K increased by equal proportion and FYM @ 5 kg every year up to an age of 16 years and doses are stabilized after that.				
	16 years or above	100	1200	1200	1200
Persimmon	Fully grown	40-50	NPK @ 100 g each/5 cm trunk diameter.		

Source: Anonymous, 2014.

The method of nutrient application is considered ideal using which applied nutrient reach the active root zone easily with minimum leaching losses. In hilly tracts on sloppy lands, fertilizer use efficiency can be increased by altering the time and placement of fertilizer application to coincide with peak periods of their utilization. Application of fertilizer too early or too late of their need by the tree and wrong application methods may result in lower nutrient recovery. Some of the strategies that can be helpful in improving the nutrient recovery of applied fertilizers in the orchards established on hill regions are as under:

- N and K fertilizers especially in sandy soils and high rainfall areas should be applied in split doses and single application should be avoided.
- Due to acidic nature of the soils, regular use of urea and ammonium fertilizers should be avoided as this may further increase the acidic conditions and make the major nutrients unavailable to the trees.
- Use of nitrate fertilizers in light textured soils may lead to the leaching of the nutrients.
- Under poor soil aeration conditions there may be loss of nutrients through denitrification resulting into formation of nitrogen gas or its oxides.
- Instead of premixed fertilizers, single nutrient fertilizers should be preferred.
- In high rainfall areas, particularly under sandy soil conditions, use of slow-release nitrogenous fertilizers should be preferred.
- Fertilizers should be placed in the rings formed at 30-45 cm away around the tree trunk in the basin, rather than broadcasting them on the surface

of the basin soil. Surface application of ammonium or urea fertilizers results in higher volatile losses of nutrients.

- Phosphorus and potash fertilizers should be applied during dormant season in January-February well advance before the initiation of flowering.
- Under acidic soil conditions, phosphorus reacts with Fe and Al and under alkaline conditions with Ca and thereby fixed in the soil and become unavailable. Potassium may be lost by leaching in high rain fall or excessive soil moisture conditions.
- Where drip irrigation system is installed in the orchard, fertigation should be done to provide nutrients in a number of split doses at critical stages of nutrient requirement to avoid nutrient losses and increase fertilizers use efficiency.
- Micro nutrients should be applied through foliar sprays. The spray solution concentration must be controlled to prevent leaf scorching.
- Some part of primary nutrient fertilizers can also be applied through foliar sprays during active growth and critical stages of nutrient requirement.
- Dry spells and prolonged drought conditions in the orchards create temporary deficiency of nutrients as fertilizer application in the soil is confined to tree basin which does not cover the active root zone of the tree. Such kind of nutritional deficiencies can be easily overcome by foliar application of fertilizers.
- In rainfed orcharding, efficiency of applied nutritional inputs can be improved by using controlled release fertilizers with low solubility coatings and natural organic materials which require microbial action for release of nutrients. This helps in reduction of nutrient losses through volatilization, hydrolysis and leaching.

Canopy Management

Canopy is the above ground portion of a tree, which includes trunk, branches and reproductive parts. Canopy management is an important orchard management practice in deciduous fruit crops grown in hilly region achieved through training and pruning of trees. The young trees are trained during the initial years mainly to obtain the desired shape and framework so that scaffold will develop strong wide angled crotches capable of supporting large yields without breakage. Training fruit trees starts at planting to set the structure of the orchard. Plant growth regulators or plant hormones control the growth and development of trees. For instance, cytokinins generated in the roots ascend to the tree and stimulate cell division and promotes bud break. However, auxin descends with gravity and is created at the tip of fresh vegetative growth. The relation of auxin to cytokinin affects vegetative growth. High ratio of auxin:cytokinin will inhibit bud break leading to apical dominance. Manipulating the branch angle will change this ratio enabling tree training (Sallato, 2022). The bearing trees are pruned with the primary objective of regulating vegetative growth, flowering and fruitfulness to obtain good yields of quality fruits. It also ensures better light interception in the canopy for quality fruit production and removal of dead, diseased branches and regulates fruiting every year.

Training

Tree training imparts a major role in developing the tree canopy and accounts for maximum sunlight interception, higher fruit yields and quality, and easier worker access. Choosing the correct scion and rootstock combination is essential to ensure that trees will have the appropriate vigour and perform properly for the training methods used. The soil type, irrigation system, tree spacing, and management experience will also contribute to the success of a rootstock and training system combination. Tree training requires basic knowledge of the structural elements of fruit plants, their development and function as well as their mutual relationships. Knowledge of the structural plan and growth laws of fruit trees helps to guide their growth in a targeted manner. Training helps in developing strong tree architecture that can support crop loads, promote precocity and improve the efficiency of many operations, while it could be costlier at the beginning as it is labor intensive. In hilly areas, generally two types of training structure such as free standing and trellis system are followed.

Free Standings Canopy Training

Vigorous trees raised on semi-vigorous and standard rootstocks are trained as freestanding modified center leader and open-center training systems. In hilly terrains on sloppy lands, the standard tree has been the most common form grown on seedling rootstock and planted far apart in low density. Such trees are less efficient, less productive and less precautious besides taking many years to reach full production. It consists of central trunk around which scaffold branches of desired number and spacing can be arranged. The branches which originate from central leader of tree are called primary branches and those arise from primary branches are secondary branches. The secondary branches are subsequently furnishing with spurs or laterals on which fruit is borne. The spurs and laterals are periodically renewed in a bearing tree. The lower scaffold branches should be longest and more vigorous than the upper branches for better light interception. It is essential to maintain more vigour of lower branches as they have a tendency of becoming weak. The main considerations about the distances between scaffold branches are to maintain sufficient distance to admit sunlight and to ensure that the branches, when grown and enlarge, do not clash with each other. A distance of 25-30 cm is sufficient for the branches growing in same direction, however, distance of 15 cm is kept for branch growing in opposite direction. System of training relies on wide angled laterals for forming the primary branches. Narrow crotch angled branches should be removed by giving cut on cone with bud facing downwards. A new branch grows pointing downwards.

Trellis Trainings

Trees raised on dwarf or semi-dwarf rootstocks are trained to spindle canopies (tall spindle, vertical axis, bi-axis *etc.*) that require tree support using trellis systems planted at closer spacing (high density planting). The Spindle is the most widespread tree form in commercial fruit growing, with different tree sizes and canopy volumes. All spindle forms have an accentuated centre with subordinate fruit branches evenly distributed in a spindle shape around the central axis. These taper in thickness and length from bottom to top. The centre axis always remains dominant in all spindle sizes. Once the final height is reached, the centre is diverted to a perennial, fruiting side branch. Ideally, only a few longer and worn, excessively long fruiting branches on cones should be removed each year, from whose buds a new shoot is grown.

General Rules of Training of Fruit Trees

Limb Positioning

Limb positioning involves securing young branches with spreaders. Bending nearly vertical limbs to an angle of 45 to 60 degrees from the vertical typically encourages earlier fruit production in the tree's lifespan. The primary objective of different training techniques is to optimize light interception and strengthen the limbs. Various methods for positioning limbs include:

Spreading

Once the young lateral branches reach a length of 3 to 6 inches, it's recommended to gently spread them outward to promote the formation of a wide crotch angle (45 to 60°). This deliberate arrangement encourages a strong union, providing essential support for future fruit production. Two effective tools for spreading branches in young trees are toothpicks or clothespins, typically requiring 4-8 clothespins per tree. Alternatively, wooden spreaders are utilized for this purpose in mature trees.

Tying and Bending

Scientific evidence confirms that main branches characterized by wide crotch angles tend to initiate fruiting earlier compared to those exhibiting upright growth patterns. These primary branches, referred to as scaffolds, not only generate secondary branches but also contribute to the rapid expansion of the tree canopy. Manual bending of scaffolds aids in reducing the juvenile period, particularly beneficial for late-bearing fruit plants. Bending the scaffolds downward promotes spur formation, leading to earlier fruiting. As branches mature, concrete weights can be attached, or branches can be loosely tied down with string or twine wrapped around the limb and fastened to the base of the tree stake or clips embedded in the soil. Spread lateral branches exhibit slower growth due to reduced apical dominance, fostering increased secondary branching and earlier initiation of flower buds, which can lead to early fruit production attributed to diminished vigour. Forshey *et al.* (1992) noted reduced vegetative vigour in bent limbs, suggesting that bending reduces gibberellin production, which is antagonistic to flowering. Thus, the formation of moderately vigorous lateral shoots and spurs creates additional sites for flower formation, resulting in increased flowering and earlier fruiting. Fruit quality is also enhanced due to improved light penetration (Ferree and Schupp, 2003).

Figure 15.4: Tall Spindle System of Training in Apple.

Trellising

In this method, trees are supported by a single leader growing vertically, while primary branches extend along the trellis system. These primary branches are carefully fastened to the trellis with strings to improve the crotch angle, offering structural reinforcement for the fruit load. Unlike traditional permanent scaffold systems, trellised fruit trees yield heavily.

Pinching

Pinching of shoots is primarily employed in one or two years-old apple trees, particularly in trees trained to Tall Spindle system (Fig. 15.4), with the main goals of stimulating fruiting and effectively controlling upper tree growth. Targeted for pinching are rapidly growing shoots, typically measuring 6 to 8 inches in length, located in the top quarter of the tree and originating from the leader. Pinching serves two main purposes: first, it weakens shoot growth while reducing competition with the leader, and second, it promotes the development of a fruit bud from June to July. This technique not only prevents the tree from developing excessive top growth but also minimizes the need for less desirable dormant pruning cuts in the future.

By removing immature apical leaves through pinching, the number of branches on the current-season growth is increased (Ouellette *et al.,* 1996).

Stripping

Stripping is a method used in training of young apple trees to protect and prioritize the central leader. When newly planted trees produce three to four buds just below a heading cut, these buds give rise to sturdy, upward-growing shoots. The goal is to select the most vertically oriented shoot to become the central leader. Once identified, the competing shoots directly beneath this chosen leader, typically numbering 2 or 3, are either stripped away with a downward pull or pinched out. It's crucial to remove these shoots promptly when they reach a length of 3-4 inches to direct growth into the central leader and prevent the formation of a multi-leader tree.

Notching and Nicking

Notching is a technique used to encourage bud sprouting, particularly in dormant buds. It involves making two wedge-shaped cuts about 5 to 10 mm apart and approximately 5 mm deep on the bark above the bud. By removing a portion of the bark along with some wood, dormant buds are stimulated to sprout. The opposite of notching is nicking, which involves making a similar cut and bark removal below the bud. Nicking aims to induce spur formation from the bud or inhibit shoot growth by accumulating carbohydrates and inhibitors. Notching promotes lateral branching from buds that would not typically grow by interrupting the downward movement of auxin from the shoot tip (Greene and Autio, 1994; Verma *et al.,* 2023).

Pruning

Pruning is a systematic and judicious removal or cutting of various plant parts, including shoots, spurs, leaves and roots, or the selective nipping away of terminal parts to maintain the balance between vegetative growth and fruiting. Pruning is done in bearing trees to correct or maintain tree structure. The principal goal of pruning is to correct or sustain the tree's structural integrity while enhancing its overall utility. The training is completed during initial 4-5 years after planting of plant, but pruning is continued after training throughout the life of tree.

Basic Rules of Pruning

- ✰ Prune young tree very lightly following the rules of minimal pruning (Robinson,2003) and mature tree more heavily, especially if they show poor growth.
- ✰ Prune the top portion of the tree more heavily than the lower portion.
- ✰ To reduce the height of an excessively tall tree, cut whole limbs at the top, making cuts flush with the bark of a lower limb. Stubs left in the top of a tree would not heal.

- Follow 3: 1 pruning rule (Cook and Strydom, 2000), the leader should be 3 times the diameter of any of the lateral branches in the upper part of the tree.
- Thinning out and heading back are two main techniques of pruning. While pruning trees by thinning, final cut should be made just beyond the branch collar and branch bark ridge. However, in heading back, cut at a slight angle above the bud, whose position on the top or underside of the branch determines the future direction of the new shoot. A cut too far from the bud will result in the untended stump drying out and a cut too close to the bud or too diagonally, harbors the risk of drying out and non-sprouting of the bud.
- Strong branches on the trunk which act as competitors to the main shoot are removed and base branches on the periphery are pruned using heading back with bud pointing upwards.
- More a tree is pruned in winter, more it will sprout again in spring. Pruning too hard can lead to excessive new growth and number of buds that do not sprout is reduced by severe pruning. Whereas, light pruning causes more short shoots with flower buds to develop on the longer remaining shoot.
- Pruned extension shoots do not grow as strongly in the vicinity of unpruned competitor shoots, if all shoots in the tree are pruned evenly.
- Intensive pruning leads to strong new growth and tends to inhibit flower formation. The proportion of old growth with productive fruit wood is reduced. Whereas, less pruning is the quickest way to produce a lot of fruiting wood and a high yield capacity.
- Pruning can be used to weaken growth for canopy, height limitation and growth control. Pruning to strong buds results in a strong growth impulse, while, pruning to weak buds results in relatively weak new growth. Pruning back to a flower bud always has the effect of weakening growth.

Time of Pruning

Pruning can be carried out at any time of the year and has different effects on shoot vigour. The strength of the desired new growth can be influenced by the time of pruning. The reaction of a tree to pruning and new shoots is also determined by the time of pruning within the year. Deciduous fruit trees can be pruned in pre-winter or even during the winter, but only if it is warmer than -4°C. Lower temperatures can damage the cut frozen wood. If there is a risk of falling temperatures, the cones of cut branches should be left longer. However, summer pruning slows down new shoots and winter pruning promotes new shoots. Timing and amount of shoot removal is important since it can impact carbohydrate supply to developing fruit and also direct sun exposure of fruit that could lead to sunburn. Summer pruning to remove fire blight requires care to make cuts 8-10 inches below the canker and it requires disinfestation of tools between cuts to prevent disease transmission.

Dormant Pruning

Dormant pruning performed during winter is a common orchard operation in deciduous fruit crops grown in hilly regions. Trees are winter pruned to establish the basic tree architecture, manipulate the tree size and better sunlight interception into the canopy. Dormant pruning is the mechanism for carbohydrates and energy production to support growth and development; enhance air circulation, increase strength of crotch angles, stimulates vegetative growth by diverting the stored energy in fewer buds, maintain the right balance between vegetative growth and fruiting and removal of dead/infested wood. While all pruning methods have a dwarfing effect, dormant pruning tends to stimulate the newest growth while simultaneously reducing trunk size and canopy volume compared to unpruned trees. Although dormant hedging or shearing has been effective in controlling tree size, it often leads to an abundance of shoots along the tree's periphery, resulting in densely packed canopies (Forshey *et al.,* 1992). Peach trees subjected to severe pruning reduce canopy volume by 21 per cent (Miller and Byers, 2000). Although dormant pruning may initially appear to stimulate growth, the total dry matter produced by pruned trees is typically less than that of unpruned trees (Myers and Savelle, 1996). To mitigate localized and invigorating effects, pruning primarily consists of thinning cuts, with only a few heading cuts (Marini *et al.,* 1993).

The best time to start dormant pruning depends upon a number of factors like size of orchard and availability of labour. However, pruning in early winter is hazardous than late winters, because extremely low temperature following pruning operation may result in serious freezing injury to the trees. Therefore, in areas, where sub-zero temperature occurs, the pruning is recommended to be done in latter part of the dormant season before bud burst. Pruning before dormancy removes previously stored reserves from the trees and frost can damage young trees more easily. Younger trees are therefore pruned later than older trees. Pruning performed towards the end of winter, favours less growth and the strength of the shoots.

Summer Pruning

Summer pruning is performed between rapid shoot growth until the end of the growing season, is a devigorating process and involves removing photosynthetic leaf area, thus reducing carbohydrate production (Sallato, 2022). Summer pruning accounts for manipulation of tree size; promote sunlight distribution within canopy, maximize photosynthesis in shaded canopy areas; reduces branch crowding and amount of foliage inside the canopy, enable better air circulation that reduces humidity and allows better coverage of plant protectants (pesticides) and improves fruit quality. However, excessive intervention in the leaf mass can result in reduced sugar and acid content of the fruit and moderate fruit colour. Summer pruning increases the photosynthetic photon flux density (PPFD) within the tree canopy

(Morgan *et al.,* 1984). Summer pruning enhances light penetration within tree canopies (Marini and Barden, 1987), although levels may decrease in the following year (Taylor and Ferree, 1984). Light interception and distribution into the canopy promoted by summer pruning enhances fruit coloration especially for varieties grown in regions where color development is suboptimal besides boosting the fruit size and weight. Summer pruning performed in early July has been found to result in a significant yield increase (Musacchi *et al.,* 2014).

Early summer pruning induces development of 2-3 new shoots from the buds just below the cut and vigour to the shoot besides prolonging succulent growth. This growth could be detrimental to flower formation and shoots need to be re-pruned for formation of flower buds in the same year. Therefore, in productive trees, pruning is recommended to be performed late in summer during mid-July to mid-August when the shoots tip is terminating in buds. Summer pruning is generally confined to one-year-old shoots or a lateral emerges from primary scaffolds. The shoots which are 5-6 inches long and terminate in flower buds are not summer pruned. Some of vigorously growing shoots are removed from the base, while few selected shoots are cut leaving 3-4 buds (4-5 inches). Summer pruning promotes the development of spurs and a better fruiting system in the older section of the branch.

REFERENCES

Anonymous, 2014. *Fruit Production and Protection.* Directorate of Extension Education, Dr. YS Parmar University of Horticulture and Forestry. 286p.

Atkinson ET, 1882. The Himalayan Gazetteer. Vol II, Cosmo Pub, Delhi, 369 p

Brown G and Schimanski L, 2002. New weapon to combat apple replant problem. *Pome Fruit Australia,* pp. 10-12.

Cook NC and Strydom DK, 2000. The South African high-density system. *Acta Horticulturae* **513**: 321-27.

Edwards L, Utkhede RS and Vrain, T, 1994. Effect of antagonistic plants on apple replant disease. *Acta Horticulturae* **363**: 135-40.

Ferree DC and Schupp JR, 2003. Pruning and training physiology. **In**: *Apples: Botany, Production and Uses* (Ferree DC and Warrington IJ Eds.) CABI Publishing, Wallingford, Oxon, United Kingdom pp. 319-44.

Forshey CG, Elfving DC and Stebbins RL, 1992. *Training and pruning of apple and pear trees.* American Society for Horticultural Science, Alexandria, Virginia, USA. 162 p

Ghosh SP, 1985. Horticulture in North-Eastern India. Associated Pub Co, New Delhi, pp 230.

Greene DW and Autio WR, 1994. Notching technique increases branching of young apple trees. *J. American Soc. Hort. Sci.* **119**: 678-82.

Isutsa DK and Merwin IA, 2000. *Malus* germplasm varies in resistance or tolerance to apple replant problem in a mixture of New York orchard soils. *HortScience* **35**: 262-68.

Ladon T, Chandel JS, Sharma NC and Verma P, 2024. Optimizing apple orchard management: Investigating the impact of planting density, training systems and fertigation levels on tree growth, yield and fruit quality. *Sci Hortic.* **334**: 113329.

Laurent S, Merwin A, Fazio IA, Thies G and Brown JE, 2010. Rootstock genotype succession influences apple replant problem and root-zone microbial community composition in an orchard soil. *Plant and Soil* **337**: 259-72.

Marini RP and Barden JA, 1987. Summer pruning of apple and peach trees. *Hort. Rev.* **9**: 351-75.

Marini RP, Barden JA and Sowers D, 1993. Growth and fruiting responses of 'Red Chief Delicious' apple trees to heading cut and scaffold limb removal. *J. American Soc. Hort. Sci.* **118**: 446-49.

Miller SS and Byers RE, 2000. Response of winter-injured peach trees to pruning. *HortScience* **35**: 421-50.

Morgan DC, Stanley CJ, Volz R and Warrington IT, 1984. Summer pruning of Gala apple: the relationship between pruning time, radiation penetration and fruit quality. *J. American Soc. Hort. Sci.* **109**: 637-42.

Musacchi S, Abeti C, Ancarani V, Bucci1 D, Gagliardi F and Serra S, 2014. Summer pruning of apricot cvs. 'Bora' and 'Carmen Top'. *Acta Horticulturae* **1058**: 85-93.

Myers SC and Savelle AT, 1996. **In**: *Tree Fruit Physiology Growth and Development a comprehensive manual for regulating deciduous tree fruit growth and development.* Maib K. M. (Ed.) Good Fruit Grower, Yakima, WA.

Ouellette DR, Unrath CR and Young E, 1996. Manual and chemical branch inducement in fall-and spring-planted 'Empire' apple on two rootstocks. *HortScience* **31**: 82-88.

Papola TS, 1996. Planning for environment and economic development in mountain areas -concepts, issues and approaches. Discussion Paper No. MEI 96/2. ICIMOD, Kathmandu, Nepal.

Robinson TL, 2003. Apple orchard planting systems. **In**: *Apples: Botany, Production and Uses* (Ferree D. C. and Warrington I. J. Eds.) CABI Publishing, Wallingford, Oxon, United Kingdom 345-408 pp.

Robinson TG, Fazio HA and Hoying S, 2004. Performance of the new Geneva apple rootstocks in trials in the US, NZ and Europe. *Compact Fruit Tree* **37**: 91-94.

Rumberger A, Merwin IA and Thies JE, 2007. Microbial community development in the rhizosphere of apple trees at a replant site. *Soil Biology and Biochemistry* **39**: 1645-54.

Sallato B, 2022. Pruning and Training Systems. Washington State University, Fruit Tree. https: //treefruit.wsu.edu.

Sharma NC, Verma P, Singh N and Babita, 2020. Causes and control measures of apple replant problem. *International Journal of Bio-resource and Stress Management* **11**: 246-52.

Singh N and Sharma DP, 2017. Soil treatments and rootstocks against apple replant problem. *International Journal Chemical Studies* **5**: 2164-70.

Soni M, Thakur M and Modgil M, 2011. *In vitro* multiplication of Merton I. 793–An apple rootstock suitable for replantation. *Indian Journal Biotechnology* **10**: 362-68.

Taylor BH and Ferree DC, 1984. The influence of summer pruning and cropping on growth and fruiting of apple. *J. American Soc. Hort. Sci.* **109**: 19-24.

Verma HS, 2002. Use of DRIS in integrated nutrient management in temperate fruits. **In**: *Enhancement of temperate fruit production- In changing climate.* Jindal, K.K. and Gautam, D.R. (Eds.), pp. 289-95.

Verma P, Sharma NC, Sharma DP and Thakur H, 2023. Induction of lateral branches in Apple (*Malus × domestica* Borkh.) by notching. *Applied Fruit Science (Erwerbs-Obstbau)* **65**: 2219-28.

Yao S, Merwin IA, Abawi GS and Thies JE, 2006. Soil fumigation and compost amendment alter soil microbial community composition but do not improve tree growth or yield in an apple replant site. *Soil Biology and Biochemistry* **38**: 587-599.

Chapter 16

Nursery Production, Orchard and Canopy Management in Grapevines in Temperate and Sub-tropics

N.K. Arora, Gagandeep Kaur and Rachna Arora

Department of Fruit Science, Punjab Agricultural University, Ludhiana–141004, Punjab

Grapevines (*Vitis vinifera* L.) belong to family Vitaceae and are generally deciduous woody vines, characterized by tendrils and the fruit is botanically a berry, appearing in clusters. Most of the domesticated grapevine species come under the genus *Vitis* native to the Mediterranean and Central Asian region. The ability of grapevine to develop deep root systems allows its easy adaptation in a wider range of agroclimatic conditions. Worldwide 27.9 million metric tons of grape were produced in 2022-23. China is the largest grape producer in the world with production volume of 12.73 million metric tons per year. India is ranked 7th in grape production with yearly production of 37,13,187 metric tons of fresh fruit.

Nursery Production

Grapes are generally propagated asexually, however, rarely, some varieties are propagated through seeds also. But the new varieties developed through seeds have to be propagated asexually to ensure that the progeny retains the desired properties of the mother plants.

Propagation Technology and Rootstocks

Propagation through stem cuttings is the most used method in grapes, particularly in the region where own-rooted plants are grown. However, in the recent past, the problem of salinity stress and moisture stress contributed to the sharp decline in the productivity of vineyards established on their own roots. In such cases, grafting is the method of choice for propagation. Therefore, the salinity and drought-tolerant rootstocks like 'Dogridge' came into use and the vines grafted onto it out yielded their own-rooted ones of the same varieties. However, both methods have their advantages and disadvantages. Initially, during vineyard establishment, the rootstock performed better. But later, the plants encountered many problems such as reduction in fruitfulness, uneven bud sprouting and decline in yield *etc.* Some other rootstocks came into use later, such as 110R. Later, it was found that the performance of rootstocks varied in different soil and climatic conditions. In northern India, most of the commercial plantations are own rooted, and propagation through hard wood cuttings is the only method of propagation. While in states like Maharashtra, Karnataka most of new grape's plantations comprise of grafted plants.

Preparation and Planting of Cuttings

The plants, which are healthy, free from diseases, and earmarked for quality bunches should preferably be used for the preparation of cuttings. The cuttings of pencil thickness and having at least 4 buds are prepared from one-year-old mature shoots (canes). The terminal shoots are avoided for preparation of cuttings, as they are very thin and may be infected with anthracnose disease, which starts from tips towards middle of shoot. The upper cut of cuttings should be away from the bud, while lower cut should be near the bud, which helps to promote better, rooting. The main criteria of the time of preparation of cuttings are the period before sprouting of dormant shoots. Keeping this in view, cuttings must be prepared before mid-February. The cuttings after preparation are tied in bundles and are buried in moist soil for 1-2 weeks. The process allows callusing on cut ends which subsequently help in better rooting of these cuttings.

After callusing the cuttings are planted in nursery beds at a distance of 15 cm from cutting to cutting and 30 cm from line to line. While planting cuttings, $2/3^{rd}$ portion should be buried in soil, and $1/3^{rd}$ above the soil. At the time of planting, chlorophyriphos treatment is must to avoid termite attack. Grape cuttings are highly susceptible to termites.

Rootstocks

In Europe, U.S.A., Australia and Russia to tackle the problems of soil borne diseases, insect-pests (phylloxera) grapes are grafted on compatible rootstocks. Recently, in commercial grape growing area of India, due to problems of soil salinity, drought, nematode and poor fruitfulness, use of rootstock tolerant to these problems is gaining popularity over own rooted plantation. Grape rootstocks have been found to impart resistance to soil borne fungi, drought and vigour to the scion. Hence, the choice of rootstock should be made with careful consideration.

Characteristics of some important rootstocks are given as below:

- **Dogridge**: It is seedling selection from *Vitis champini,* which is recommended for use with wine and raisin varieties on light sandy soils. It is highly resistant to nematode infestations and produce most vigorous grafted vines. Cuttings of this rootstock root with great difficulty and once rooted they bud and graft readily.
- **St. George**: It is also known as Rupestris St. George. This rootstock is a pure seedling selection chosen from the wild type species *Vitis rupestris.* It has been planted more extensively than any other rootstock in most of the grape growing countries. It is vigorous, roots-well from cuttings, grafts well and is highly resistant to phylloxera and drought. This rootstock produces only staminate flowers.
- **Teleki 5-A:** It is a hybrid between *Vitis berlandieri* and *Vitis riparia* which show moderate degree of resistance to nematodes and highly resistant to phylloxera.
- **1616** (Syn. Couderc 1616, Solonis Riparia 1616): It is hybrid between *Vitis solonis x Vitis riparia,* having moderate resistance to phylloxera and is highly resistant to nematodes. In sandy loam soils, it produces scions of moderate vigour that yield good crops. Cuttings of this variety easing to root and graft readily.
- **1613** (Syn. Couderc 1613, Solonis x Othello 1613, Solonis-othello): This rootstock is a hybrid between the species *Vitis solonis* and the fruiting variety Othello. It produces moderately vigorous scion and is resistant to nematodes and moderately resistant to phylloxera.
- **Salt Creek** (*Vitis champini*): This rootstock imparts great vigour to its scion variety and is resistant to nematodes. It has performed well with wine and raisin varieties of California in light sandy soil of low fertility. Rooting of cutting is poor but bud take is higher than Dogridge.
- **110 R** (Richter 110): It is reported to be moderately vigorous to vigorous. The initial vine growth is slow because it first develops the root system. It is not appropriate for varieties with irregular fruit set. It is reported to have a very long vegetative cycle, which delays maturity. Its roots are not as deep growing as those of Richter 99 or Rupestris du Lot. Richter 110 is well suited to all kinds of soils, including acid soils. It is an excellent rootstock in warm grape-growing areas with an arid climate. There is no single ideal rootstock for a particular cultivar or agro-climatic region. The choice depends on some other important factors, such as, soil profile, availability, amount and quality of irrigation water, climatic conditions, vineyard nutritional management *etc.*

Orchard and Canopy Management

Irrigation

Water management is very important for optimum grape production from a vineyard. The irrigation during dormant season should be restricted, as it results in early sprouting of vines. Vines would not resume normal growth in spring unless irrigation is applied to replenish the soil moisture. Any water stress condition hampers growth and productivity of the vine *viz.*; the shoot growth decreases and the internodes become shorter, the tendrils drop, the leaf margins curl and the older leaves turn yellow. The irrigation during sprouting or initiation of new growth is must to have uniform vine growth and initiation of flowering, while during flowering should again be restricted, as the excessive irrigation during flowering diverts food energy towards vegetative growth and thereby leading to imbalanced flowering and vegetative growth. Irrigation during fruit development is must, as the berries are constituent of more than 90 per cent of water. During maturity and ripening, the irrigation should again be restricted to increase TSS/acid ratio and results in better fruit quality. The irrigation after harvesting should again be restricted as rainy season approaches during this period.

Under Punjab conditions, one irrigation should be done after pruning in the first fortnight of February. The second irrigation is applied in first week of March. After fruit set in April till first week of May, irrigation is recommended at 10 days intervals. During the rest of May, at weekly interval, during June at 3 or 4 days interval and during July to October, irrigate when prolonged dry spell prevails or rainfall is insufficient. One irrigation is recommended from November to January, if soil gets extremely dry.

Manuring and Fertilization

Adequate nutrition supply to the grapevines is an essential pre-requisite from the time of vineyard establishment as it determines the plant vigour and overall health of the vineyard. Grapevines show slow growth during initial few months of planting due to very less carbohydrate reserves in the transplanted rooted cuttings or grafted vines. Optimum fruit yield, quality and maximum returns can be achieved from a vineyard only if nutrition is maintained from planting stage.

As vineyard soils are either sandy loams or heavy clays, the usage of organic manure has assumed high importance in India. A standard dose of 60 g urea and 125 g muriate of potash is applied per plant in April and repeat application is given in the month of June. In older vines planted at 3 x 3 m in Bower system, following fertilizer schedule is recommended:

Age (Years)	*Farmyard Manure (kg/tree)*	*Dose per Tree (g)*		
		Urea	*Super-phosphate*	*Muriate of Potash*
1	20	400	1500	250
2	35	500	2500	350
3	50	600	3500	500
4	65	800	4000	650
5 and above	80	1000	4500	800

Apply whole Farm yard manure and superphosphate as well as half N and K fertilizers after pruning. Apply remaining N and K after fruit set in April. Use of high doses of fertilizers may result in excessive vegetative growth and barrenness. Complete shade under bower is an indication of excessive growth. In that case, the nutrient content of soil should be checked with a soil test and the fertilizer dose be adjusted, if necessary. Give two sprays of urea @ 1.0 per cent first at full bloom and second at fruit-set to get higher yield and better fruit quality in the Perlette grapes.

The diagnosis of nutrient deficiency should be made on the basis of soil and plant tissue analysis.

Training and Pruning

Being indeterminate in growth, grapevines need proper support to develop canopy architecture. Properly trained vines will have increased sunlight distribution within the vine canopy. It will ensure better photosynthesis, more fruiting as well as better quality fruits. Pruning of the vine ensures balanced distribution of bunches on vines and increase in fruiting area on vines. It helps to maintain the vitality of the vine to get consistent productivity. During the first two to three years of growth, pruning is done to shape and train the vine. This is achieved by tying the growing vine along with a strong support to develop a strong trunk with well-placed laterals to form a permanent framework. The grape bears on growth of one-year-old shoots called cane. Pruning is done to regulate yield, size and quality of fruit. Regular and right pruned vines bear the good quality fruit for longer time. The technique of training and pruning is explained below:

Training

An ideal support/training system is one on which quality fruits/bunches can be produced, involves ease in carrying out cultural operations, there is less disease incidence, can accommodate more number of plants per unit area and is durable over the years. The choice of training system depends upon the bearing habit and vigour of a cultivar. For instance, Anab-e-Shahi cultivar of grape which is very vigorous should be trained on Bower system. Some cultivars with medium vigour (Perlette, Flame Seedless, Beauty Seedless, Thompson seedless and Punjab MACS Purple) may be trained on head, kniffin, bower, Y- trellis or telephone system.

Training System of Grapes

Head System

This system is suited to those cultivars which produce fruitful shoots close to the base of the canes. The vines are planted at the distance of 2.5 to 3.0 meters. After planting the rooted cutting, vine grows luxuriantly. The side branches should be removed as soon as they appear and the main trunk should be cut back at a height of about one meter in July. This would encourage the growth of a few laterals below the cut. During the next January, when the vines are dormant, these lateral branches should be shortened to 1 or 2 bud spurs. When growth starts in spring these spurs will produce a number of shoots which are retained as framework arm. During the third dormant season, 8-10 arms growing around the head with one or two spurs are retained. This training system is the cheapest and very easy to adopt and maintain. Although the yield per vine is low, it is compensated by having more number of vines per unit area. This system is suitable for spur bearing variety like Flame Seedless, Perlette, and Beauty Seedless.

Kniffin System

This system is named after William Kniffen of New York who first developed it in 1850. This system of training of grape vines is good for cultivars bearing on spurs as well as on canes. Trellising of Kniffin system should be completed before planting. The angle iron posts are securely fixed in the ground at 6 m from row to row and 3 m from plant to plant. Two or three wires (10 gauges) are stretched along these posts; the lower wire should not be less than 75 cm from the ground and the second wire about 50 cm above the first and so on. In the beginning, the vine is trained on this system just as it is done for head system. A strong trunk is developed which is then cut-back at the level of the top wire of the trellis. This would firstly encourage the growth of canes, the two of which are selected and tied firmly to the top wire. Similarly, two canes growing nearest to the lower wire are tied to it, one on each side of the trunk removing all other canes. Two renewal buds are kept near the base to produce canes for the next year's crop. This system is comparatively more expensive to establish than head system.

Telephone System

This system is constructed on a horizontal plane like that of wires on telephone poles. This system combines most of the advantages of Kniffin and bower system while eliminating many of their disadvantages. Angle-iron posts, standing two meters above the ground, are fixed in the soil at a distance of 6 m from each other in the row and 3 m from plant to plant. Two angle-iron arms, about 13 m long, one on each side of the post are shouldered on the post at 45° angle at the height of one meter. Three wires (l0 mm gauge) are then stretched, one through the main post and one through each of the side arms. The side arms are strongly braced.

Bower System

This system is suited for training very vigorously growing cultivars like Anab-e-Shahi and others. The 'Perlette' commercial cultivar of grapes in Punjab is also trained on this system. Bower may be constructed with angle-iron, concrete or brick pillers. Brick pillers are best as they withstand the stress. The brick pillars 2.10 m tall, 37cm thick are constructed at 3-5 metres in the row and from row to row. Angle-irons are fitted at the top of the pillars and 10–12-gauge wires are stretched both ways at 50 cm through these angle-irons.

The vines are planted at distance of 3.0 m in the row and double the distance between the rows in case of cane pruning cultivars and at distance of 3mX3m in case of spur pruning cultivars like Perlette. The vines are trained during the first year just to develop a good root system. When the vines start growth during the second year, they are tied in a fan shaped manner to the front wires of both sides. The arbour should be covered as early as possible and developing shoots should be regularly tied loosely to the wires. This system is more costly to establish than the kniffin and Telephone system. Other disadvantages are (1) Difficulty in spraying the vines against insect, pests and diseases (ii) pruning is very tedious and time consuming. However, this system gives higher return than the other systems of training and is suitable for medium as well as vigorously growing cultivars of grapes.

Figure 16.1: Bower Training System.

Y-trellis System

Conventionally bower/pandal system is used for training vines. The bower system has several disadvantages like difficult and labour-intensive cultural operations, high disease incidence and poor fruit quality. Due to shade under the bower, there is enhanced incidence of fungal diseases like anthracnose and powdery mildew. The manual sprays of fungicide and insecticides for disease and insect pests control virtually fall on the field worker, thus making it a health hazard. As an attractive alternative a new training system (Y-trellis system) has been evaluated and recommended by PAU, which has several advantages over conventional bower system.

Advantages of Y-trellis system over bower system:

a) *Early fruit maturity*: The fruit on the vines trained on Y-system mature 2-3 days earlier than on the bower system.
b) *Better fruit quality*: The TSS of the fruits on Y-system is 17-18 per cent as compared to 16-17 per cent TSS obtained on bower system.
c) *Low incidence of diseases*: As the bunches are exposed to sun, there is lower humidity around the bunches as prevalent under bower system, due to shade of bower itself, the incidence of fungal diseases like anthracnose and powdery mildew is relatively low in Y-system.
d) *Ease in cultural operations*: As the vines are at the average height of field worker, it is easy and economical to carry out fruit quality improvement technologies like flower bud thinning, GA_3 dips and girdling.
e) *Less health hazard*: The health hazards while using fungicides and insecticides are relatively less as compared to conventional bower system.

Training of Vines on Y-trellis

- *First year:* The planting should be done in the months of Jan- Feb at 5 x 12 feet, plant-to-plant and row-to-row, respectively. The side growing shoots should be regularly pinched off, so as to maintain a single main shoot (a) up to 4 feet. Then the main shoot should also be pinched off, so that side shoots emerge. Two side shoots (primary arms) (b and c) in each opposite direction should be developed, up to the distance of lowest wire.
- *Second year:* The side shoots (b and c) should again be pinched, so that the two secondary shoots (b1, b2 and c1, c2) in each horizontal direction of a wire are developed. A sort of H is thus formed after this training *i.e* one main shoot, two primary shoots and four secondary shoots.
- *Third year:* The shoots developed on secondary shoots should be allowed to grow vertically on both the sides of Y system, so that the maximum canes (one year old shoots) are developed on secondary shoots.

The pruning in the 4th and the subsequent years should be aimed at maintaining 40-50 equally distributed canes per plant and 4 buds per cane. In general, the shoots crossing the upper most wire of a Y should be pinched off after fruit harvest.

The Y system of training is superior as compared to conventional bower system in respect of early fruit maturity, better fruit quality and low cost on cultural operations and has minimal health hazards.

Figure 16.2: Y-trellis System.

Flat Roof Gable System

This system is particularly followed for vigorous vines (vines grafted on rootstocks) and has the advantage of both bower and the extended Y systems. In this system an inter-connected Y trellis forming a flat roof gable is being adopted. This system has been developed at IIHR, Bangalore and is gaining popularity among the growers due to certain advantages like ease in management; clusters are protected from direct sunlight and well exposed to sprays of pesticides and the clusters hang within the reach of the worker of an average height.

Pruning

Grapevine branches show acropetal growth habit and continue climbing with tendril. This behaviour is more pronounced in tropical conditions and if branches left as such proximal buds become unsprout or bear small size cluster. Hence, grapevine needs regular pruning every year. During the process, almost 50 per cent of the shoots developed in a year are removed. The aim of pruning is to evenly distribute the bearing branches all over the vine and to regulate the so that it bear crop for longer period. The one-year-old mature shoots are called canes. Pruning severity depends upon the bearing habit of a cultivar. The number of canes and renewal spurs to be kept on each vine at the time of pruning depends upon vigor of the cultivars, training system and cultural practices. There are two systems of pruning:

Spur Pruning

Spur pruning is followed in the cultivars which bear at the basal portion of the cane and each spur possesses upto 4 buds *e.g.*, Beauty Seedless, Perlette, Flame Seedless and Punjab MACS Purple *etc.*

Cane Pruning

This type of pruning is followed in cultivars which bear on their distal buds, *i.e.* 5-10 buds *e.g.* Thompson Seedless and Anab-e-Shahi.

Table 16.1: Pruning Schedule of Important Grapes Cultivars Grown in India

Sl.No.	*Cultivar*	*Distance (m)*	*Training System*	*No. of Canes per Vine*	*No. of Buds per Cane*
1	Perlette, Beauty Seedless, Flame Seedless, Punjab MACS Purple	3 x 3	Bower	60-80	4
2	Thompson Seedless	3 x 3	Bower	40-50	5-12
3	Anab-e-Shahi	3 x 6	Bower	80-100	6-8

Pruning Time and Intensity

In north India, pruning is generally done during dormant season before sprouting i.e from second fortnight of January to second fortnight of February. Severity of pruning depends upon variety and training system. At present maximum area under grapes in this region is occupied by Perlette, Flame Seedless or other spur bearing varieties. Only one crop is taken in these varieties under northern India. The spur pruning is done, retaining 3-4 buds/canes. These are termed as fruiting canes. About 60-80 canes on bower system are retained per plant, every year after pruning. It is important to keep canes uniformly placed on the vine. Along with fruiting canes, almost an equal number of shoots having only 1-2 buds (spurs) are also retained on a vine. In a particular growing season, the fruiting canes will bear bunches and the spurs will develop into vegetative shoots, which in the next season act as fruiting canes. In this way, the vine potential to yield uniform fruit over the years is retained. Furthermore, pruning is necessary to keep the fruiting wood near the center of vine, otherwise if pruning is not done regularly every year, the central portion of vine near main stem becomes barren. It is also easy to manage the vineyards by practicing necessary regular pruning.

Fruit Thinning

Fruit thinning in terms of flower bud thinning or removal of some clusters are important practices in grape cultivation helpful for quality improvement. The management of optimum load on vine also maintain the health of vineyard for longer time. Different grape varieties do not set the fruit equally well with normal thinning. Some tend to set straggly clusters, others nearly perfect clusters while some like Perlette over compact clusters. To obtain maximum improvement in fruit quality in these varieties, a set of different methods of thinning have been recommended.

Flower Cluster Thinning

This practice consists of reducing the number of clusters without reducing the number of leaves. The purpose is to balance the crop load and increase the leaf area: cluster ratio. With an increase in leaf cluster ratio, the flowers on the retained clusters are better supplied with food materials. The flower parts develop more perfectly, fruit set will be more and there will be a large percentage of normal berries. Flower cluster thinning is followed in cultivars which bear more number of clusters than desired. Flower cluster thinning can also be practised to rejuvenate weakened vines. The flower cluster thinning in Flame Seedless cultivar under northern India beneficial for uniform colour development of cluster.

a) Un-thinned Panicle

b) Panicle after thinning

c) Girdling

Flower Bud Thinning

In this process, almost $1/3^{rd}$ of unopened flowers are retained and the rest are removed using a special brush having distantly placed bristles. Flower buds are thinned carefully by holding a bunch with one hand and brushing slowly from base of bunch towards the tip (narrow side of bunch). After brushing, uniform flower buds should be seen on the bunch. It is necessary to irrigate the field after brushing, so as to achieve cent percent retention of unbrushed flower buds. Flower bud thinning is done one week before flowering, when flower buds are still closed. Flower bud thinning after flowering damages the ovary and leads to deformities in developing berries at later stages. This methods is very famous in cultivar Perlette which tend to set heavily and bear compact cluster.

Cluster Thinning

It consists of removal of entire clusters after the fruit set. This method has no effect on fruit set or cluster length. In this practice misshapen, small sized and oversized clusters are removed. This provide more favourable conditions of growth for the remaining clusters, giving larger berries. Cluster thinning is easiest and best means of reducing crop on overload vines of highly productive wine grape or raisin grape vineyards. Cluster thinning, however, is a time consuming and costly affair. In case of Perlette, thinning of bunches is necessary so that the remaining bunches develops properly. It is recommended that not more than 100 bunches be left on each vine (trained on bower), planted at a distance of 3 m × 3 m. The thinning should be done soon after the berry set and all small and under developed clusters should be removed.

Berry Thinning

Berry thinning consists of removing parts of clusters or individual berries after fruit set. The rachis of the cluster can be cut back to retain the derived number of berries. In addition to its removal of some of the laterals it provides more space for berry development and avoids compactness of the clusters. Berry thinning should be done soon after fruit set especially in seeded cultivars. The removal of individual berries is very time consuming and uneconomic. Berry thinning is not as effective in increasing the berry size in seedless cultivars. It is, however, very important in preventing over compactness of the clusters when berry size is either increased by girdling or use of growth regulators.

Quality Improvements Practices in Grapes followed in Northern India

In northern India, Perlette occupy around 90 per cent area under grapes. This cultivar has compact clusters and bear low quality small berry size fruit, if kept as such. Hence, in Perlette grape and others cultivars bearing compact clusters, the following practices should be adopted to get optimum yield and quality:

1. Thin flower buds one week before flowering by leaving 100—120 flower buds per panicle.
2. When berry size is 4 mm, girdle the vine by removing a 2 mm wide ring of bark from the main stern and dip the clusters in 40 ppm GA_3. While girdling, there should be no injury to the wood and no piece of bark should remain attached with the wood.
3. One week after the first GA_3 + treatment, give a second dipping in 40 ppm GA_3.

4. The field should have sufficient moisture during girdling and at least 3 weeks after that.
5. Harvest the crop when attained the requisite TSS content.
6. Two foliar sprays of potassium sulphate @ 1.5 per cent, first one week after fruit set and second at colour break stage, improve the quality and colour of Perlette grapes.

Without Spray

Potassium Sulphate Sprayed

Girdling

Girdling consists of removing a narrow ring of bark entirely around the trunk, arm or cane. The common width is 4-5 mm. It is an old practice in grapes and was introduced into Greece accidentally as a means of improving the set of Block Corinth in 1833. It is essential that the ring be completely removed and even if a small portion of the bark is left, there is no effect. The immediate effect of a complete girdle is to interrupt the supply of carbohydrates and hormones so that their level increases in the parts above the girdle. The width of the girdle should be such that the wound heals in a short time. It is especially true when the trunk is girdled otherwise restricted food supply to the roots results in death of the vine. Similarly removal of wood below the bark injures the vine. Vines are girdled to accomplish one or more of the three objectives *viz.* to improve fruit set, to increase berry size and to advance maturation.

Fruit set in variety like Perlette can be improved by girdling the vine before blooming. Sizing of the Seedless cultivar like Thompson Seedless and Perlette can be hastened and their ultimate size can be appreciably increased by girdling timed to be effective during the period of rapid berry enlargement. If girdling is done to improve berry colour or enhance ripening, the girdle must be open and effective during the early part of the ripening period.

REFERENCES

Anonymous, 1985. Grape Bulletin. Maharashtra Grape Grower Association, Pune, pp.16.

Anonymous, 2012. Manual on good agricultural marketing practices for grapes Ministry of agriculture (department of agriculture and cooperation) Directorate of marketing and inspection branch, Nagpur (agmarknet.nic.in/grapes.pdf).

Bal JS, 2007. Grapes. In Bal JS (ed.) Fruit growing. Kalyani Publishers; 2nd revised edition. pp. 213-242.

Chadha KL, 1999. The Grape-Improvement, production and post-harvest management. Malhotra Publishing house New, Delhi.

Gill MIS, Arora NK, Kumar K, Karibasappa GS, Tetali S, Karkamkar SP, Misra S and Ghosh SN, 2014. Gprape. *In*: S. N. Ghosh (ed.) *Tropical and Sub-tropical Fruit Crops: Crop Improvement and Wealth*. Jaya Publishing House, New Delhi. pp. 293-334.

Keller M, 2010. The Science of Grapevines: Anatomy and Physiology. Academic Press, Burlington, Massachusetts.

Pearson RC and Goheen AC, 1988. Compendium of Grape Diseases. APS Press, St. Paul MN.

Pandey RM and Pandey SN, 1990. The grape in India. ICAR, New Delhi, pp. 115.

Singh SK and Dhillon WS, 2013. Grape In: Dhillon WS (ed.) Fruit Production in India. Narendra Publishing House, New Delhi. pp. 235-261.

Winkler AJ, 1965. General Viticulture, Univ of California Press, Berkeley and Los Angeles.

Chapter 17

Nursery Production, Orchard and Canopy Management in Dragon Fruit, Raspberry and Fig

Jaswinder Singh Brar and Trina Adhikary

Department of Fruit Science, Punjab Agricultural University, Ludhiana–141004, Punjab

Dragon Fruit (*Hylocereus* spp. and *Selenicereus* spp.)

Dragon fruit (*Hylocereus* spp. and *Selenicereus* spp.), commonly known as pitaya is now emerging as a 'super fruit' internationally. Due to the numerous health benefits and medicinal properties of this fruit, it is being considered a potential high-value fruit crop for diversification under the present scenario of climate change and changing dietary habits. The origin of dragon fruit is probably the tropical forest regions of Mexico and Central and South America. The commercial production of dragon fruit had started in South and Central America. Numerous other countries like Vietnam, Taiwan, Indonesia, Thailand, Philippines, Australia, USA, Sri Lanka, Cambodia, Israel, *etc.* are now cultivating dragon fruit on commercial scale. Dragon fruit introduced as a superfruit in India is expected to be a promising and profitable fruit crop. The large scale commercial cultivation of this exotic fruit crop first began in Karnataka, Gujarat, Maharashtra, Tamil Nadu, Kerala, West Bengal, Odisha, Telangana, and Andhra Pradesh states. Though it is a cactus plant of tropical areas, owing to its wide adaptability and tolerance for various soil and climatic conditions, it is growing successfully in sub-tropical regions too. Currently, its cultivation has extended to other states such as Punjab, Madhya Pradesh, Haryana, Rajasthan, Uttar Pradesh, and the North Eastern States of India.

Dragon fruit has received global appreciation; initially as an ornamental cactus and then as a fruit crop. In recent years, the popularity of dragon fruit has impressively augmented worldwide due to its appealing fruit and diverse pulp color, as well as the pleasant taste it offers. The fruits of this cactus been acknowledged as of supreme beauty in the Cactaceae family. The fruits can be consumed as fresh fruits or can be processed into many products such as juice, jams, jellies, wine, marmalades and beverages. It is rich in nutrients and low in calories. Dragon fruit contains antioxidants called betanins which are present in red pulped varieties as well as the peel of white fleshed dragon fruit and have radical scavenging ability. Dragon fruit contains phyto-chemicals such as polyphenols, flavonoids and vitamin C which confer robust antioxidant potential to this fruit.

Moreover, the proteins, minerals, and vitamins available in fruits adorn the nutritional value of fruits. The betalains; deep red pigments responsible for imparting the red colour to the fruit pulp in red pulped fruits and are very helpful in reducing the 'bad' LDL cholesterol. The diverse antioxidants present in fruits are linked to prevent many human ailments related to brain health, heart diseases, cancer, *etc.* In addition to this, the oil extracted from dragon fruit seeds contain a high proportion of essential fatty acids; especially linoleic acid. These essential fatty acids help in maintenance of brain functions, cell membranes and nerve impulses transmission. Moreover, these fatty acids are usually outsourced by humans as a source of essential fatty acids; which raises the value of dragon fruit from this standpoint as well.

Dragon fruit is a tropical fruit native to southern Mexico, Costa Rica, and Guatemala. At present, it is being cultivated in many countries around the world including Vietnam, China, Indonesia, Thailand, Taiwan, Malaysia, Australia, and others. Vietnam, China, and Indonesia account for more than 90 per cent of global dragon fruit production. Since its introduction in South Asian countries in the 1990s, there has been a significant expansion in the area under this fruit. In India, dragon fruit cultivation has been gradually increasing since the 1990s and by 2020, it had reached about 4000 hectares. Up to the year 2025, the area under dragon fruit in India is estimated to reach 50,000 hectare. Currently, more than 70 per cent of the dragon fruit area is located in Gujarat, Karnataka, and Maharashtra alone.

Species

Dragon fruit belongs to the family Cactaceae and genus *Hylocereus.* It is also nicknamed pitaya due to the presence of scales on its fruits. It has 14 species of which four species *viz. H. undatus, H. costaricencis, H. monacanthus* (Lem.) Britton and Rose (formerly known as *H. polyrhizus*) and *H. megalanthus* (formerly known as *Selenicereus megalanthus*) are the most widely cultivated species. *Hylocereus* species are recognized chiefly by their fruit morphology, pulp colour, areole characteristics, and the number and form of spines. The fruits of *H. undatus* have a white pulp and red peel, whereas the fruits of *H. monacanthus* and *H. costaricencis* have a red pulp and red peel and. The fruits of *H. megalanthus* are differentiated by white pulp and yellow peel.

This exotic fruit crop is new to our country and has different type of plant growth and behavior hence; requires unique production and management practices, so, the technical information for its successful cultivation in different regions of India is of utmost importance. Proper nursery management, orchard establishment, and canopy management are critical for optimizing productivity and fruit quality in dragon fruit cultivation.

Climatic Requirements

The ideal temperature for dragon fruit cultivation range is 25-35°C, though it can tolerate higher temperatures in semi-arid regions also. The well distributed rainfall around 1000mm with moderate temperature and high relative humidity favors the growth of vines. However, under sub-tropical conditions the extremes of temperature during winters and summers adversely impact the plantation; hence protective measures are required to minimize the temperature stress conditions. The proper and regular care and management of dragon fruit orchards is very necessary for higher and better quality fruit production. Wind breaks must be established around the dragon fruit plantation. Establishment of wind breaks by planting tall and fast growing trees such as barma dek, jamun, seedling mango, arjun, eucalyptus *etc.* especially of north and western side can be very beneficial for commercial plantation. The site should also be near the road for easy transportation and marketing of fruits

Selection of Site and Soil

Selection of site for planting dragon fruit is very crucial as heavy initial investment is involved for its plantation. Wrong selection of site for commercial planting may cause heavy economic loss especially under sub-tropical conditions. If possible, select the site towards the area of already exited tall trees to minimize the loss due to extremes of temperature. Never plant in the areas prone to floods or water stagnation. Dragon fruit plants thrive very well in soils with good drainage having pH around 7.0. The plant can also grow in marginal or degraded soils with some amendments, but it perform very well in sandy-loamy soils rich in organic matter.

Planting System and Canopy Management

Planting system and canopy management in Dragon fruit cultivation is vital for better exposure to light, air circulation, and overall productivity and fruit quality. The main aim of pruning and training is to maintain a balance between vegetative growth and fruit production along with minimizing the biotic and abiotic stress. Planting of dragon fruit can be done in the month of February -March and July-

September. As the dragon fruit is a climbing cactus, it requires proper training system. The rooted cutting should be planted at raised beds so that it may remain protected from excessive soil moisture or water stagnation during rainy season. Planting should be done using proper training system. Generally two types of training systems can be used for planting of Dragon fruit *i.e.* single pole or umbrella system and trellis system.

Single Pole or Umbrella System

Single pole system with concrete ring at the top is most preferred training system for dragon fruit. The height of the concrete pole should be 7.0-7.5 feet and it should be about 5 feet above the ground level. For longevity of the training system, the breadth of pole at the base should be 5-6 inches and at the top it should be 4-5 inches. The concrete pole should be erected at 2 feet depth at required distance. The concrete ring of 2 feet diameter should be fixed on the top of the pole to enable the dragon fruit vines to climb and hang downwards in such a way that a unit of 4 plants can attain an umbrella like shape. The reinforced concrete ring should be at least 2 inches wide for better strength. Four plants are planted on each side of the pole in such a way that these can climb up on the pole and develop aerial roots in the pole. In this training system about 500 poles can be installed and 2000 plants can be planted in one acre land area.

Figure 17.1: Single Pole Training System.

Figure 17.2: Trellis Training System.

Trellis System

For high density planting of Dragon fruit, trellis system can also be established. In this system about 2-3 time more number of plants can be planted as compared to single pole system. For the establishment of trellis system concrete poles or iron-angles can be erected at 10-12 feet distance in rows. On the top of the poles, the galvanized iron or steel wires can be fixed like a telephone system. Six wires should be fixed in such a way that 3 wires at 6 inches distance from each other should be erected on both side of the poles. The iron frame with 6 holes can be used for fixing of these wires. Care should be taken that the trellis system should be strong and durable enough to bear heavy weight in future. The plants can be planted singly in alternate fashion with two rows or in pairs at 1.5 to 2.0 feet distance between the poles and staked with bamboo stick to train them on to the wires. In this system, 4000-6000 plants can be accommodated per acre land area.

Pruning

Pruning in dragon fruit involves the removal of old, damaged, or excess cladodes to encourage the development of new fruitful branches. Pruning should be initiated right from the planting to force the plant to climb over the poles or trellis properly. The lateral shoots must be restricted and 1-2 main stems should be allowed to grow from each plant. The plants of dragon fruit are quick growing vines and during initial stage, these develop dense branches. The excessive lateral branches or lateral buds should be pruned regularly so that there may be no excessive growth below the rings and the plants should be tied with the help of plastic string. Likewise, in trellis training system, the support with bamboo sticks can be given to train the plants on the trellis-wires. Once vines climb up to the top of the poles through circular rings, the then lateral branches should be allowed to grow further. To spread the lateral growth, the tips of the cladodes above the rings can be removed. In this way, in single pole system, plants can attain an umbrella like structure to bear flowers and fruits. Properly developed and trained vine may produce 30-50 branches in a year and more than 100 branches in four years. In addition to pruning of vines for the purpose of training, it also comprises of removing all the damaged, diseased, intermingled stems. The cladodes affected with heat and frost under sub-tropical regions should also be pruned. Generally, the pruning is carried out during dormancy period to promote new growth before the next fruiting cycle. However, under sub-tropical conditions pruning can be done in first fortnight of February as well as in end of June to remove frost affected and heat stressed cladodes, respectively. For regular pruning, retains 100-120 cladodes per pole, leading to improved airflow and reduced sunburn. For rejuvenation of canopy, 60-80 cladodes can be retained per pole.

Propagation

The major hurdle in the area expansion in the different states is non-availability of true to type, high quality planting material. Dragon fruit is generally propagated through cuttings, seeds and grafting methods, however, the most successful method of plant multiplication is planting of cuttings.

Propagation through Cuttings

For multiplication of dragon fruit plants, select mature and healthy cladodes free of any symptom of disease or damage. The size of cutting should be at least 15-20 cm having multiple buds or nodes. Always use a sharp, sterilized knife or pruning shears for proper cutting of the cladodes with clean cut. These cutting should be cured by placing under shaded and dry place for 3-4 days. This practice will help to develop the callous on the cut end which will be helpful for prevention against infection. Always place the cuttings in flat or in upright position for even drying. For better rooting, IBA can be applied at 1000 ppm rate as a quick dip. The cut end of the calloused cutting can be placed in the IBA solution for 2-3 minutes. Cutting can be prepared after harvesting of fruits as well as before the onset of new vegetative flush. Under north Indian conditions, the cutting can be prepared in the month of February-March after winters as well as in the month of June-July. The propagation time may be adjusted according the suitability of planting time as the overgrowing nursery plants may be difficult to handle. Do not use fresh growth for preparation of cuttings; the multiplication of dragon fruit should be done using cuttings of previous year's growth of cladodes. As the time of preparation of cuttings, discard weak, distorted, damaged as well as disease infected cladodes. To avoid mixing of two or more varieties, extra precautions should be taken during preparation and planting of cuttings of different varieties of dragon fruit.

Figure 17.3: Propagation through Cuttings.

For planting of cuttings in the polybags, the soil mixture containing garden soil: well rotten Farm Yard Manure in 1:1 ratio can be used. For planting of cuttings in bed for the sale of bare rooted plants, a soil mixture of garden soil: FYM: Cocopeat in 1:1:1 ratio can be used. The cuttings will develop roots in few weeks. Insert the cut end of the cutting about 8-10 cm deep into the soil and apply light irrigation after planting, ensuring the soil is moist but not soggy. Sprinkling of water may be the better option. Partially shady conditions will be suitable for better growth and development of plants; however, for winter planting, cutting should be planting under playhouse conditions. There should be sufficient light penetration in the planning area. Check the soil moisture regularly and allow the top inch to dry out before watering again. The plants can become saleable when the newly emerged cladode attains about 20-30 cm length and there are sufficient fibrous roots.

Propagation through Seeds

Dragon fruit can also be propagated through seeds. The viability of seeds varies among different *Hylocereus* species as well as within the same species. The seeds can be extracted from fully ripened fruits by slicing, pulping, rinsing under tap water repeatedly to separate out the seed. After separating the seeds, place the seed on clean paper or muslin cloth the remove the mucilage properly. The seeds can be sown in plastic trays or polybags filled with Cocopeat: FYM: Soil mixture in 1:1:1 ratio. Spread the seed evenly and cover it with fine mixture of potting material. Do not sow the seed deep inside the potting mixture. The seed germination stars within a week and very slow growth occur in seed propagation method. Moreover, the plants propagated through seeds are not true to type and exhibit lot of variability among the plants. Hence; seed propagation is not recommended for propagation of plants of dragon fruit for commercial purposes.

Figure 17.4: Propagation through Seeds.

Irrigation Requirements

As the crop belongs to Cactaceae family, it requires very less amount of water except during most critical period. As the dragon fruit plants have shallow root system, so drip irrigation is of utmost importance for its commercial cultivation. The application of irrigation water varies with soil type, crop status and prevailing weather conditions. Under dry condition application of irrigation can be done at weekly interval, whereas in humid climate, it may be done at 15 days interval. The soil should remain moist but not soggy during vegetative and reproductive growth period. Increase in temperature lead to higher evaporation rates; therefore, increase watering frequency. Likewise in winters, reduce watering meaningfully, as the plant enters a dormancy phase and requires less moisture. In the areas with high rainfall, adjust irrigation to prevent waterlogging. Always plant the dragon fruit on the raised beds to prevent damage due to excess water. Under sub-tropical conditions of north Indian states, the irrigation during the summer months *i.e.* March to June is very critical for the development of the cladodes. The flowering and fruit setting starts from the end of June and continues till October. Drip irrigation system must be used to maintain soil moisture during fruit development. Four drippers of 2 lph capacity can be used for each pole for light irrigation. Always apply irrigation water during late evening hours. Avoid over irrigation during summer months and during dormancy period. Likewise, the irrigations in winter months should also be limited under such conditions. To reduce the evaporation from the plantation area, mulching with polypropylene woven sheet or weed mats can be done. The sheet of 1.2 m width can be spread on the beds to reduce evaporation from the soil as well as to check the weed growth. Mulching with paddy straw can also be done for this purpose. Spread 10 cm thick layer of paddy straw on the beds of dragon fruit plantation in the month of March after application of organic manures. Always apply irrigation water free of excessive salts for better growth and development.

Manuring and Fertilization

Organic manures play a key role in dragon fruit growth and development. Each pole should be applied with 10 to 15 kg of organic manure 15 days before planting. Application of inorganic fertilizers should be practiced 2 to 3 months after planting; the period of commencement of fast growth of dragon fruit plants. Application of urea, single super phosphate and muriate of potash should be done at 50, 100 and 50 g per year per pole in four equal splits with the increment of same quantity of fertilizers every year up to 4 years. After 4 years onwards, the chemical fertilizers should be applied with urea 250 g, SSP 500g and MOP 250 g in the same number of splits and at same time. Splits should be applied in February, May, August and September. Well rotten farm yard manure should also be applied at the rate of 5-7 kg per pillar per year age of the plants. After 4th years onward 20-25 kg of well rotten FYM can be added in the month of January-February. Fertigation is a highly effective way to ensure proper nutrition for dragon fruit plants. Through this method, plants will receive the optimum quantity of nutrients at the right time resulting promotion of healthy growth and optimal fruit production. Moreover, it minimizes the risk of runoff and leaching compared to traditional broadcasting methods. Regular and balanced nutrient supply supports vigorous growth and enhances fruit quality. Fertigation can be done every 2-4 weeks during the growing season. Adjust frequency based on plant growth and soil nutrient levels. Apply fertigation early in the morning or late afternoon to reduce evaporation and enhance nutrient uptake. Dragon fruit has specific nutrient needs that can vary based on growth stage. Studies suggest that in the fully grown dragon fruit orchards, the application of nitrogen (N) can be done at 100-120 kg/ha/year. Application of phosphorus (P) can be done around 40-50 kg/ha/year. Potassium (K), the crucial element for fruit development and quality can be applied at 80-100 kg/ha/year. However, the application rates may vary with soil type, region of planting, age of the plants *etc.* Likewise, elements like magnesium, calcium, iron, and zinc are also important, often supplied in smaller amounts through chelated forms.

Figure 17.5: Flowering Stage.

Figure 17.6: Fruit Bearing Stage.

Figure 17.7: Fruit at Maturity Stage.

Harvesting

The harvesting time for dragon fruit prevails between July-November under north Indian conditions, which bears in 4-6 flushes, however in southern and central states of India, fruiting flushes are higher. The fruit matures 30-55 days after flowering depending up on the months of the crop flush. The fruits of early flush mature in short period as compared to the late flushes due to prevailing climatic conditions. The fruit can be harvested 3-4 days after the peel colour changes from green to red or pink for domestic market., while for far markets/distant marketing purposes, it can be harvested a day after colour break. The harvesting of fruits can be done by using sharp knife in such a way that small fruit stalk may remain intact with the fruits and ensure that there is no damage to the peel of fruits on the stalk end side during harvesting. Dragon fruit start bearing fruits from second year of planting while the potential average yield from fully grown orchards may be 8-20 kg per pole depending up on the area of planting, variety and care and management of orchards.

Raspberry (*Rubus idaeus*)

Raspberry belongs to genus *Rubus* and the family *Rosaceae*. Wild plants are small shrubs commonly grown under tropical highlands and temperate forests. The raspberry is primarily cultivated in Russia, Europe and the Pacific Coast of North America. Most of the produce is used for processing. The crop harvesting operations are mostly carried out mechanically. Walk in tunnel are also being used for raspberries production in Europe for extended period of crop production and now it is becoming popular in North America also.

Nursery Production

Nursery production lays the foundation for raspberry plants, determining future productivity and quality. Key nursery practices include propagation method selection, disease control, and optimizing plant hardiness and size for transplanting. Micro-propagation of raspberry is an advanced technique enables the production of disease-free plants and provides the consistency required for large-scale commercial cultivation. Micro-propagated plants (*e.g.*, stage-IV types) show strong growth and productivity. It can also be propagated through suckers and root cuttings as it is less costly and widely used method. This method maintains genetic traits but is more susceptible to soil-borne pathogens. For nursery production, irrigation and fertigation is essential to ensuring healthy growth, consistent irrigation is crucial, particularly for tissue-cultured plants that are more sensitive during initial establishment. Controlled drip irrigation systems allow precise water and nutrient delivery. Isolation of nursery plants from production areas minimizes disease transfer risks, especially those causing crumbly fruit.

Planting and Layout

Proper orchard setup is critical for efficient raspberry production, affecting plant health, maintenance practices, and harvesting ease. Raspberries thrive in slightly acidic, well-drained soils with high organic content. Raised beds or ridged rows improve drainage and root aeration, benefiting growth, especially in humid or rainy climates. For planting, row spacing should accommodate airflow and sunlight penetration, typically 2-3 feet between plants within rows and 8-10 feet between rows. This layout supports healthier canes and reduces disease potential.

Training and Pruning

Various trellis systems support raspberry growth, reducing cane damage and improving yield. Systems like T-trellis and V-trellis structures offer support, while allowing easy access for harvesting and maintaining canopy openness. Pruning practices such as cane thinning and branch length adjustments are vital for maximizing yield. Practices vary depending on cane type (floricane or primocane), with floricanes removed post-harvest and primocanes pruned to promote lateral branching for increased fruit set. Effective canopy management practices enhance sunlight exposure, airflow, and overall plant health, reducing disease incidence and improving fruit quality.

For summer-bearing varieties, remove spent floricanes after the harvest to reduce disease risk and make room for new primocanes. Retaining only strong primocanes in each hill allows sufficient light exposure and minimizes overcrowding. In ever bearing or primocane-fruiting varieties, tipping stimulates branching, resulting in higher yield potential as more fruit-bearing laterals develop. However, excessive tipping can increase disease susceptibility, such as cane blight. Maintaining optimal canopy density ensures adequate sunlight reaches the plant, promoting fruit ripening and reducing humidity within the canopy, which can cause fungal diseases like Botrytis.

Harvesting

The high quality raspberries produce should be free from decay, injury and sunscald. The uniformly ripened and colored fruits with glossy appearance and still firm fruits are good for fetching better market returns and processing. The shelf life of raspberry fruit is very short and the fruits intends to decline rapidly in the field and even quicker after harvesting. Hence; for the maintaining post-harvest fruit quality, special consideration to pre-harvest and post-harvest factors are of utmost importance. Moreover, the fruit pickers must be trained to recognize the appropriate ripening stage and berry appearance for harvesting. For the wholesale marketing of fresh raspberry should be done on the onset of appearance of fully red color. The fruits can be easily pulled from the receptacle, and should be still firm. Before harvesting, the berries should not be touched as they are very fragile and prone to be damaged easily during harvest.

Fig (*Ficus carica* L.)

Fig (*Ficus carica* L.) is a sub-tropical fruit known for its nutritional value and unique properties as both a fresh and dried fruit. This fruit is commonly known as 'Anjeer' in India. The fig tree has been cultivated for centuries, with regions like the Mediterranean, South West Asia, and parts of India being primary centers for its growth. It thrives in dry, warm climates and is highly adaptable to drought conditions, making it a popular fruit in arid and semi-arid regions. As a fresh fruit, it is very delicious in taste. Over the centuries, figs have been esteemed for their medicinal and nutritional properties. The figs can be consumed fresh and processed into dried figs, candied, canned or used for jam making. The dried figs are one of the richest sources of energy and are known for premium preserved fruit throughout the world.

The tree being deciduous it can tolerate frost and low temperature without any injury. Very high temperatures in summer may cause scorching of terminal leaves. Heavy rains at the time of fruit ripening cause splitting of ripe fruits. It is grown on a wide variety of soils and grows well on heavy clays, rich loam and light sandy soils with good drainage.

Propagation

The successful cultivation of figs depends greatly on efficient propagation methods and well-managed orchards. Techniques such as hardwood cuttings, air layering, and grafting ensure the establishment of vigorous nursery plants. Generally, the figs are propagated through hardwood cuttings, typically taken during dormancy periods. The ideal cuttings are from 3-year-old wood, approximately 25-30 cm in length and 1.25-1.50 cm in diameter. They are selected from the base of shoots or the lower part of the crown, as these areas tend to have better rooting potential. Under northern Indian conditions, cuttings are collected in January-February, while in southern regions; they are gathered during the rainy season (August-September). The cuttings are planted either directly in the soil or in nursery beds. The cuttings should be planted on well prepared flat beds in the nursery. Keep one third portions of the cuttings above ground and bury the remaining two third in the soil. Root development generally occurs within a few weeks, making this method reliable for mass propagation. In addition to the cuttings, figs can be propagated using air layering. This technique involves inducing root formation on aerial parts of the plant before severing it from the parent plant. In addition to this grafting and budding can also be performed. Patch budding and side grafting are effective methods for producing high-quality plants, especially in commercial settings. Using a rootstock such as *Ficus glomerata* offers resistance to pests like root-knot nematodes. Propagation success is enhanced by maintaining ideal conditions in the nursery. Substrate selection is crucial, with peat-perlite mixtures being popular for their superior root aeration and moisture retention. In commercial nurseries, a trough culture method is also employed, wherein fig cuttings are grown in troughs to prevent contamination from soil-borne diseases.

Orchard Establishment

The best time to plant fig trees is either during the rainy season or early spring, depending on regional climate patterns. However, under sub-tropical climate, the fig plants can be planted in the month of Januray-Februray; before the emergence new vegetative buds. A square or hexagonal planting system is recommended, with spacing between trees depending on soil fertility and the variety being grown. In most cases, a spacing of 5-7 meters is recommended. The dwarf varieties can be planted at 3x3m distance also.

Training and Pruning

The height and shape of fig trees should be controlled to facilitate orchard operations such as harvesting and spraying. Canopy management ensures light penetration, which is crucial for fruit quality and overall tree health. Fig trees can be trained to modified leader system. The training is completed in 3-4 years from planting. It bears fruits individually in the axils of leaves on current season's growth. Properly trained trees can yield around 10-60 kg fruits per tree, depending on the variety and management practices. Pruning is a critical practice in fig orchards as it helps to manage the tree's canopy, promote fruiting wood, and maintain tree health. Fig trees produce two crops a year: the first on the wood from the previous season, and the second on the current season's wood. Pruning is, therefore, essential to encourage the growth of productive branches. Fig plants can also be trained to central leader system to maintain a strong structure and ease of management. To stimulate lateral branch formation, notching can be performed by making small cuts in the bark above a node, promoting better canopy development. Pruning is also necessary with a goal is to remove weak, diseased, or dead branches while retaining strong branches that will bear fruit. The timing of pruning should be adjusted according to local conditions to maximize fruit production. The fig varieties 'Poona' and 'Deanna' can be planted in high density planting system. About 1000 plants can be planted in one hectare area with proper training and pruning of plants. Initially, the plants are trained to open center and then annual pruning of previous season's shoots is done to the basal six nodes regularly. This system results in significant improvement in fruit yield as compared to normal planting system of 500 plants per hectare. Under sub-tropical conditions and in the winter dormant varieties, the ideal time of pruning is the middle of February under Indian conditions, when the plants are still in dormant phase.

Irrigation and Fertilization

While figs are drought-tolerant, proper irrigation practices can significantly boost yield. Irrigation during the summer months, especially once every 10-12 days, is recommended to maintain optimal soil moisture. However, irrigation must be carefully managed during the fruit-ripening stage, as excessive moisture can lead to fruit splitting and reduced quality.

Fertilization involves applying organic manure (such as 20 kg of farmyard manure per tree) and balanced doses of nitrogen and phosphorus to support growth. Nitrogen application enhances vegetative growth, while phosphorus promotes root and fruit development. To attain higher yield of quality fruits, suitable plant nutrition must be guaranteed in young as well as in mature orchards. The requirements of different plant nutrients differ with the age of plant, variety and soil type. The application of nitrogen can be done in two splits; one half can be applied after pruning and the other half can be applied two months after fruit set under North-West –Indian conditions.

Fruit Maturity and Harvesting

Fruits can be harvested picked when they become fully ripe but still firm. At ripening stage, fruit start to curve at the neck. Moreover, sudden increment in fruit size is another indicator of fruit maturity. The fruits should be harvested manually by twisting and pulling. Fruit pickers should also wear gloves and protective cloths as the latex oozing from the detached end of the fruit can cause skin irritation. The packing of fruits should be given due care to evade latex drops staining the surface of picked fruit. The freshly harvested figs are very delicate, so additional care is needed during harvesting and post-harvest handling operations. The fruits should not be packed in the carton in more than two layers.

REFERENCES

Anonymous, 2023. Dragon fruit. Package of practices for cultivation of fruits. Commination centre, PAU, Ludhiana. Ed. Brar, J.S. p. 142-144.

Doke A, Kakade VD, Patil RA, Morade AS, Chavan SB, Salunkhe VN, Nangare DD, Boraiah KM, Thorat KS and Reddy KS, 2024. Enhancing plant growth and yield in dragon fruit (*Hylocereus undatus*) through strategic pruning: A comprehensive approach for sunburn and disease management. *Scientia Horticulturae,* **337**: 113562.

Karunakaran G, Arivalagan M and Subbaraman S, 2019. Dragon fruit country report from India. *Dragon Fruit Network (DFNet): Marketing and Whole Value Chain and Steering Committee Meeting* (2019): 105-112.

Wakchaure GC, Satish Kumar, Meena KK, Rane J and Pathak H, 2021. Dragon Fruit Cultivation in India: Scope, Constraints and Policy Issues. Technical Bulletin No. 27. ICAR–National Institute of Abiotic Stress Management, Baramati, Pune, Maharashtra, India. pp. 1-44.

Nangare DD, Taware PB, Singh Y, Kumar PS, Bal SK, Ali S and Pathak H, 2020. Dragon Fruit: A Potential Crop for Abiotic Stressed Areas. Technical Bulletin No. 47. ICAR National Institute of Abiotic Stress Management, Baramati, Pune, Maharashtra, India, pp. 1-19.

Ramadan MF, 2023. Introduction to Fig (*Ficus carica*): Production, Processing, and Properties. In: Ramadan, M.F. (eds) Springer, Cham.

Sarkhosh A, Yavari A, Ferguson L, 2022. The fig: botany, production and uses. CAB International, Wallingford.

Crisosto H, Ferguson L, Bremer V, Stover E, Colelli G, 2020. Fig (*Ficus carica* L.),by Elhadi M. Yahia, In Woodhead Publishing Series in Food Science, Technology and Nutrition, Postharvest Biology and Technology of Tropical and Sub-tropical Fruits, Woodhead Publishing, pp. 134-160e.

Popa Roxana-Gabriela, Emil Cãtãlin 'chiopu, Aurelia Pãtra'cu, Aniela Bãlãcescu, and Florentina Alina Toader, 2024. Raspberry Production Opportunity to Develop an Agricultural Business in the Context of the Circular Economy: Case Study in South-West Romania. *Agriculture* **14**(10): 1822.

Kempler C, Hall H, Finn CE, 2012. Raspberry. In: Badenes, M., Byrne, D. (eds) Fruit Breeding. Handbook of Plant Breeding, Vol 8. Springer, Boston, MA.

Graham J, Hein I, Powell W, 2007. Raspberry. In: Kole, C. (eds) Fruits and Nuts. Genome Mapping and Molecular Breeding in Plants, vol 4. Springer, Berlin, Heidelberg.

Chapter 18

Orchard and Canopy Management in Fruit Crops of North-East India

T.K. Hazarika

Department of Horticulture, Aromatic and Medicinal Plants, School of Earth Sciences and Natural Resources Management, Mizoram University, Aizawl–796004, Mizoram

The North-Eastern region of India comprising eight states namely Arunachal Pradesh, Assam, Manipur, Meghalaya, Mizoram, Nagaland, Tripura and Sikkim lies between 21°.51 'and 29°.5' latitude and 85°.5' and 97°.5' E longitude. Nearly two thirds of the region is under hill terrain. The dense forest covers across 67.05 per cent of the total geographical area generates large quantities of humus that add to the fertility of the soil. Out of approximately 15,000 species of flowering plants available in India, the north-east region alone accounts for about 8,000 species and six out of the nine important vegetation types found in India. The great altitudinal differences in the region coupled with varied physiography contribute to extreme climate variations in the entire region, having definite pockets representing temperate, sub-tropical and tropical areas. The region has its own unique combination of living species, habitats and eco-systems, which together make up its diversity rich resource. The region is one of the richest reservoirs of genetic variability and diversity of different crops *i.e.* various kinds of fruits, different vegetables, spices, ornamental plants and also medicinal and aromatic plants.

Primarily, indigenous farming practices are followed in the North-East region which are chiefly organic in nature and depend on the indigenous technical knowledge systems ubiquitous in the region for centuries. The agro-climate of the region is generally rainfed, which is most conducive for organic farming. Low use of synthetic fertilizers (<12.0 kg/ha) and chemicals, abundant availability of biomass and animal manure are typical to the region.

The entire North-Eastern region is bountiful in horticultural diversity, yet research and development have remained at a very low level. Due to the lack of suitable planting material, lack of skilled labour, lack of trained personnel, lack of demonstration orchards, language barrier, non-existence of suitable transport facilities, non-adoption of suitable plant protection measures, non-existence of suitable marketing agency for the produce, non-adoption of suitable agro-techniques, the horticulture industry have not been exploited properly and still it remains in a neglected state. The increase in area and production of horticultural crops, conservation, management and proper utilization of the bio- resources will not only strengthen the nutritional security of the region, but also assist in boosting the local economies and sustainable livelihoods.

Fruit Crops of North-East India

North-East India is considered as the centre of origin of many citrus species. Khasi mandarin (*Citrus reticulata*) and sweet orange (*Citrus sinensis*) is cultivated commercially in the region. Among the citrus crops available in North-Eastern Region, Khasi mandarin is the most economically important one and plays a vital role in the socio-economic development of the people in this region. Khasi mandarin is well known for its quality, fruit colour, unique sugar acid blend and shelf life which make it the most popular citrus cultivar in North-Eastern region of the country. It covers the largest area in the region due to its commercial value. Assam lemon (*Citrus limon*) is one of the most important citrus crops of Assam. This exotic citrus has a unique aroma and is seedless. The fruits are produced year round, having two peak seasons in a year (Barua and Bharadwaj, 2017). Moreover, a number of other species *viz. Citrus indica* Tanaka (Indian wild orange), *C. latipes* (Swingle), *C. ichangensis* Swingle (Ichang Papeda), *C. medica*, *Citrus assamensis*, *Citrus macroptera* and *C. hystrix* are also recorded wild in the sub-tropical forests of North-East India and the foothills of the eastern Himalayas.

The highest genetic variability of *Musa acuminata* and *M. balbisiana* is documented in NE India. *M. flaviflora* is limited to Manipur and Meghalaya. The north-eastern states of India are major hotspots of banana diversity (Roy *et al.*, 2015). Although there are around 15 to 20 varieties of banana available, the famous ones are: Jahaji, Bor Jahaji, Chenichampa, Malbhog, Amrit sagar. Due to the favourable climate, edaphic and market prospective in the region, there is great potential for increasing the production of banana in the region (Sarma *et al.*, 2020). Goalpara district of Assam is well-known for its commercial cultivation of banana and has tremendous potential for further expansion of areas under banana cultivation (Baruah and Sahoo, 2018).

Table 18.1: State-wise Area and Production of Fruits Cultivated in North-East India

State	*Area (' 000 ha)*	*Production ('000 MT)*
Arunachal Pradesh	48.13	125.70
Assam	147.26	2123.62
Manipur	47.61	455.59
Meghalaya	32.81	316.51
Mizoram	63.19	340.51
Nagaland	39.50	380.52
Sikkim	19.36	54.90
Tripura	53.75	547.52
Total NER	451.61	4344.87

Source: Horticulture Statistics, Division, Department of Agriculture, Cooperation and Farmers Welfare, Ministry of Agriculture and Farmers Welfare, 2017-18.

The unique climatic and soil conditions favour pineapple cultivation in almost all the states of north-eastern region of India. In Cachar district of Assam, famous for Queen variety of pineapple, there is a good potential for expansion of area under pineapple crop. Out of the major fruits in Manipur, pineapple contributes the highest production to the total fruit production of the state. In Nagaland, Pineapple grown is of excellent quality in terms of size, appearance, TSS and other aspects, but commercial cultivation is still in its infancy.

Table 18.2: Major Fruit Crops of North-East Region

State	*Fruits*
Arunachal Pradesh	Apple, orange, kiwi
Assam	Banana, orange, pineapple, arecanut, coconut, Assam lemon
Manipur	Pineapple, papaya, banana, lemon, orange, passion fruit
Meghalaya	Pineapple, banana, orange, passion fruit, strawberry
Mizoram	Orange, banana, papaya, grapes, passion fruit
Nagaland	Pineapple, banana, orange, passion fruit
Sikkim	Orange, passion fruit, kiwi fruit
Tripura	Pineapple, banana, litchi, jackfruit, mangoes

Dragon fruit cultivation had been started from past few decades in Mizoram, Assam, Sikkim and Nagaland so that there can be availability of high value fruits in local market and with income generation to the farmers. The climatic condition of North-Eastern states is suitable for the cultivation of dragon fruit on a large scale, production and processing of this fruit crop improves nutritional and livelihood security. Dragon fruit provides fast return with economic production in the first year after planting (1.5 t/acre accommodated 448 poles), and full production is attained within 3-4 years. The life expectancy of the crop is about 20 years. Initial investment for dragon fruit cultivation is comparatively high, and is mainly due to training system establishment. Marketability of Dragon fruit in India is expected to be very high because of limited number of commercial producers.

In Meghalaya strawberry cultivation was recently grown in Sohlya village of Ri-Bhoi district which is now known as Strawberry village. Though the strawberry cultivation was newly practice in the state but gives has a great impact to the livelihood of the people (Banerjee, 2010). For the reason of its promising high net income, farmers from various villages of the state now are moving forward for cultivating strawberry. However, at present scenario Ri-Bhoi district (61.46%) was the leading producer of strawberry, followed by East Khasi Hills district (37.23%) and East Garo Hills district (1.29%) (GoM, 2014).

Avocado is a very popular fruit crop in Mizoram; its fruits are in high demand due to its buttery texture and nutritive value. But the availability of the fruits is very less in the markets due to less area and very few commercial plantations.Most of the avocado orchards in North-Eastern region are rainfed and are seedling progenies which generally takes 5-6 years to attend bearing period. Presently, plantations are not well organized and they are scattered. There is a quantum jump in the area and production of avocado because of its growing awareness as a diet fruit among the consumers. Realising the advantages of avocado, serious attempts have been made during the last two decades to develop trait-specific superior types, popularise the cultivation of underutilised plants. Varietal improvement in Avocado has so far been limited to selection of regular bearing and high yielding genotypes with improved fruit quality.

Kiwi is a temperate fruit crop, which has shown tremendous potential in recent years, owing to its preference amongst the growers as well as the consumers due to its favourable properties for easy maintenance and marketing as well as having high nutritional as well as health-promoting attributes. In Mizoram, Hnahlan village, near the Indo-Myanmar border, noted for its extensive grape cultivation, is now striving to become the largest grape producing village in India. Around 80 per cent of the villagers are engaging in grape-cultivation.

More than 300 wild edible plant species are reported from NE India and some are consumed by different ethnic groups of the region. Two *Elaegnus* species, namely, *E. latifolia* and *E. pyriformis* are reported frequently in the region. It is quite common in Sibsagar, Dikho valley, Naga Hills, Khasi and Jaintia hills. Soh-Shang (*Elaeagnus latifolia*) is among the significant potential fruit crops extensively spread in North-East India. *Docynia indica* and *D. hookeriana* are regularly observed in the region. *Pyrus pashia*, a medium sized deciduous fruit tree and two types of tree tomato (*Cyphopmandra betaceae*) (dark red and yellow fruit type cultivars) are also reported.

This region enjoys the unique advantage of growing temperate fruits. Temperate fruits *viz.* apple, pear, plum, peach, kiwifruit, walnut, chestnut, strawberries *etc.* are grown in semi commercially at small scale in Arunachal Pradesh. In Arunachal Pradesh, young apple orchards are now coming up in few places above 1600 m. Plum (cv. Santa Rosa), Asiatic pear and peach are common backyard plants. High hills of West Kameng, Tawang, Ziro, Anini, Upper reaches of Anjaw, Upper Subansiri, West Siang, Upper Siang and Tirap are some potential areas suitable for promotion of temperate fruit production. In Meghalaya, temperate fruits cultivation is limited to homestead only. Peach and plum are common kitchen garden plants. Kiwifruit is becoming popular with few farmers.

Besides, major fruit species, the region is a treasure trove for many underutilized temperate fruit crops such as *Persea americana, Pyrus pashia, P. pyrifolia, P. indica, Prunus nepalensis, P. jenkinsii, P. avium, P. cerasus, P. serotina, P. undulata, P.cerasoides, Rubus ellipticus, Actinidia callosa, A. stragosa, etc.* Many of these fruits are rich in nutrients and medicinal properties and have immense potential to contribute location specific food production, as they are well adapted to existing and adverse environmental conditions. Apart from providing diversification of the food base, different indigenous fruit species produce at different times of the years, ensure year round supply of nutrition. Furthermore, these crops have long been a traditional part of cropping systems.

Traditional Orchard and Canopy Management Practices of Fruit Crops of North-East India

Traditional farming methods are prevalent and used by the different indigenous tribal folks of the North-East region. Farming methods such as shifting cultivation and terrace (bun) agriculture are the two popular ones. Farmers also practise the method of tree-based farming practice, where the crops are grown together with trees such as alder, areca nut, coconut, bamboo, and Khasi pine.

Shifting Cultivation

Shifting cultivation or Jhum is the most primitive and popular agroforestry system practiced across the entire NEH region. Local farmers first select a piece of land in the forest. Then trees or bushes are cut down, left to dry and burnt. Then in the cleared area, the seeds are broadcast without the use of ploughs. When the crop yields begin to decrease after a few years, the farmers move to a new patch of forest area and repeat the same process. Farmers abandon the land to let it recuperate and after a period of 2 to 20 years they come back to the same location. This primitive method of cultivation has been used for thousands of years. But, due to its adverse ecological effect on forested areas, it is now only practised in a few communities of north-east states.

This system of farming is a transition between food gathering and hunting to food production. The folks who practice Jhum cultivation are called Jhummias. *Jhum* is a complex system with wide variation that depends upon the ecological variation in the area and cultural diversity among various tribal clans. However, the basic cropping practice is quite similar. Usually all the essential crops such as paddy, maize, tapioca, colocasia, millets, sweet potato, ginger *etc.* are grown on the same piece of land as mixed crop. *Jhum* in its most traditional form is not a very unsustainable land use practice particularly when the *Jhum* cycle is more than 20 years. The soils get enough time to rejuvenate and restore their health and productive capacity. However, with increase in population pressure on land resources, the *Jhum* cycle is getting reduced very fast and reached at 4 years at present. This makes the system unstable and lead to severe land degradation as a result of soil erosion and associated factors such as reduction in soil organic matter, nutrients *etc.* There is decline of forest cover due to shifting cultivation in the NEH region although the degree varies from one state to the other. Total area under *Jhum* also varies among the different hill states. The maximum area under shifting cultivation is reported in Mizoram followed by Arunachal Pradesh (Table 18.3). In terms of percentage of the total geographical area, Nagaland (36.15%) and Tripura (30.49%) are the most severely affected by *jhum* cultivation.

Table 18.3: Area under Shifting Cultivation in different States of the NEH Region

State	*Geographical Area (sq km)*	*Area under Jhum (% of geographical area)*	*Abandoned Jhum (> 10 yrs) (% of geographical area)*	*Current Jhum (5-10 yrs) (% of geographical area)*
Arunachal Pradesh	8374	6.25	0.77	5.58
Manipur	2233	18.10	3.51	14.59
Meghalaya	2243	14.42	2.58	11.84
Mizoram	2108	51.90	15.75	36.15
Nagaland	1658	12.22	0.69	11.53
Tripura	1049	32.43	1.94	30.49

Source: Roy *et al.* (2002), Rao and Bhattacharyya (2005).

Bun or Terrace Cultivation

Bun or terrace cultivation is practised widely in slopes and valleys and in Meghalaya. This cultivation method has been prevalent for the last 3 decades. It provides an improved production system, helps conserve soil moisture, and prevents land degradation and soil erosion. In this system, bench terraces are built on the hill slopes. The gap between each bun is levelled using the cut-hill method. The vertical break between each terrace is one meter. Thus, preventing erosion and maintaining a balanced water holding capacity within the lopes. It also helps to safely dispose-off the additional runoff from the slopes to the lower areas.

Agri-horticulture System

NEH region has ample potential for horticulture based agroforestry systems. For the development of suitable agro-forestry system for the region, an experiment was conducted at I CAR Research Complex for NEH Regopn, Umiam in the year 1987 where various fruit trees were grown with different combinations of agricultural crops. The results of different tree crop combinations have been summarized in Table 18.4. Khasi Mandarin was planted initially at a tree density of 800 trees per hectare. However, it was observed that 400 trees per hectare produced maximum yield. Average yield of mandarin was 12.8 kg per tree after 7 years of plantation which increased up to 57.3 kg per tree after 12 years of plantation.

Table 18.4: Performance of Fruit Tree Based Agri-Horti Systems in the NEH Region

Tree Crop	*Field Crop*	*Variety of Field Crop*	*Trees ha^{-1}*	*Net Return (Rs ha^{-1})*
Khasi Mandarin	Groundnut	JL-24	400	4541
	Soybean	Alankar	400	19625
	Turmeric	RCT-1	400	30375
	Ginger	Nadia	400	33416
	Taro	Local	400	18583
Guava	Groundnut	JL-24	400	3000
	Soybean	Alankar	400	916
	Turmeric	RCT-1	400	2750
	Ginger	Nadia	400	15791
	Chillies	Local	400	1125
Assam Lemon	Soybean	Alankar	400	2583
	Turmeric	RCT-1	400	1916
	Ginger	Nadia	400	36625
	Radish	Japanese White	400	2583

Source: Deb (2020).

With the increase in age of the tree crop and gradual closure of the over storey canopy, yield of groundnut and soybean started to decline. This opened scope to grow crops like ginger and turmeric and replace these oilseed and pulse crops in the mandarin based system. Among the field crops, ginger was the most remunerative. However, major share of income in these systems was generated by *Khasi mandarin* which was about 80 per cent of the total income irrespective of the associated field crops.

In the guava based system, cv *Allahabad safeda* was planted in association with five field crops namely groundnut, chilli, soybean, turmeric and ginger. Fruit yield of guava increased upto 8th year after which it started to decline because of fruit borer infestation. After two years of plantation, fruit yield was about 56 q ha^{-1} yr^{-1}. Average yield of groundnut, chilli, soybean, turmeric, and ginger were 17.7, 9.8, 6.5, 63.7 and 47.5 q ha^{-1}, respectively. Majority of the income was from ginger crop as the market price of guava is low as compared to mandarin.

Assam lemon based agri-horti system was relatively more profitable than the guava based system. Maximum yield of the fruit trees was attained after 7 years of plantation. Different intercrops such as ginger, turmeric, soybean and radish were cultivated in the inter row spaces of these tree crops. The net return was maximum in Assam lemon + ginger followed by Assam lemon + radish.

Agri-Horti-Silvi-Piscicultural System

This is an age-old agroforestry practice adopted by the Kalita community of Assam. This is one of the productive systems, where different agroforestry components are cultivated on the same land management unit. Farming is the mainstay of their economy. The farmers choose the crops and crop combinations based on their own wisdom and perceptions acquired over generations of experiences, the criterion being their day-to-day requirements of food, fuel, fodder and timber. The farmers generally plough their field with bullocks. The bullocks, however, incur high cost of feed and fodder. Hybrid varieties of paddy (Ranjit, Bora, Sonalika, Ijong), vegetables (cabbage, carrot, cauliflower, chilli, pumpkin), specific trees, shrubs, and palms are deliberately planted on the cultivated lands. Recently, horticulture has become very profitable and popular because of the improvement in the production potential of edible fruits and timber species like *Musa* sp., *Citrus* sp., *Ananas comosus, Bauhinia* sp., *Erythrina indica, Gmelina arborea, etc.* From the socio-economic and cultural viewpoint, some species were maintained and utilized as cash crops. For example, arecanut tree plays a vital role in the economy of the local society. The labour input for managing the crop is less than that for many other crops, which makes it an ideal crop for the people engaged in other occupations. The economic advantage of the system is derived from the cash-sale of the agricultural products. The family income is greatly improved, as the farmers save the cash that otherwise would have been spent on food. The system also provides a more or less full-time employment to most participants who have no other source of income. The farmers also maintain fish ponds in the farmyard. Trees are planted surrounding the fish pond, and crops inter-planted forming an integrated biological production system (Deb *et al.,* 2009). Common carps, silver carps and grass carps are generally preferred by the traditional society. The litter of many leguminous plants like *Leucaena* sp., *Moringa oliefera, etc.* has been found to serve as a good fish-feed when offered as pellets and improved the fish production. Further, the trees and shrubs in the traditional systems play an important role in regulating the microenvironment of the system. They are the principal source of rural energy and provide countless medicinal products used in the households (Deb, 2020).

Agri-Horti-Silvicultural System

This is a common and age-old agricultural system practiced by the Nyishi community of Arunachal Pradesh. The farmers grow traditional crop species such as paddy, *Eleusine coracana, Zea mays, etc.* in their field. Among the tuber crops, *Manihot esculenta, Dioscorea* sp., *Colocasia* sp. *etc.* are the most widespread and chief subsidiary food crops. Vegetables grown in the system are *Capsicum, Colocasia, Solanum tuberosum, Solanum melongena, etc.* Pineapple is a common floor crop grown along with the vegetables in the homesteads. A number of cultivars of banana are also cultivated. The farmers cultivate tree species on the boundary and agricultural crops in the middle. So, this farming practice is predominantly subsistence-oriented. The farmers have poor resource base, their landholdings are small and fragmented, and they have diverse requirements of food, fodder, fuel and timber. The farmers have a rich tradition of cultivating medicinal plants in their traditional agroforestry systems. In addition, other species that provide leaves, spices and condiments (*Betel vine, Piper longum, Zingiber* sp. *Oscimum sanctum, Azadirachta indica, etc.*) are also cultivated. In practical terms, the main expectation from an intercropping system in a perennial plantation cropping system is that the overall return from a unit piece of land is increased without adversely affecting either the current or the long-term productivity of the perennial crop. At the same time, the returns from the additional crops should justify the adoption of the intercropping practice and should contribute to the long-term productivity of the system (Deb, 2020).

Horti-Silvi-Pastoral Systems

This system is practiced by the Nyishis in Arunachal Pradesh. Timber species such as *Terminalia myriocarpa, Gmelina arborea, Mesua ferrea* are grown in combination with the non-timber yielding species like bamboo. *Livistonia jenkinsiana* (locally called 'Toko') cultivation is the unique feature of this tribe. This species is having a greater potential in agroforestry due to its shorter gestation period and recurring economic returns, and the Toko leaves are used as roofing material. The fruit trees like *Mangifera indica, Psidium guajava, Artocarpus heterophyllus, etc.* are common in this system. Crop-animal systems, in which livestock play a multi-purpose role yielded various products and performed different functions, such as recycling of nutrients and energy and helped to achieve self-sufficiency, stability and sustainability of the system, are common among the Nyishis. The horti-silvi-pastoral land use brings stability in total biomass production through fruit, fodder and fuel wood. During initial period, pasture would contribute most of the income and later the fruit trees. The cost incurred during the preliminary stages would be substantially higher due to warranted performances of all operations like rubbing, ploughing, sowing, planting and use of materials like manure, seeds *etc.* However, the cost reduces substantially in the next year. The expenditure incurred was mainly on collection of grass, seeds and harvesting of grasses and maintenance of pasture. Horti-silvi-pastoral system has proved to be quite beneficial in generating more income for the farmer especially during off-season when crops are not cultivated.

The success of the system largely depends upon the proper selection of fruit tree species under different agroclimatic conditions to meet various objectives (Deb, 2020).

Intensive Integrated Farming System

Intensive integrated farming system (IIFS) is based on the concept that there is no waste and waste is only a misplaced resource which can become a valuable material for another product (Edward *et al.,* 1986). It is a more refined and holistic approach of land use system through practices in which a number of production components are integrated with the primary objective of developing a self-sustainable system. In IIFS all the components of agriculture like crop, fish, forestry, horticulture are integrated in a complementary way. The sloppy land was cleaned and contour bunds were prepared for gradual conversion of the slope to bench terraces at fixed vertical intervals of 3 m. Hedge row of *Tephrosia candida, Flemingia macrophylla, Indigofera tinctoria, Desmodium rensonii, Crotolaria tetragona* and *Cajanus cajan* were raised on contour bunds for soil and water conservation and soil fertility build up. One year old seedlings of multi-purpose tree species (*Gmelina arborea, Alnus nepalensis, Chukrasia tabularis, Michelia champaca, Bauhinia variegata, Symingtonia populnea* and *Morus alba*) and fruit trees (*Psidium guajava, Citrus reticulata, C. lemon, C. sinensis, Pyrus communis, Prunus persica* and *Artocarpus heterophyllus*) were planted. The area at the lowest elevation of the farm (about 3.31 ha) was marshy where crop cultivation was not possible. Small water harvesting earthen ponds were created in a small area and large marshy land were brought under cultivation of high value crops with assured irrigation facilities. Fish fingerlings were introduced in each ponds with species composition of catla (*Catla catla*)- 20%, rohu (*Labeo rohita*)- 10%, mrigal (*Cirrhinus mrigala*)- 20%, silver carp (*Hypopthalmichthys molitrix*)- 2%, grass carp (*Ctenopharyungodon idella*)-20% and gonius (*Labeo gonius*)- 20%. Duck (Indian Runner and Khaki Campbell), pig (Large Black), layer birds (White Leghorn), goat (Black Bengal) and cow (*Holstein*) were reared and integrated with fishery. One pond was kept as control to compare the fish growth without integration of livestock/poultry/ducks. Vermicompost, liquid manure and mushroom cultivation was started in IIFS.

Bari System – The Home Garden of Assam

The *bari* system of farming has evolved over the years in the North-East India and has had great significance from the point of conservation, consumption and management of biodiversity. Bari in Assam is a traditional land-use practice, mostly around the dwelling house and sometimes in isolated distant locations, where numerous plant species including fruits, vegetables, plantation crops, timbers, firewood, spices, herbs, ornamentals along with some other components of farming systems, *viz.*, livestock, poultry, fishery, sericulture, bee-keeping, *etc.* maintained by the members of the farm family with the primary objective to fulfill various household needs. At present, there are about 6.4 million home gardens in the state having enormous significance. Like any other part of the world, the concept of home gardens is gaining importance in the state of Assam in the present critical scenario of lesser availability of agricultural land, due to increased urbanization, fragmentation of land holdings, and conversion of agricultural land to non-agricultural purposes. The *baris* have been designed to allow optimal harvest of solar energy through the strategy of fitting phenological classes and life forms together in space and time, and through niche diversification techniques. Multiple crops are present in a multi-tier canopy configuration. The leaf canopies of the components are arranged in such a way that they occupy different vertical layers with the tallest components having foliage tolerant to strong light and high evaporation demand and the shorter components having foliage requiring or tolerating shade and high humidity. Although the *baris* exhibit a general pattern, each garden is unique in its spatial and temporal structure, crop mix and arrangement, and overall design.

During scientific planning for growing varied fruit crops in *Bari*, it is of utmost importance to consider nutritional security of the farm family, priority of the house hold and conservation of biodiversity through use. The lion's share of produce in fruit zone is often consumed in the house hold as in case of vegetable crops; the surplus amount is either shared within the neighborhood or sold in the local market. Usually mixed plantation of several species is preferred to be grown in *Bari* for fulfilling nutritional balance of the farm family round the year. However, large scale plantation of *Citrus reticulata* cv: Khasi mandarin and *Areca catechu*, in *Bari* is traditionally popular throughout the state. Monoculture of *Carica papaya*, *Citrus sp.* cv. *Citrus lemon*, *Cocos nucifera*, *Fragaria vesca*, *Musa sp.*, *Zizyphus mauritiana*, *etc.* are also becoming popular. Large scale cultivation of other fruit species is yet to gain momentum. Among the fruit species *Artocarpus heterophyllus*, *Artocarpus lakoocha*, *Baccaurea sapida*, *Citrus grandis*, *Citrus medica*, *Flacourtia cataphracta*, *Mangifera indica*, *Myrica nagi*, *Prunus communis*, *Prunus jenkinsii*, *Prunus persica*, *Spondias mangifera Syzygium cuminii*, are the most common (Baruah *et al.*, 2019).

Recommended Orchard and Canopy Management Practices for Fruit Crops of North-East India

Use of quality planting material: In north-east India, most of the farmers cultivating the fruit trees used the planting materials from their own and majority of them uses seeds as planting materials. Vegetatively propagated materials result in uniform early bearing trees giving rise to superior productivity and better quality fruits. These trees are mostly free from transmittable diseases and insect-pests. Trees arising from vegetatively propagated materials are dwarf in size and hence more number of trees can be accommodated resulting higher output per unit area in the fruit zone.

Different recommended propagation techniques needed for generation of quality planting materials are mentioned in Table 18.5.

Table 18.5: Types of Vegetative Propagules to be Used for some Important Fruit Crops

Sl.No.	*Crop*	*Methods of Propagation*
1.	*Achras zapota*	Wedge grafting
2.	*Aegle marmelos*	Wedge grafting
3.	*Anacardium occidentale*	Soft-wood grafting
4.	*Ananas comosus*	Sucker, slip
5.	*Annona reticulate*	Wedge grafting
6.	*ArtocarpusHeterophylaaus*	Soft-wood grafting, patch budding
7.	*Citrus limon/Citrus aurantifolia*	Cutting, layering, budding
8.	*Citrus sinensis*	Wedge grafting, T-budding
9.	*Citrus reticulate*	Wedge grafting, T-budding
10.	*Emblica officinalis*	Wedge grafting, patch budding
11.	*Fragaria vesca*	Runner
13.	*Litchi chinensis*	Air-layering, wedge grafting
14	*Mangifera indica*	Grafting (soft-wood, wedge and veneer)
15.	*Musa* sp.	Sucker, corm
16.	*Prunus domestica*	Grafting (wedge and tongue), T-Budding
17.	*Prunus persica*	Grafting (wedge and tongue), T-Budding
18.	*Psidium guajava*	Wedge grafting
19.	*Punica granatum*	Wedge grafting, air layering
20.	*Pyrus communis*	Grafting (wedge and tongue), T-Budding
21.	*Syzygium cuminii*	Soft-wood grafting

Source: Singh, 2012.

Recommended cultivars suitable for the fruit orchards of North-East India: In north-east India, most of the farmers use the locally available cultivars for most of the fruit crops which are unproductive and prone to a number of diseases and pests. The farmers should use the recommended cultivars for the fruit orchards in order to increase the productivity of the fruits.

Table 18.6: Recommended Cultivars of some Major Fruit Crops for North-East India

Crop	*Cultivars*
Achras zapota	Cricket Ball, Barmasi, Oval, Co.1
Ananas comosus	Kew, Queen, Mauritius
Areca catechu	Local Selection
Artocarpus heterophyllus	Singapore Jack, Rudrakshi, Local Selection
Carica papaya	Dioecious type: Co.1, Co.2, Washington, Ranchi, Haflong Hermaphrodite type:

Crop	Cultivars
Cocos nucifera	Assam tall, TxD hybrid (Chandrasankar), Bengal Hazari, Kamrupa, Kahikuchi Hybrid-1 Solo, Coorg Honey Dew, Hybrid (RL, Sapna)
Litchi chinensis	Muzaffarpur, Dehradun, Calcutta, Rose Scented
Mangifera indica	Langra, Malda, Gulabkhas, Krishnabhog, Dashehari, Fazli, Mallika, Amrapali
Musa sp.	Dwarf: Dwarf Cavendish Medium: Chenichampa, Malbhog, Bar jahaji Tall: Kachkal, Jatikal, Bhimkal, Manoharkal
Psidium guajava	Lucknow-49, Allahabad Safeda, Chittidar, Seedless

Source: Barua *et al.*, 2019.

Canopy Management for Fruit Crops of North-East India

Canopy management in crops deals with the development and maintenance of their structure in relation to the size and shape for maximizing quality product without increasing production cost. As the canopy design and shape influence light interception with assured higher monetary returns, therefore, early height control along with canopy management are important techniques which should be practiced for fruit crops. Pruning, an integral part of management of canopy architecture, begins at an early stage of plant growth and is continued till plants starts providing optimum fruiting.

Canopy management practices for some of the important fruit species of north-east India is discussed briefly (Singh, 2010).

Citrus Limon cv. Assam Lemon

- ✰ In the nursery stage, pruning should begin soon after the saplings start sprouting from the stem cutting or leaf bud cutting or air layering;
- ✰ After planting in the main field, all branches should be removed keeping only single main trunk up to 60 cm from the ground level and above 60 cm, pruning is confined to develop the canopy
- ✰ During winter when bearing of tree is low, the branches touching the ground should be cut close to the laterals without leaving any stub besides pruning to reduce the height and breadth;
- ✰ All diseased, injured, criss-cross branches and water sprouts should be removed so that light interception to all the portions of the plant is not be hindered.

Psidium guajava

- ✰ Branches are allowed to flourish from the main stem only at the height of 60-70 cm from the ground level;
- ✰ 3-4 sprouts present almost at equidistance and at the upper portion of the main stem are allowed to elongate up to 40-50 cm;
- ✰ Afterwards, pruning is carried out on these selected sprouts at the mid-point so that numerous branches emerge just below this point. These branches are allowed to elongate to 40-50 cm followed by pruning as earlier. After several pruning in such manner will result the satisfactory shape of the tree;
- ✰ Pruning in the similar manner should be continued in the second year after planting, finally the height of plants should be around 2.5 m with a canopy confined within 2 m width with a satisfactory shape. Pruning should be done each year during the month of May-June.

Litchi chinensis

- ✰ Plants grown from air layering are usually single up to the height of 40-50 cm
- ✰ Sprouts or branches emerging from the ground are to be removed;
- ✰ Equally distributed sprouts at the height of 40-50 cm should be trained to become main branch; pruning should be carried out to excessive branches regularly to give the satisfactory shape to the plants;
- ✰ After harvesting from the fruit bearing sprouts, pruning (25-30 cm) should be done 2-3 times to encourage fruiting in the next year;
- ✰ Pruning should be done in all the 4-5 m long branches of the low productive trees; this practice will encourage fruiting after 2-3 years from the new sprouts.

Mangifera indica

- ✰ Pruning should be done before one year of age during October-November;
- ✰ During the month of May, excess branches should be removed allowing only four sprouts to exist in different directions of the main stem at equal distance;
- ✰ If the angle of the branches to the stem is very less, the branches should be trained with the help of the jute string/rope;
- ✰ Pruning of the sprouts should be done when the colour of the sprouts changes from green to brown *i.e.* become matured;
- ✰ Pruning for the second time should be done during October-November;
- ✰ Excess sprouts from the primary branches are to be removed allowing 2-3 sprouts to exist.

Passiflora edulis

- Vines should be allowed to trail on strong trellis preferably with two arm kniffin system placed in north-south direction
- Pruning should be done to the lateral branches nearby active buds.

Artocarpus heterophyllus

- Encourage the top portion of the tree to be needle-shaped during the first year;
- Tree should be established on a single stem at the height of 70-80 cm above the ground level and 3-4 branches at different sides of the main stem at equal distance should be encouraged to flourish
- Excess branches or sprouts should be removed carefully and regularly.

Averrhoa carambola

- Pruning some branches at the top portion of the tree during February-March should be carried out;
- Excess sprouts should be removed so that the branches are only about 40-50 cm in length during first 1-2 years after planting;
- Pruning some of the selected branches should be carried out so that height of the mature tree will not be more than 1.5-3.5 m. Some of the upper branches which obstruct light interception to the lower branches should also be removed each year with utmost care;
- Fruits can be harvested during October when pruning is done during August; however, if pruning is carried out in November-December, harvesting season becomes April to June.

Elaeocarpus floribundus

- Remove all the sprouts and lateral branches up to 45 cm allowing to flourish only a single stem; one lateral branch at right side and another one at left side should be allowed and trained with a wire downwards just above 45 cm during first year;
- During second year, two more lateral branches are allowed to grow and are to be trained in the similar manner;
- During third year similar operation should be done with two more lateral branches; besides branchlets of these branches also should be allowed;
- Sprouts from main stem and lateral branches should not be allowed to be dense. When the height of the tree reaches about 4 m, pruning should be done in weak lateral branches. Such practices should be continued up to the bearing stage.

Emblica officinalis

- ✰ In the first year, major branches should be allowed at 0.75-1.00 m from the ground level;
- ✰ In the early year 2-4 branches, present in opposite direction of the stem and forming obtuse angle each, should be allowed to grow; excess branches should be removed during March-April. Dead, weak, diseased and criss-cross branches should also be removed;
- ✰ In subsequent years, 4-6 branches are allowed to grow;
- ✰ Middle portion of the canopy are encouraged to grow to maximum 2.5-3.0 m.

Aegle marmelos

- ✰ Severe pruning, at the height of 1.5-2.0 meter from the ground, should be done on seedling plants during winter and it should be carried out on excessive shoots during May-Jun;
- ✰ To reduce the height of tree, long branches should be cut at the mid portion;
- ✰ Only 3-4 branches grown at equal interval around the main stem should be allowed to grow for 6-7 months followed by heading back at mid-point of their length.

Syzygium cuminii

- ✰ Pruning should be carried out at the height of 60-90 cm from the ground level on the main stem; the branches or sprouts below this height should be removed;
- ✰ New sprouts emerging just below this point should be removed keeping only 3-4 branches intact. When the length of these branches become 50-60 cm pruning should be done at the mid-point;
- ✰ All the diseased, weak, dry and criss-cross branches should be removed from time to time.

Intercropping in Commercially Grown Fruit Crops

As already mentioned, the most popular market driven fruit crops in Assam are *Areca catechu, Citrus reticulata, Citrus limon* cv. Assam lemon, *Musa sp., Zizyphus mauritiana, Carica papaya, Fragaria vesca, Psidium guajava, Cocos nucifera, etc.* Most of these species provide economic yield after 5-6 years of planning; however, only a few species like *Musa* sp., *Carica papaya, etc.* are relatively short duration fruit crops. To get additional income from the same area under commercially grown fruits, it is extremely important to maintain the expenditure incurred for planting material and intercultural operations.

Table 18.7: Suitable Intercrops in Fruit Orchards of North-East India

Fruit Species	*Intercrop*
Areca catechu	☆ Assam lemon ☆ Black pepper ☆ Banana, chilli, elephant foot yam, pineapple, guinea grass, *etc.* ☆ Betel vine ☆ Elephant foot yam, black pepper
Artocarpus heterophyllus	☆ Sesame – toria (up to 4 years) ☆ Citrus sp. cv. Assam lemon ☆ Green manuring crops, fodder
Carica papaya	☆ Chilli, onion, potato, beans ☆ Chilli, onion, cabbage, radish, knolkhol, tomato ☆ Tomato, chilli, beans, water melon ☆ Green manuring crops, fodder
Citru limon	☆ Onion, chilli, green gram, cowpea (up to 3-4 years) ☆ Pea, turnip, cauliflower, carrot, radish (up to 3-4 years) ☆ Pea, cowpea, tomato, okra, potato (up to 3-4 years) ☆ Pea, cowpea, beans (up to 3-4 years)
Cocos nucifera	☆ Black pepper, banana, Assam lemon, pineapple, ginger ☆ Betel vine, banana, turmeric, Assam lemon, colocasia ☆ Ginger ☆ Turmeric ☆ Banana ☆ Assam lemon ☆ Pineapple
Musa sp.	☆ Brinjal, colocasia, yam, chilli, okra ☆ Ginger, colocasia, turmeric ☆ Onion, tomato, capsicum, potato ☆ Turmeric, ginger, pineapple, beans ☆ Turmeric ☆ Cowpea, sweet potato, maize ☆ Green manuring crops, fodder
Psidium guajava	☆ Cowpea, French bean ☆ Okra, onion, turmeric, garlic, cauliflower, cabbage, chilli, leafy vegetables ☆ Green manuring crops, fodder

Growing of suitable intercrops on space available between rows of fruit trees at least for initial few years is the only answer of this hurdle. Based on relevant research works in different parts of India and similar agro-climatic situation of Assam, following suitable intercrops may be encouraged:

Technological Interventions in Orchard and Canopy Management for Fruit Crops of North-East India

Rejuvenation of Old Khasi Mandarin Orchards

Mandarin orange is one of the most widely cultivated fruit crops in the state since long. However, during the recent past, in whole North-East India, mandarin orange production was crippled by citrus decline, bringing down a large area under orange orchard to be unproductive. In a bid to revamp the mandarin orange cultivation and production scenario in the region, the state horticulture department of every state took up several steps such as implementation of rejuvenation of senile orchards, canopy management, and human resource development including infrastructure development for production of quality planting materials. Few state developed scion bank to boost the production of quality disease free planting materials. With these efforts, the declined orchards become productive once again, and new productive orchards are also established. Owing to the use of quality planting materials budded on resistant rootstocks with short gestation period coupled with the intervention of the aforesaid schemes, the level of production and productivity has increased considerably over the years, and cultivation of mandarin orange is now serving as a means of livelihood to many farmers.

The following points should be considered while, adopting rejuvenation programme (De and Patel, 2019)[

1. Preparation of half-moon terrace and cleaning of tree basin and mulching with straw, dry leaves.
2. Pruning of old declined Khasi mandarin and acid lime at primary branches and 1.5 m produced fruits ranging from 100-220 and 185-237 respectively, after two years. In Khasi mandarin, pruned plants produced fruits with higher quality.
3. Periodical removal of water shoots, dry and diseased twigs, lichen, mosses and phanerogamic parasites.
4. FYM @ 30 kg in the month of March-April along with 1.3 kg urea + 3 kg SSP and 1.2 kg MOP/tree/year in three split doses *i.e.* March-April, June-July and SeptOct.
5. Dolomite lime @ 3 kg/tree once in three year in the month of February is applied.
6. Multiplex a mixture of micro-nutrients @ 2.5 ml/litre of water or mixture of 0.5 per cent zinc sulphate, 0.4 per cent copper sulphate and 0.5 per cent magnesium sulphate is sprayed when new flush comes.

7. Trunk borer (damage 20-30%): Collection and destruction of trunk borer adults by shaking the branches 2-3 times at 10 days interval in the month of May. Injection of 3 to 5 ml of petrol or monocrotophos in to the bored hole through syringe and plugging the hole with wet soil. Apply Beauveria bassiana (1020 spores/ml of water on the trunk).
8. Gummosis or foot rot: Proper drainage to avoid water stagnation around the plants. Scraping of affected portions and pasting of Bordeaux paste.
9. Root rot: Removing of soil around the trunk and drenching the root with 1 per cent potassium permanganate solution followed by 1 per cent Bordeaux mixture.
10. Scab: Spraying of Bavistin (0.05%) or Blitox (0.25%) in April, June and September.

Rejuvenation of Old/Senile Peach Trees

For rejuvenation of old/senile peach trees, old/senile peach trees of more than 20 years old should be selected and heading back of primary branches leaving 50 cm on main trunk used to be carried out during November-December. The cut portion should be pasted with copper oxychloride and main trunk with Bordeaux paste, mixing 25 kg FYM and fertilizer, mulching of basin area with black polythene. After heading back, 10-15 new shoots will emerge 30-45 days. The 2-3 new shoot/ branch should be selected and remaining will be removed during March-April. It is then followed by terminal pruning of selected shoot up to 50 per cent of its length. The reproductive pruning of the shoots should be performed in October-November.

Early Maturity and Quality Fruit Production in Peach cv. Flordasun

Profuse bearing of peach cv. Flordasun, resulted in to excessive crop load of undersized fruits with impaired fruit quality and limb breakage. In addition, the ripening of this cultivar coincides with rains in late April which hampers fruit quality and marketability. To get rid of these problem fruit/fruitlets of cv. Flordasun should be thinned on 20 days after full bloom with spacing of fruits at 10 to 15 cm on shoot on whole tree canopy. This advanced the fruit maturity by 12- 14 days with an average fruit yield of 28.10 kg/tree.

Increase the Production of Grafted/Budded Plants of Khasi Mandarin

To increase the production of budded/grafted plants, division of Horticulture ICAR, Umiam, Meghalaya has standardized the wedge grafting technique, which is done in July-August on 6 to 7 month old rootstock for the production of grafted plants with more than 80-85 per cent plant survival. Thus it is suggested to do wedge grafting during July-August on 6-7 months old rootstock. Rootstock left after grafting can be utilized for T-budding during February to extend the grafting/ budding period up to 3 months with the use of available resources.

Off -season Production of Strawberry under Low Tunnels

Strawberry can be produced 30-35 days earlier than normal period, when planted in low tunnels of 50 per cent shade net in the month of July or August and the period of fruit availability may be extended to 47 days from normal when planted in the month of November under UVS polythene tunnels. Normal fruiting period under open condition was 2nd fortnight of January to 2nd fortnight of March.

Stage-wise Requirement of N and K in Banana

Department of Horticulture, Assam Agricultural University standardized the Stage wise requirement of N and K in banana by applying 60 per cent of N at planting to five month stage, 20 per cent of N at shooting, 20 per cent of N at last hand opening to one month before harvest and potassium fertilizer: 40 per cent of K at shooting to last hand opening and 60 per cent of K at last hand opening to one month before harvest stage.

Low Cost Structure for Organic Kiwifruit Multiplication through Cuttings

Semi-hardwood cuttings of 0.5-1.0 cm thickness with relatively short internodes and about 15-20 cm in length were most ideal and showed 70 per cent rooting under low cost structure with the media composition of sterilized growing medium composed of perlite, vermiculite, cocopeat and vermicompost (1: 1: 1: 1 ratio). Jute bags should be removed after 90 days planting of cuttings.

High-Density Orcharding of Apple, Kiwi, Persimmon

Apple, kiwi and persimmon are one of the potential crops to grow under high density in NE region. These crops are very potential for the region, due ideal climatic conditions. In apple, Gala varieties grafted on clonal rootstocks have been proved very profitable in Arunachal Pradesh. Likewise, the kiwi and persimmon are another fruit crops are yielding significantly higher yield of quality fruits as compared to other regions of the country.

Low Chilling Cultivars Suitable for Sub-tropical Conditions

Selection of cultivars of the temperate fruits to be grown under sub-tropical climate is one of the key requirements for their successful cultivation in the sub-tropical regions. Only those cultivars which have low chilling requirement and ability to tolerate high summer temperature should be selected. Breeding programs have provided new cultivars selected for their short chilling requirements and tolerance to high temperature. One of the most successful cultivars is 'Anna' apple, developed by A. Stein in Israel. This is now extensively grown in areas with limited chilling in Assam, Manipur and Mizoram.

Strategies and Measures for Development of Fruit Orchard Industry in NE Region

Orange is a unique horticultural crop of NE region that need priority research attention for increasing its productivity and quality of the fruits. Organic cultivation of oranges can pave the way for successful venture. Indiscriminate introduction of quality planting materials from other region has to be avoided and indexing and certification programme should be implemented properly to ensure availability of quality planting material. Production technologies in fruit crops like pineapple, banana, orange and rejuvenation technologies for declined mandarin orange orchard, *etc.*, are some useful technologies for dissemination among farmers so as to increase the productivity and quality of produce. Area should be increased under papaya, pomegranate, and jackfruit at low to mid altitude and walnut, chestnut, cherry, almond and kiwifruit at higher altitude of the region. Few fruit crops like banana, citrus, pineapple, jackfruit, coconut are well adopted in larger parts of the region. These crops need priority research attention for increasing the productivity and quality of the produce. In recent past, throughout the world, efforts have been made to control the fruit pests and diseases by use of natural enemies, parasite, predators besides host specific insect viruses and other entomo-pathogenic microorganisms and success has been achieved to a considerable extent. This technology should be adapted in the region to control the diseases and pest in vegetables crops.

Conclusion

Till recently, for most farmers, fruit orcharding has been mainly a backyard activity as they are generally busy throughout the year in cultivation of food crops and have little time for development of horticultural crops on a commercial basis. Besides, due to the long gestation period involved in fruit crops, the cultivation of these crops has been generally confined to small backyard gardens developed by almost every household. It is only in the past decade that there has been a more focused attention to the development of fruit sector in the region. There is need to create awareness and make the farmers receptive to the new technology through farmers participating demonstration and training. There is a need for establishing a sound system with forward and backward linkage so that vast potential of fruit crops can be exploited through adoption of improved production technology. Evolving of the new varieties, production, protection, post-harvest technologies and value additions are essential to keep up the pace with the current growth of fruit crops for self-sufficiency. Besides, clean planting material production, horticulture cluster development, digital horticulture, organic horticulture, and natural farming also assume importance. The efforts for boosting productivity must continue, through the introduction of climate resilient varieties and quality planting material production for area expansion.

REFERENCES

Banerjee R, 2010. The Economics Times. Shillong creates eco-tourism packages around strawberry fields.

Barua BC and Bharadwaj S, 2017. Assam lemon: A prospective NPD initiative aimed at global market positioning. *Int. J. Res.*, **4**: 727–734.

Barua JP, Ahmed AA, Gogoi S, Pathak S and Hatibaruah P, 2019. *Baree* – The Home Garden of Assam. Horticultural Research Station, Assam Agricultural University, Kahikuchi, Guwahati- 781017, Assam.

Baruah P and Sahoo C, 2018. A study on Daranggiri- one of Asia's largest banana markets, its problems and prospects. *Int. J. Basic Appl. Res.* **8**: 620–625.

De LC and Patel RK, 2019. Rejuvenation of Citrus; c2019. DOI: 10.13140/RG. 2.2.23364.99208.

Deb S, 2020. Traditional Agroforestry Systems of Northeast India. In: Roy, N., Roychoudhury, S., Nautiyal, S., Agarwal, S. K., and Baksi, S. (eds.), *Socio-economic and Eco-biological Dimensions in Resource use and Conservation.* Springer Nature Switzerland AG 2020. 10.1007/978-3-030-32463-6_5. Pp. 103-115

Deb S, Arunachalam A and Das AK, 2009. Indigenous knowledge of Nyishi tribes on traditional agroforestry systems. *Ind. J Trad. Knowl.* **8**: 41–46.

Edwards P, Kaewpaitoon K, McCoy EW and Chantachaeng C, 1986. Pilot scale crop/ livestock/fish integrated farm. AIT research report, 184, Bangkok, Thailand, 131p.

GoM, 2014. Directorate of Horticulture Meghalaya, Shillong. *Report on area, production and yield of Horticulture crops* due on 2013-14.

Rao Nageswar PP and Bhattacharyya KC, 2005. Application of geographic information systems and remote sensing in assessment and management of natural resources. *In*. Agroforestry in North-Eastern India: Opportunities and Challenges. (*Eds*) B.P. Bhatt and K.M. Bujarbaruah. ICAR Research Complex for NEH Region, Umiam, Meghalaya. pp. 15-31.

Roy A, Das SK, Tripathi AK, Singh NU and Barman HK, 2015. Biodiversity in North East India and their conservation. *Progr. Agri.* **15**: 182–189.

Roy PS, Singh S, Aggrawal S, Gupta S, Joshi PK, Aung S, Singh IJ and Shukla Y, 2002. TRESS II-Tropical forest assessment in India and north Myanmar. In: Forest cover assessment in Asia: Proceedings of the international workshop on Asia forest cover assessment and conservation issues. Feb 12-14, Dehrandun, India, pp 61-100.

Sarma B, Choudhury M, Sarma R and Nath RK, 2020. Production constraints of banana cultivation in western district of Assam. *J. Pharmacog. Phytochem.* **9**: 1829–1830.

Singh G, 2010. Canopy Management in Fruit Crops - a Practical Manual. Publ. by Dept. of Agri. and Cooperation, New Delhi. p. 75.

Singh G, 2012. Protocols and Standards for Vegetative Propagation of Fruit Crops. Dept. of Agri. and Cooperation, New Delhi. p. 50.

Chapter 19

Nursery Production, Orchard and Canopy Management in Litchi

Ashok Kumar Singh and Kuldeep

Department of Horticulture, G.B. Pant University of Agriculture and Technology, Pantnagar–263145, Uttarakhand

Nursery management forms the cornerstone of successful litchi cultivation. The quality of planting material is vital for establishing a healthy orchard. By selecting disease-free, genetically uniform saplings and utilizing appropriate propagation techniques, growers can promote robust orchard development. This stage requires precise control of factors such as soil health, irrigation, and pest management, which directly influence the vigor of young plants (Sahni *et al.*, 2024). Orchard management is essential for the long-term success of litchi cultivation. It encompasses land preparation, planting, irrigation, fertilization, and pest control strategies tailored to the specific requirements of litchi trees (Xie *et al.*, 2022). Proper orchard management not only maximizes fruit yield but also extends the productive lifespan of the trees, thereby enhancing the economic sustainability of the orchard. Canopy management plays a critical role in optimizing tree performance (Singh *et al.*, 2023). This involves careful pruning and shaping of the trees to improve light penetration, air circulation, and fruit quality. Effective canopy management aids in disease prevention, facilitates efficient harvesting, and helps maintain tree health and productivity over time. Additionally, it ensures a balance between vegetative growth and fruit production, which is essential for achieving consistent yields.The primary objective of this chapter is to provide a comprehensive overview of best practices for nursery, orchard, and canopy management in litchi cultivation. It aims to equip growers, horticulturists, and researchers with the essential knowledge to optimize every stage of litchi production, from the selection and propagation of planting material to the effective maintenance of mature orchards. The chapter will also explore common challenges in litchi cultivation and present evidence-based strategies to address them. By integrating insights from scientific research with practical experience, this chapter aspires to promote sustainable, efficient, and profitable litchi farming practices.

Nursery Production and Management

Selection of Nursery Site

Several factors are responsible for the selection of a suitable nursery site for litchi. Some important consideration are as follows.

a) ***Location***: A nursery must be located in a pollution-free environment-away from brick kilns, smoke-emitting industries and rough motorized roads as dirt and dust settle on plants, covering the foliage, which not only reduce the photosynthetic efficiently of plants but also gives them a dull look. It must be ensured that the nursery site gets adequate sunlight However, care must be taken that the plants are protected against severe heat.

b_ ***Topography of land***: The topography of land at the nursery site must be even. If it is undulating, it must be leveled. In hilly area, it may be dived in to leveled terraces.

c) ***Soil***: The soil must preferably be loam or sandy loam with large quantity organic matter. The pH of the soil must be near neutral (6.5-7.5). It must be having adequate water retention capacity and aeration.

Propagation Techniques

Litchi is predominantly propagated through vegetative methods (mainly through air layering for commercial purposes. Seed-propagated plants exhibit slow growth, an extended juvenile period of 10–12 years, and fail to produce true-to-type fruits. Therefore, it is strongly recommended to use only vegetative propagated, authentic planting material for commercial cultivation. The techniques employed for plant propagation are outlined below:

Sexual (Seed) Propagation

Seeds are primarily used to raise rootstocks for grafting purposes. Propagation by seed is not common in litchi because the plants raised from seed take 7-10 years to come into bearing and plants normally do not produce true-to-type plants but may prove good as rootstock and such plants create variability for selection. Seeds are sown generally to raise seedlings for rootstock immediately after extraction from the fruit, as they lose their viability If seeds remain in the fruit allowed to dry, they can be kept viable for 3-4 weeks. For germination, seeds soaked in water for 18-20 h should be placed horizontally about 1.5 for below the surface of a well-drained soil. The growth of seedlings is slow leading to more time lag in nursery, development of hard wood at matrix height and thus poor graft uptake, hence the seedling growth needs to be improved by the use of mycorrhizal soil, microbial consortia, organic/inorganic manure in the nursery bag/bed (Saúco *et al.,* 2018).

Vegetative Propagation

Vegetative propagated plants are true to type in nature and start early bearing. Litchi is propagated commercially by air layering, which is the most common and best method adopted all over the world. Grafting (cleft and approach) is another vegetative propagation method for litchi but the rate of success depends on rootstock growth and management. An appreciable success of 75-80 per cent has been noticed in cleft grafting under controlled condition (Das *et al.,* 1999). Budding is not commonly practiced in litchi. The commercial method of vegetative propagation in litchi are given below:

Air Layering/Marcottage

Among the various vegetative propagation methods, air layering is the most widely used for producing high-quality planting material in litchi. Air layering is typically performed after fruit harvest (mid-June to August) when the plants are in an active growth phase, with the monsoon season being the most favorable period. In mother trees maintained exclusively for propagation, layering can also be conducted during spring with appropriate water and nutrient management. Sphagnum moss is commonly used as the rooting medium for air layering (Mohsin *et al.,* 2019).

Steps/Procedure for Air Layering

- ☆ ***Selection of branches***: Healthy terminal branches exposed to ample sunlight, with a thickness of 1.25–1.50 cm, are ideal for air layering. Branches on the periphery of the tree are preferred for ease of handling. Shoots that recently flowered or fruited, having depleted food reserves, should be avoided as they generally produce poor roots.
- ☆ ***Preparation of the layer***: A 2.5 cm ring of bark is removed approximately 45–60 cm below the shoot tip. The cambium layer is carefully scraped off to expose the woody portion. To promote early and proper rooting, a 1000 ppm IBA paste can be applied immediately to the ring. Alternatively, applying 5000 ppm IBA in lanolin paste at the upper edge of the ring enhances root formation. This practice promotes more effective root formation.
- ☆ ***Application of rooting medium***: Rooting media such as sphagnum moss are widely used due to their excellent moisture retention properties. A layer of moist sphagnum moss is wrapped around the prepared area and enclosed in a 20 x 25 cm piece of 200-gauge polyethylene sheet,

secured tightly at both ends to maintain moisture and encourage root development. In cases where sphagnum moss is unavailable, alternative media like a mixture of clay soil and farmyard manure (FYM) in a 1:1 ratio, combined with a moist gunny bag, can effectively retain sufficient moisture to support root development. Other materials such as wet sawdust, vermiculite, and wood chips can also be used to cover the girdled portion of a branch (Ali, 2016).

Alternatively, an enriched rooting medium can be used, prepared with pond silt (40 kg), Farmyard manure (40 kg), Decomposed jute bag or leaf mold (10 kg), Neem/castorcake (2 kg), Urea (200 g), Single superphosphate or bone meal (200 g), Chlorpyrifos dust (100 g). The mixture is heaped, thoroughly mixed, and covered with UV-stabilized polyethylene for 45–60 days under sunlight to decompose. For layering, a moist ball (~200 g) of this medium is placed over the ringed area, ensuring it covers 2 cm above the ring.

- ***Root development in layers***: Roots in layers develop within 50–60 days and become visible through the polyethylene. Once 6–8 roots are clearly seen, turning from white to creamy brown, the air layer is detached by making a sharp cut 5 cm below the ring.
- ***Planting the air layers***: The detached layers are transplanted into a partially shaded area. Excess vegetative growth is pruned to maintain balance between the shoot and root system. Each plant is potted in individual 15 x 15 x 25 cm bags containing a well-prepared mix of soil, vermicompost, sand, and vermiculite. After 2–3 vegetative flushes, the plants are hardened off in open air before their final planting.
- ***Rooting media for filling in root trainers, bags, pots and nursery bed to promote proper rooting in the layers***: A mixture of soil, sand, and farmyard manure (FYM) in a 1:1:1 ratio is the most commonly used rooting medium. Alternatively, a combination of riverbed soil and vermicompost in a 1:2 ratio is also recommended for effective root development. For enhanced results, the addition of vermiculite (50 g/kg), coco peat (50 g/kg), and DAP (5 g/kg) to the recommended medium can promote the production of more secondary roots and vigorous planting materials, ultimately leading to higher survival rates in the nursery (Bal, 2016).
- ***Use of "Root Trainer" for better root development and survival in the field***: Root trainers of various volumes and geometric designs are readily available in the market and can be effectively used for raising litchi for the initial establishment of air layers in the nursery. Among these, root trainers with a 500 ml capacity arranged in trays with 12 conical cavities have proven to be particularly suitable for litchi propagation. These root trainers, when mounted on iron stands of 30–45 cm in height and appropriate width, provide optimal conditions for plant growth. The

planting medium used should be lightweight, with excellent water-holding capacity and sufficient nutrients to support healthy development. The design of the root trainers ensures that developing roots are directed downward toward the base of the cone. When placed on a stand, the roots emerging from the bottom are naturally pruned by air exposure, promoting the development of a dense network of active secondary and tertiary roots throughout the medium within the trainer. After one or two vegetative flushes, a robust root system forms, creating a complete web of roots within the saplings. At this stage, the plants can be transplanted into larger containers/polythene bags with additional substrates to facilitate further growth. Root trainers also offer the advantage of accommodating a greater number of plants in a mist chamber or greenhouse, optimizing space utilization and improving nursery efficiency. As per the study conducted at GBPUAT, Pantnagar, the litchi saplings produced by use of root trainers has very less mortality at the farmers field and it is because of properly developed and deeper roots system through intervention of trainers. The step-by-step procedure for use of root trainers can be visualized by Figure 19.1.

Bubbler System

Figure 19.1: Mother Block under Protected Conditions and Use of Root Trainer for Production of Quality Planting Material in Litchi.

Source: Singh *et al.*, 2011 (Horticulture Research Center at GBPUA&T, Pantnagar, Uttarakhand).

Orchard Management

Manure and Fertilizer

Among the various factors influencing litchi production, balanced nutrition is considered the most crucial determinant of productivity and fruit quality. Litchi responds positively to the application of manure and fertilizers, though the response varies depending on the cultivar, climatic conditions, and soil types. A survey conducted in the Doon Valley revealed that 80 per cent of orchards are deficient in nitrogen (N) and phosphorus (P), necessitating their application. In West Bengal, NPK levels were reported to be below the optimum, while in Punjab, deficits in nitrogen (N), potassium (K), magnesium (Mg), zinc (Zn), and manganese (Mn) were identified. In India, litchi is primarily cultivated in fertile soils. A graded dose of NPK combined with organic manure, micro-nutrients, and biofertilizers has been shown to be effective; however, farmers rarely adopt these practices. Severe shortages of N, P, and K, stunt litchi growth and hinder floral initiation, while micronutrient deficiencies negatively impact fruit quality. The use of biofertilizers alongside organic manures enhances nutrient uptake (Nayak *et al.*, 2024). A generalized recommended nutrient schedule for litchi in the North Indian plains is provided in Tables 19.1 and 19.2.

Table 19.1: Manure and Fertilizers Schedule for Litchi in India

Age of Plant	*Manure and Fertilizers/Plant/Year (kg)*			
	Farmyard Manure	*Urea*	*Super Phosphate*	*Muriate of Potash*
1-3 years	10-20	0.15-0.45	0.3-.9	0.15-0.50
4-6 years	25-40	0.60-0.90	1.2-1.8	0.60-0.90
7-10 years	40-50	1.00-1.50	2.1-3.0	1.00 -1.50
Above 10 years	60	1.50	3.0	1.5

Table 19.2: Manure and Fertilizers Schedule Recommended for Bihar

Manure/ Fertilizers	*First-Year*	*Increasing Amount Every Year (up to 10 years)*	*Fertilizer Dose of Full-Bearing Tree (kg)*
Compost (kg)	10	5	60.0
Neem/caster/other cake (kg)	1.0	0.5	5.0
Single super phosphate (kg)	0.50	0.30	3.0
Murate of potash (kg)	0.15	0.15	1.5
Urea (kg)	0.15	0.15	1.5
Zinc (kg)	0.10	0.02	0.25
Boron (kg)	0.10	0.02	0.25

The field experiments conducted on different cultivars at different locations have clearly demonstrated the effect of a graded dose of NPK on growth, yield and quantity of fruits. Application of 600-800 g N, 200-300 g P2O5 and 400-600 g K2O per plant is recommended for 12-15-year-old trees. Excessive application of nitrogenous fertilizer before flowering should be avoided. Phosphorus application at the time of flower bud differentiation improves flowering and fruiting. Application of cakes and manure during the month of September and October is generally practiced to get better quality fruits. In general, litchi orchards maintained with higher doses of organic manure have better yield and quality as compared to orchards maintained with chemical fertilizers. Additional application of Ca, Zn, B, Cu and Mn is recommended. Application of 0.6 per cent Ca as calcium chloride improves fruit weight and quality. Zn is applied in the form of 0.5 per cent zinc sulphate hydrated with lime, which helps in reducing fruit drop and enhancing fruit yield and quality. Boron @ 0.1 per cent enhances fruit setting and reduces fruit cracking. In acidic soil, application of 10-15 kg lime/tree once in 3 years has been found to increase the yield. Other micro-nutrients like Fe (Ferrous sulphate), Cu (Copper sulphate), Mg (Magnesium sulphate) *etc.* are applied if deficiency symptoms are observed in the plant (Shihua, 2011).

In general, application of FYM, potassic and phosphoric fertilizers in major litchi growing areas of the country is done during June-July, just after harvesting of the crop. However, in heavy rainfall areas like West Bengal, Uttarakhand and Uttar Pradesh, manure and fertilizers are applied in the month of September-October just before the end of monsoon. However, the nitrogen is applied in two equal spilt doses, the first dose after fruit set, in the month of March-April while the remaining half dose immediately after harvesting of the crop. Standard method of manuring by making circular trench of about 1.5 to 2.0 m away from tree trunk, filling with fertilizers mixture and leveling with soil similar to that of other fruit crops is followed for proper uptake of the nutrients by the plants. Proper soil moisture is essential at the time of fertilizer application (An *et al.*, 2022). The total requirement of nitrogen, phosphorus and potash is applied through basal application. However, 2.0 per cent urea spray may be practiced once or twice during the period of fruit growth and development as per the need.

Foliar application of the plant bio-regulator, NAA (20 ppm) at an interval of 10-15 days during the period of fruit growth and development, once or twice, has been found to be useful for minimizing the fruit drop.

Irrigation, Mulching and Water Conservation

Litchi being an evergreen plant, the maintenance of optimum soil moisture is critical for growth, development and fruit production. Litchi is grown successfully, if the rainfall is evenly distributed and supplementary water requirement depends upon cultivar and evaporation demand. Water requirement ranges from 600-800 mm. Investigations carried out to determine the irrigation needs have clearly indicated that irrigation is critical at the fruit development stage to get better yield and quality of fruits. Interestingly, differential management of water in the vegetative phase and reproductive phase is also suggested. To achieve faster growth of the plant no water stress should be permitted, while in the reproductive phase water stress is beneficial at the time of fruit bud differentiation. Light irrigation during summer and winter months and cleaning of the basin is advocated. Irrigation at the intervals of 2-3 days during the initial stage of plant establishment is considered essential. Further, the young plants should be irrigated during dry periods and winter months at intervals of 3-5 days. For young plants mulching with dry leaves or residues in the basin help in better moisture conservation. Certain physiological disorders like poor sex ratio, poor fruit set, heavy fruit drop and high fruit cracking, besides sunburn of the fruits can be minimized with proper water management. The basin or flood method of irrigation is normally practiced. However, adoption of drip/bubbler irrigation (Figure 19.2b–c) has been found to be effective in the economic use of water and enhanced growth, especially in an area where water availability is not satisfactory. As per the study conducted at GBPUAT, Pantnagar, spraying plain water three to four times in the early morning hours of the day during the advanced stage of fruit growth and development have been found to be highly effective for achieving better growth with minimized fruit cracking (Figure 19.2c).

Moisture conservation through mulching using dried weeds or black polythene sheet has been found useful. Trials have also been conducted to conserve moisture using farm residues and polythene sheets. Through adoption of mulching, frequency of irrigation is reduced. To check fruit cracking mulching with 3-4 irrigations during fruit growth has been found to be satisfactory.

Weeds Management

Weeds compete for water and nutrients during the initial year of growth. Depending upon location, more than 25 species of weeds is reported to be present in litchi orchards. Inter-culture operation and mulching suppress the weeds. Herbicides, although found effective for the control, are rarely applied. In mature orchards the weed problem is not encountered due to heavy shade and natural mulching by falling leaves.

Filler Plant and Intercropping

Litchi is a slow growing plant and takes about 15-16 years to develop canopy and cover the area. During the initial period of establishment, the space between the plants can be utilized for planting of filler plants/intercrops. The planting of guava, custard apple, lime/lemon in the centre, between and within the rows of litchi have been found to give additional income in the initial stage of planting without competing with the main crop. Papaya may be also planted as filler plant at the spacing of 2.5 x 2.5 m. In between the plants in the initial stage, cowpea, french bean, okra, brinjal or other suitable crops of the regions are grown as intercrops. In the mature lychee orchards, cultivation of partial shade loving plants (ginger, turmeric, elephant foot yam) is practiced successfully, which provides additional income

(a) Bubbler System (b) Drip System

(c) Micro Sprinkler System
*Av. Yield 189 kg/plant (Max- 240 kg/plant)
*Less than 2 per cent fruit cracking
(As observed at GBPUAT, Pantnagar)

Figure 19.2a-c: View of Micro Irrigation (Bubbler, Drip and Micro Sprinkler) in Bearing Litchi Orchard.

Source: Singh *et al.,* 2011 (Horticulture Research Center at GBPUA&T, Pantnagar, Uttarakhand).

Ringing or Girdling

Ringing is a well-known method for stimulating fruit bud formation in plants. Girdling, or ringing, is a horticultural technique where a strip of bark is removed from a tree's trunk or branches, blocking the flow of nutrients and sugars. This causes sugars to accumulate above the girdle, stimulating growth and flowering, especially in fruit trees. When performing girdling on plants, it's important to follow a few key guidelines to ensure the health of the plant while achieving the desired effect. First, select primary branches with a diameter of more than 10 cm. The best time for girdling is between the last week of August and the first week of September. To begin the process, remove a 3-4 mm strip of bark in a circle around the selected branch, ensuring that it is done carefully. It's crucial to leave at least one main limb or branch un-girdled to avoid starving the roots and risking the plant's death. Additionally, never girdle the main trunk of the plant, and take care to avoid damaging the xylem during the operation. Afterward, apply Bordeaux paste or copper oxychloride to the wound to prevent fungal infection. For subsequent girdling in the following year, make sure to perform the cut at least 5-6 cm above the previous girdle to promote healthy growth.

As per the research trial conducted at GBPUAT, Pantnagar and recommendation made under AICRP on fruits (2017), girdling of 50 per cent of primary branches with 6 mm width was found to be promotive for improved fruit weight, yield and quality. About 41.0 per cent more yield and good fruit quality attributes were obtained as compared to non-girdled tree, It has significantly overcome the problem of alternate bearing in litchi cv. Late Bedana. The Method of gridling and its ffect on bearing habit of litchi can be visualized by Figure 19.3.

Figure 19.3: Method of Gridling and its Effect on Bearing Habit of Litchi.

Source: Singh *et al.*, 2011 (Horticulture Research Center at GBPUA&T, Pantnagar, Uttarakhand).

Sun Burning Cracking in Litchi

Symptoms

Sun-burning and cracking of developing fruits of litchi (*Litchi chinesis* Son Suisba serious problem which accounts for a considerable loss to the growers every year. Losses may be up to 50 per cent under adverse weather conditions. Fruit growth in litchi is accomplished in 2 stages. In the first stage, the fruit grows rapidly along the longitudinal axis, mainly because of rapid increase in seed length. In the second phase, flesh growth takes place rapidly in May and synchronizes with the period of highest temperature and lowest humidity. High temperature and low humidity result in localized light-brown blotches on portions of the fruit-skin facing direct sun-rays. The blotches become intense in a few days, and the blotchy areas dry up. Simultaneously, a small vertical rupture appears in the dried blotchy area on account of the internal pressure because of rapid flesh growth. Occasionally, this rupture is a transverse feature also. The cracks become more pronounced with the rapid increase in flesh growth (Lal *et al.,* 2022).

Conditions for Development

The temperature is high, and relative humidity and rainfall low during the fruit-development period. Close observations on fruit growth of litchi have shown that flesh growth extends from the second week of May to the third week of June in most of the varieties. Weather data for this period indicate that temperature higher than 38°C in combination with relative humidity lower than 60 per cent are very favorable for cracking. Low relative humidity as a result of high temperature and lack of adequate moisture supply during the active period of fruit growth are the limiting factors for normal fruit growth in litchi. Thus, due to a sharp vapor-pressure gradient, the rate of water loss from leaf and fruit surface exceeds the rate at which water is absorbed by the roots. As a result, fruit skin becomes hard and inelastic. When such hard skin is subjected to increased internal pressure as a result of rapid flesh growth following irrigation, it cracks (Marboh *et al.,* 2017).

Corrective Measures

It is important that high humidity be maintained through regular and copious irrigation throughout the fruiting season. To reduce cracking of litchi fruits, two irrigations per week, between second week of May to the end of June are recommended.

Bagging of Fruits for Color Development and Quality Production

Litchi is good source of sugar, ascorbic acid, phenolics and flavonoids. There are some major problems in litchi production like seed and fruit borer, fruit drop, pericarp browning, fruit cracking and sun burn. Among these, sun burn, fruit cracking and fruit borer can be managed with intervention of bagging to the fruit bunches in litchi. Bagging of fruit is becoming most popular in litchi as it not only reduces sun burn, fruit cracking and fruit borer but it also improves colour of fruit and reducing blemishes which is one of the most important characters of litchi to attract the consumers. Development of fruits inside bags on the tree avoids the infestation of fruit borer because the bags served as a successful physical barrier against the fruit borer. Thus, pesticides load can be drastically reduced in litchi.

Types of Bagging

The bags act as a barrier to protect the fruit against sun burn and fruit cracking and attack by seed and fruit borer. Individual litchi fruit bunches are bagged with different types of bags at pea stage (March- April) when remain bagged till harvest (May to June). The stage of bagging to fruit bunches also affect the level of control of borer, sun burn and fruit cracking. The pesticide or fungicides pressure can be reduced once the bags are placed on the fruit. There are different types of bag available such as: Kraft paper, Maslin cloth, Poly ethylene, Butter paper, Light-yellow colored bag, Black and blue bags, Nylon bags, Cellophane or fabric bag, White coated bag, news paper bag, brown paper, pink paper bag *etc.* which can be used to protect the fruits. The efficacy of bags depends on the type of materials and colors.

Benefits of bagging of litchi fruits:

1. It increases in fruit growth, size and weight.
2. It reduces the incidence or attack of birds and insect pests.
3. It is a good technique to maintain a physical separation between the environment and the produce.
4. It reduces the direct penetration of light and improves the fruit colour. Fruit colour is the fundamental feature that attracts consumers.
5. It also reduces the incidence of sun burn and fruit cracking.
6. It reduces the number of damaged fruits.

As per the research trial conducted at GBPUAT, Pantnagar and recommendation made under AICRP on fruits (2020) the bagging of litchi fruit bunches with non-woven polypropylene pink bags at 25-30 days after fruit set, checks fruit cracking and sunburn, improves fruit quality (colour, size, TSS.), gives desirable consumer appeal. It reduces the incidence of insect pests and diseases, especially fruit borer incidence and protects from hail storms and bats. The effect of bagging of litchi bunches with non-woven polypropylene bags on fruit quality can be visualized by Figure 19.4.

Canopy Management

Litchi is an evergreen spreading tree, which attains the height about of 10-12 m at its full growth and development. Light penetration of its canopy is also desirable for proper fruiting; hence planting in square system at a distance of 10-12m has been practiced. However, due to shrinking land to man ratio and mounting demand for quality litchi fruits have made it imperative to achieve higher productivity from per unit area of litchi orchards. Therefore, it was soon realized that high density planting with some modification suited to the litchi tree canopy could greatly enhances productivity of litchi orchards. In this regard a spacing trial at Horticultural Research Centre of GBPUA&T, Pantnagar was conducted.

Figure 19.4: Bagging of Litchi Bunches with Non-Woven Polypropylene Bags for Quality Fruit Production.

Source: Singh *et al.*, 2011 (Horticulture Research Center at GBPUA&T, Pantnagar, Uttarakhand).

Amongst various planting systems, planting of litchi in a double hedge row system accommodating 222 tree/ha has been found to be the best and gave higher yield of equally good quality fruits up to certain years of its plantation, which later become over crowded. Due to overcrowding quality and productivity both diminished. As it is well known fact that litchi commences 3-4 growth flushes in a year which result in a buildup of dense canopy very soon thus, causes shading effect to its lower counterpart. Excessive vegetative growth also competes for nutrition, light and air which badly affects productivity and quality of the produce. The fact that the flowering and fruiting in litchi occurs at the periphery of canopy suggests the importance of light and air regimes in litchi fruits production (Jackson, 1980). Instead of thinning out the selective branches of a tree in high density planting systems, pruning of overcrowding branch was employed to make the tree more productive. However, different strategy was employed for canopy management of 60-year-old litchi cv. Rose Scented under square system of planting accommodating 100 plants/ha. It is well known fact that flowers in litchi are borne on current year's growth and old shoot rarely produces flowers. Thus, an efficient canopy management in litchi which calls for judicious removal of excessive vegetative growth and diverting photo synthetic energy for optimum reproductive growth seems to be the key factor for improving productivity and quality of litchi fruits. In India low and high density both kind plantings exist and both demands different canopy management strategy.

Training and Pruning (Canopy Architecture)

Training and pruning is very essential orchard operation in litchi. Since commercial harvest in litchi tree is obtained from two third of the lower canopy from the outer periphery, boosting more fruiting branches in this region is the main concern of training and pruning operation. Management of optimum stature of litchi tree with compact and stereo bearing canopy is an important aspect of orchard management. Generally, litchi tree grows continuously which results in larger canopy. In view of the present shortage Training and pruning is very essential orchard operation in litchi. Since commercial harvest in litchi tree is obtained from two third of the lower canopy from the outer periphery, boosting more fruiting branches in this region is the main concern of training and pruning operation. Management of optimum stature of litchi tree with compact and stereo bearing canopy is an important aspect of orchard management. Generally, litchi tree grows continuously which results in larger canopy. In view of the present shortage of space and requirement of high density orcharding, promotion of large bearing area, accommodation of a greater number of plants/unit areas, management of tree height, canopy concentration to provide more fruiting branches and penetration of more light to inner portion is more relevant. Hence, proper shaping of tree from initial stage and pruning of branches after fruit harvest in young plants, shaping is quite convenient at this stage. However, removal of 20 cm shoots after harvesting should be done every year to promote healthy vigorous shoots for the next year crop. Reiterative pruning is practiced to rejuvenate the old orchards whereas, corrective pruning by removing undesired, unfruitful and centre leader branches to ensure light and aeration inside the canopy should be adopted to enhance the stereo fruiting and thereby the productivity of the orchard. Training is an important operation and should be initiated at the nursery stage itself. Single stem air layered plants should be raised in bags and allowed to grow up to 40-50 cm. The layered plants have strong tendency to produce branches at the ground level which should be pinched or pruned. Further, strong, well-spaced, out facing with wider crotch shoot should be allowed to form the main branches. It is necessary to continue shaping by removing all the branches forming crotches with main branches as and when grow (Nath *et al.,* 2016).

The main aim of frame working is to develop sufficient fruiting terminals. Compact and stereo-bearing canopy can only be achieved by well thought out pruning at early stage and shaping of young tree. The canopy should have well illuminated and sufficient number of strong and healthy branches for stereo bearing. To develop good and compact canopy, 25-30 cm fruit bearing shoots at the time of harvesting should be removed. In this way, 2-3 new terminals will develop which will consequently develop into fruiting branches next season. As principle, young trees are lightly pruned to provide larger area for Photosynthesis. Severe pruning at young stage should however be avoided as it hinders the tree development and increases the period of juvenility. Besides, removal of dead and diseased shoots, small internal branches which prevent the sun rays to penetrate inside the canopy should be removed. Although, removal of branches along with the fruit bunch had shown its promise, but culturing compact canopy seems to be more useful in other litchi growing countries. The techniques include, more pruning at top (open window) and outer parts of the tree canopy and less at bottom and inner side of the canopy which promotes stereo fruiting. In Israel however, dwarf stature litchi tree has been maintained by topping and hedging.

The main objective of annual pruning is to control growth of the tree as in case of high-density planting and promoting the flowering. The recent findings advocates pruning of litchi is facilitate healthy growth and better yield. However, in the past litchi trees were never pruned. Only at the time of fruit harvesting, a small portion of branches were used to be lopped off along with fruit bunch. This practice was considered to be helpful in improving subsequent fruiting by encouraging terminal shoot production. Sometimes root pruning and exposure also improves the flowering in adult trees. Time of annual pruning is very much important for the next year cropping. It should be done just after fruit harvesting. As soon as the fruit is harvested, the plants put forth new growth, therefore pruning must be completed before the initiation of new growth. Pruning after the initiation of new flush results the production of unhealthy shoots which do not mature for flower bud differentiation for the next fruiting season. During pruning tree vigour should be taken into consideration. Winter pruning should be strictly avoided. However, the unproductive shoots arising from the main branches under the canopy should be removed as and when it is appearing. Rational supplementary pruning (light pruning) can be done after flower bud formation in spring depending upon the requirement. Reiterative pruning is done to one-year old bearing mother branches by removing 1/3 or 4/5 of the main branches. The damaged shoots due to severe cold, also needs to be removed just before the new growth initiation. After pruning, the leaves should be allowed to decompose in the basin of each tree. This can be done by removing the woody branches after drying of leaves followed by mixing of leaves in the soil before application of regular fertilizer dose.

Deep Pruning/Rejuvenation of Old and Senile Orchard

The old and senile litchi orchards are not able to boost the productivity of crop vise. Poor selection of planting material, indiscriminate plantation and poor management over the year has made many orchards unproductive or less economical. The term 'Rejuvenation' means renewal or young again. Perennial fruit trees notice decline both in quality and quantity of produce after some period of time due to improper canopy and orchard management. As a result of which orcharding becomes economically non- viable and non-remunerative. It has been observed in general that litchi orchards after attaining the age of 50-60 years, turns dense with compact top canopy covering most of the branches at the bottom and bearing fruits only on the high-tops. Such plants yield less vis-à-vis pose problems in proper management such as annual pruning, pest control and harvesting. To overcome the problem of unproductive or uneconomic of litchi orchards which is existing in abundance particularly in traditional litchi producing states like

Bihar, Bengal, Uttarakhand, Jharkhand the farmers left with options of large scale uprooting and replacement with new plantations which will be a long term and expensive enterprise or go for rejuvenation of senile trees and utilize the existing established root system and capacity of plants to rebuild the canopy by activating latent buds present on main limbs and trunk which is comparatively cheaper and environmental friendly too (Nath *et al.*, 2019).

Objectives of Rejuvenation

1. To convert the low-yielding and inferior trees in to high yielding and superior trees with or without top working of improved cultivars.
2. To exploit the better root system of an established plant which has survived over the year in exiting soil and climatic conditions.
3. To increase the orchard income through sale of bi-products (wood) and inter cropping in inter-space.
4. To increase the orchard productivity and economical age of plant.
5. To ease farm operations on managed tree canopy and cut down production costs.

Technology for Rejuvenating Old Litchi Orchard

The rejuvenation technology embodies heading back (topping) of branches during August-September at a height of 1.0 to 1.5 m from the ground level depending on the structure of individual tree in the orchard. About 3 to 4 main branches with outward growth from the base are marked for pruning at the required height, with a plan of developing umbrella like or semicircular frame work of canopy to enhance bearing surface area. Pruning can be done either with manual saw or power operated saw in phased manner from the top. Care should be taken to avoid bark splitting or debarking at the cut end due to falling of heavy branches at the time of pruning. To avoid any external infection at the cut portion, it should be pasted with Bordeaux mixture or Copper-oxy-chloride (Blitox) immediately after pruning. Steps in Rejuvenation of old Litchi orchard. The major steps in rejuvenation of litchi plants include:

i. *Selection of litchi orchard*: Selection of unproductive litchi orchard through discussion with farmers and making farmer convinced for adoption of technology.
ii. *Training of the farmer/staff*: Training of the farmers who can execute the operations. Procurement and knowledge of tools and implements for rejuvenation must be ensured before execution of program. Now a day's custom hire options are also available in many areas.
iii. *Selecting right time* (Aug- Sep under Bihar condition): When there is near normal temperature and light environmental humidity which favors emergence of new sprouts and protects the tree trunk from splitting due to scorching sun or chilling winter.

iv. *Execution of rejuvenation operation:* Cutting of main branches which are growing outwards at a height of 1.0 – 1.5 m leaving 3- 4 main limb as base which helps open center canopy development.

v. *Pasting of cut ends and entire trunk*: Mud and fresh cow dung paste or blue copper or Bordeaux paste is applied to avoid any infection.

vi. *Fertigation and floor orchard management*: Ploughing of field and application of manures and fertilizers to the plant as per recommendation (1000 g N + 800 g P + 800 g K per plant) along with neem cake (2-4 kg/ plant), farm yard manure (50-60 kg/plant) micro nutrients and bio-fertilizers in the tree basin. Regularly irrigate the plants and maintain the orchard floor weed free. If possible suitable intercrops preferably legumes or high value crops can be grown.

REFERENCES

Ali M, 2016. The Influence of Different Media on Rooting of Litchi Plant through Air Layering. *New media and mass communication*, 47, 12-15. https: //doi.org/10.7176/NMMC.VOL4712-15.

An X, Jiang S, Xie C, Xu Y, Dong C and Shen Q, 2022. [Effects of reducing chemical fertilizers combined with organic fertilizers on soil microbial community in litchi orchards]. *Ying yong sheng tai xue bao = The journal of applied ecology*, **33**(4): 1099-1108. https: //doi.org/10.13287/j.1001-9332.202204.031.

Bal JS, 2006. Propagation and nursery production. In: Fruit growing. 2nd revised ed., Kalyani publishers, Ludhiana, New Delhi. pp. 27-53.

Das D, Prakash N and Bhalla-Sarin N, 1999. Multiple shoot induction and plant regeneration in litchi (Litchi chinensis Sonn.). *Plant Cell Reports*, **18**: 691-695. https: //doi.org/10.1007/s002990050644.

Lal N, Sahu N, Kumar A and Pandey S, 2022. Effect of rainfall and temperature on sun burn and fruit cracking in litchi. *Journal of Agrometeorology*. https: //doi.org/10.54386/jam.v24i2.1153.

Marboh E, Singh S, Pandey S, Nath V, Gupta A and Pongener A, 2017. Fruit cracking in litchi (Litchi chinensis): An overview. *The Indian Journal Agricultural Sciences*. https: //doi.org/10.56093/ijas.v87i1.66954.

Marboh E, Singh S, Pandey S, Nath V, Gupta A and Pongener A, 2017. Fruit cracking in litchi (Litchi chinensis): An overview. *The Indian Journal Agricultural Sciences*. https: //doi.org/10.56093/ijas.v87i1.66954.

Nath V, Kumar G, Pandey SD and Gupta AK, 2016, May. Status of litchi in India. In *V International Symposium on Lychee, Longan and Other Sapindaceae Fruits* **1211**: 153-160.

Nath V, Marboh ES and Gupta AK, 2019, June. Canopy management for improving productivity in litchi. In *VI International Symposium on Lychee, Longan and Other Sapindaceae Fruits* **1293**: 129-142.

Nayak S, Tarai R, Swain S, Samal S, Pradhan S, Sethy B, Behera S, Vali D, Ranjith P, Mandal P and Selvi S, 2024. Effect of Foliar Feeding with Nutrients and Bioregulators on Yield and Quality Attributes of Litchi cv. Bombai. *Horticulturae*. https: //doi.org/10.3390/horticulturae10020188.

Sahni R, Kumari S, Kumar M, Kumar A, Singh S and Kumar M, 2024. Mechanization of Litchi (Litchi chinensis S.) Cultivation: Current Scenario and Future Needs of India. *Journal Scientific Research and Reports*. https: //doi.org/10.9734/jsrr/2024/v30i122721.

Saúco V, Jahiel M, Modesto P, Huang X, Mitra S, Yamanishi O and Andrade R, 2018. Litchi (*Litchi chinensis* Sonn) propagation. New Technologies and innovations. *Revista Brasileira de Fruticultura*. https: //doi.org/10.1590/0100-29452018575.

Shihua T, 2011. Effect of Fertilizer Application Strategy on the Growth, Quality and Yield of Litchi. *Chinese Journal Tropical Crops*, **33**: 98-103.

Singh CP, Singh PK, Mishra DS, Joshi J and Chauhan P, 2011. Litchi: Integrated Orchard Management. Directorate of Research at GBPUA&T, Pantnagar, Uttarakhand.

Singh J, Pandey S, Lal N, Diwan G and Nath V, 2023. Effects of Canopy Architecture and Planting Density Management on Yield and Quality Attributes of Litchi (Litchi Chinensis Sonn.). *Bangladesh Journal Botany*. https: //doi.org/10.3329/bjb.v52i1.65247.

Xie J, Chen Y, Gao P, Sun D, Xue X, Yin D, Han Y and Wang W, 2022. Smart fuzzy irrigation system for litchi orchards. *Comput. Electron. Agric.*, **201**: 107287. https: //doi.org/10.1016/j.compag.2022.107287.

Chapter 20

Citrus-Based Cropping Systems in North-East India

Evening Stone Marboh and Darshan M. Kadam

ICAR–Central Citrus Research Institute, Nagpur–440033, Maharashtra

The North-Eastern region (NER) of India is recognized as a centre of genetic diversity for various citrus species. Total 17 citrus species, 52 cultivars and 7 probable natural hybrids have originated in this region, reflecting its rich genetic reservoir. This unique biodiversity makes North-East India the natural home for a wide range of citrus species, including *Citrus indica, C. assamensis, C. latipes, C. ichangensis, C. macroptera, C. aurantium, C. reticulata, C. megaloxycarpa, C. jambhiri, C. aurantifolia, C. grandis, C. limon* and *C. karna* (Barbhuiya *et al.,* 2022). However, only two types of citrus are cultivated on commercial scale *i.e.,* Khasi mandarin and Assam lemon. Trees of the rest of the indigenous citrus species are mostly confined to home-stead gardens, kitchen gardens, community land and wild habitats including forests in NER. However, despite its significant genetic diversity, citrus cultivation in NER is constrained by several challenges *viz.,* low productivity, citrus decline, inappropriate choices of intercrops, low adoption of technologies *etc.* The Khasi mandarin (*C. reticulata* Blanco), a variety of high commercial quality is grown across all North-Eastern states, covering the largest area among citrus crops. In Assam, it is known locally as 'Kamala', is widely cultivated in regions like Kamrup, Tinsukia, Dibrugarh, Karbi Anglong and North Cachar. In Meghalaya, the southern slopes of the Khasi hills, especially Shella, Dawki and Cherrapunji produce some of the highest quality Khasi mandarins, with cultivation extending into the Jaintia and Garo hills. Tripura, too, is notable for mandarin production, especially in the

Jampui Hills, where conditions support high-quality yields. Arunachal Pradesh, particularly the Alang-Basar area in Siang district, has ideal conditions for citrus, while in Sikkim, mandarin orange is predominantly grown in the East, South and West districts. The citrus-growing areas in Manipur include Tamenglong, Tipaimukh, Jiribam and Ukhrul, while Nagaland's citrus belts cover Mokokchung and Tuensang. Assam also produces a unique seedless lemon, the Assam lemon (*Citrus limon* Burm.), known for fruiting throughout the year, and distributed across districts like Nowgaon, Jorhat and Sonitpur. Mizoram and Nagaland contribute significantly to regional citrus diversity with cultivars such as the orange, lemon, and local mandarins grown across districts like Kolasib, Longnok and Wokha.

The NER is characterized by challenging terrain, varied slopes, altitudes and diverse cultivation practices influenced by unique land tenure systems. Agriculture, horticulture and allied land-based activities form the primary means of livelihood for the majority of the population. The region's agriculture system is largely rainfed, mix cropped and subsistence-based and with shifting cultivation or 'jhum' traditionally practiced on steep slopes. Due to increased human population pressures and changes in land use, the fallow cycles in 'jhum' cultivation have been shortened to two or three years from the previous 15-20 year. This shift has led to soil erosion, nutrient depletion and an overall decline in biodiversity, contributing to a degradation of land productivity. The shifting cultivation has resulted in annual losses of 88,346 thousand tonnes of soil and 10.66, 0.372 and 6.05 thousand tones of nitrogen, phosphorus and potassium, respectively in NER (Sharma and Prasad, 1995). Additionally, significant amounts of calcium and magnesium are also depleted, leading to noticeable deficiencies in crops grown within these areas. This gradual degradation of essential production resources has become a critical concern, necessitating location-specific strategies for conservation, utilization and management to optimize sustainable production without compromising quality. In this context, intercropping within orchards holds considerable promise for the Northeast region.

Farmers of NER followed an indigenous way of orchard management, with few practicing an ecologically sustainable multi-cropping system and orchard intercropping. Intercropping in citrus orchards is gaining recognition across the region as a means to diversify income, reduce dependency on chemical fertilizers and address the risks of monocropping, which is vulnerable to pest and diseases. Incorporating compatible crops, particularly legumes like groundnut and soybean, into citrus orchards enhances soil nitrogen, suppresses weeds and provides interim income during the early years of mandarin cultivation before the fruit trees reach full production. Integrated citrus-based systems enable the continuous use of land, generating additional income streams, optimizing land use, and allowing for year-round employment. In addition to intercropping, agroforestry systems (AFS) that combine 'tree + crop' practices have shown potential in soil conservation and eco-restoration. In citrus-based AFS, combining these practices has proven beneficial for the region, supporting both crop and citrus yield stability. This diversification is essential in ensuring long-term productivity and a stable livelihood for farmers in North-East India.

Most citrus-based intercropping studies to date have primarily focused on issues related to cropping systems in pre-bearing orchards (Singh *et al.,* 1999). However, there is limited research on the evaluation of intercrops in bearing citrus orchards, resulting in conflicting findings regarding their impact on the health of the main crop and fruit yield (Santinoni and Silva, 1995). These discrepancies are influenced by factors such as whether intercrops are grown under rainfed or irrigated conditions, the environmental context, crop duration and the specific nutrient needs of the intercrops (Reddy *et al.,* 2003). Moreover, limited scientific information is available with regards to intercrops practiced in NER. This chapter discusses various citrus-based cropping systems in northeast India, criteria for selecting appropriate intercrops and the advantages and disadvantages of different intercrop choices in citrus orchards.

General Significance of Intercropping in Orchards

The growing population has compelled agricultural planners to recheck the role of intercropping, the application of growing more than one crop in a field at the same time, as a tool to enhance agricultural production and to obtain efficient land use. The monoculture system faces the risk of price fluctuation, vulnerability to certain major pests and diseases, yield uncertainty which singly or jointly can results in a great economic loss. Cropping of several plant species together reduces negative effects of a monoculture and thus is commonly employed in ecological agricultural systems. Agricultural practices like intercropping are pro ecological; supporting bio-diversity and is compatible with the principles of balanced agriculture. Andrews and Kassam (1976) defined the intercropping as the agricultural practice of growing two or more species in the same land at the same time. 'Ecological intensification' with tools like intercrops will be necessary to boost food production for growing global populations while simultaneously enhancing crop protection, improving resilience to climate change, and reducing the carbon footprint of agriculture (Morugan-Coronado *et al.,* 2020).

Intercropping is considered to be one of the most significant cropping techniques in sustainable agriculture. Besides diversifying agricultural output, intercropping also improves nutrient status and the physical properties of the soil. The long juvenile period of fruit trees, poor income from initial harvest, insecurity against diseases and pests, poor marketing infrastructures are the major constraints faced by perennial fruit growers'. Effective utilization of interspaces between the citrus plants in the orchard would be a viable and profitable intercropping system for sustainable profit. It also lowers the need of external inputs and improves the stability and diversity if ecological niches are kept in the mind while selecting intercrops. Due to wide spacing and developing root patterns, during initial years (up to 6-7 years) of citrus orchards establishment, large unutilized interspace can be exploited for growing inter and mixed crops successfully. This enables farmers to raise additional income during the years when the main crop yields no or low returns.

Citrus is a nutrient responsive crop and inclusion of legumes and leguminous cover crops in intercropping of citrus orchards will helps in fixing the nitrogen in the soil biologically and restoring soil fertility (Aziz *et al.,* 2008: Srivastava *et al.,* 2007). Citrus trees are medium to large and generally planted at a spacing of 6 m × 6 m. These interspaces can be efficiently used to raise intercrops during the pre-bearing phase with a view to obtain some extra yield without sacrificing the main crop or base crop yield. Intercropping of citrus with ground nut, cotton or soyabean in kharif season and with wheat or gram in rabi season can be successful. Intercropping leguminous crops like groundnut in improves the nitrogen concentration in the orchard soil and gave maximum fruits/tree and net return (Dhyani and Tripathi, 1998). To enhance soil fertility and improve the physical condition of orchard soils, leguminous crops such as pea, cowpea, jack bean, rice bean, French bean, green gram, black gram, and stylosanthes (used as fodder) can be cultivated as intercrops. Additionally, shallow-rooted, low-nutrient-demand seasonal vegetables, along with pineapple and papaya, are suitable intercrops for orchards. On hill slopes, two rows of pineapple may be planted just below contour bunds, with a spacing of 60 cm from the bund and 30 cm between plants (Majumder *et al.,* 2011).

Types of Intercropping System

Intercropping systems involve two or more crop species or genotypes that grow together and coexist for long periods of time. The latter criterion distinguishes intercropping from mixed monocropping and rotation cropping. The intercropping of fruit orchards is a common practice in many countries. The success of intercropping systems over sole cropping can be achieved through agronomic manipulation. These manipulations included plant density, planting time, available resources, and intercropping patterns (Mousavi and Eskandari, 2011). Spatial arrangements, planting, and harvest times of crops should be considered in intercropping systems. Supplemental effects in resource use models should be considered to obtain better yield and quality in intercropping systems. Cultivars to be used in intercropping systems should improve the complementary effects between species (Baumann *et al.,* 2001). There are several intercropping systems, such as mixed, strip and row intercropping (Oad, *et al.,* 2007).

1. **Mixed cropping** – growing two or more crops together in no distinct row arrangement.
2. **Strip intercropping**: growing two or more crops together in strips wide enough to separate crop production using machines, but close enough to interact.
3. **Row intercropping**: growing two or more crops together at the same time with at least one crop planted in rows.
4. **Relay intercropping**: Plant a second crop into a standing crop at a time when the standing crop is at its reproductive stage, but before harvesting.

Beside this different AFS are traditionally practiced in the NER. These AFS varies with geographical location and socio-cultural behaviour of the people. The following systems has edge over other systems of agroforestry in NE region of India- Agri-Silviculture: (crops + trees), Agri-Horticulture: (crops + fruit trees), Silvi-Pasture: (trees + fodder crops). Horti-Pasture: (fruit trees + fodder crops), Agri-Horti-Silviculture: (crops + fruit trees + trees) and Homestead Agroforestry: (may be the mixtures of crops, vegetables, fruit trees, fodder crops and trees).

Criteria for Selection of Intercrops

The cropping system should be planned in such a way that, under unfavourable conditions, at least one crop should have competence to survive and produce economic yield. Any cropping system aims to produce higher yields per unit area through judicious use of available natural resources that would otherwise not be utilized by a single crop, offers greater stability in production under any weather condition, meets the domestic needs of growers, offers opportunities for employment generation, effective weed, pest and disease suppression. Compatibility of crops in cropping systems is an important factor before setting up any cropping system. The crops selected for intercropping should meet the criteria required for the intended system. Soil type, rainfall pattern, irrigation facilities, and climatic conditions of the area, besides market demand, are important governing factors for the selection of intercrops. Component crops with different harvest times should be preferred to reduce competition for similar resources simultaneously. Moreover, different roots and aboveground parts of plants used in intercropping systems can be used for the effective utilization of moisture and light. The component crop should not have the same rooting system and should not be as tall as the main crop. Crops with the same insect pests and diseases those as of the main crop should be avoided. The crops should be selected according to shade tolerance and the amount of solar radiation available. Intercrops have synergistic or complementary effects on the main crop (Kumar *et al.,* 2010). Each should have adequate space to maximize cooperation and reduce competition between them. Most importantly, while selecting the crop combination, care should be taken to ensure that the selected crops are compatible with the systems. Many annual, biennial, and perennial crops have been studied for their suitability as component crops in citrus-based intercropping (Table 20.1). Variations in cultivar influence the productivity of the intercropping system. For instance, in mandarin, the total rhizobia count of the mandarin tree rhizosphere and biological N-fixation ability varied among soybean cultivars. Cultivars with narrow leaves had higher light penetration and transmission within other parts of the plant, and thereby higher biological N fixation ability under interplanting conditions. Further late-maturing cultivars had higher leaf dry weight per plant which was probably related to their higher biological N fixation ability during vegetative growth and development. Accordingly, it is expected that nodule ability in the roots of late-maturing soybean cultivars is greater than that in early cultivars. (Selim *et al.,* 2020). The distinct aspects of intercropping are as follows:

Table 20.1: Different Types of Intercrops in Citrus Based Intercropping System

Crop Types	*Component Crops*	*Citrus Species*	*Effects*	*References*
Legumes	Green gram-pea-green gram	*Citrus spp.*	Increase bacterial population (21.3 x 10^6 cfu/g), actinomycetes (8.4 x 10^3 cfu/g) and basal respiration rate (193 mg/kg of CO_2-C)	Sarathambal *et al.*, 2015
	Cluster bean + wheat	Kinnow (*C. reticulata*)	Increase plant height and gross income (Rs. 41180/-) compared to Rs. 31605/- in monoculture system	Gill *et al.*, 2018
	Soybean and cowpea	Washington Navel Orange (*Citrus sinensis*)	Increased available N:P:K: (mg/kg) content in soil @ 65-66:15-21:486-533.39 as compared to 53.5:17:486 in sole orange	El-Mehy-Amira and El-Badawy, 2017
	Pea	Kinnow mandarin (*C. reticulata*)	Maximum BC ratio (2.41) and highest net returns (90,000 Rs/ha)	Singh and Kaur, 2017
	Green gram (*kharif*) - gram (*rabi*)	Sweet orange (*Citrus sinensis*)	High B:C ratio (4.22)	Dalal and Gohil, 2009
	Fenugreek-cluster bean	Kinnow (*C. reticulata*)	More vigorous plant growth as compared to sole plantations	Soni *et al.*, 2024
	Soybean	Nagpur mandarin (*C. reticulata*)	Maximum canopy volume (28.81 m^3) compared to sole Nagpur Mandarin (19.52 m^3)	Shirgure 2012
	Soybean	*Citrus spp.*	Reduces dry weight of grassy weeds (42.89 g/m^2) and broad weeds (21.47 g/m^2) compared to 82.40 g/m^2 and 113.77 g/m^2 in Sole crop, respectively	El-Mehy-Amira and El-Badawy 2017
	Soybean	*Citrus spp.*	Improves P recovery rate in soybean and citrus when P was applied in upper and deeper soil layer, respectively	Zhou *et al.*, 2009
	Soybean	Mandarin (*C. reticulata*)	Increase total count of rhizobia in rhizosphere, minimized fruit drop and increase fruit yield	Selim *et al.*, 2020
	Groundnut	Mosambi (*Citrus sinensis*)	Improved net return (Rs. 35,820.0/ha)	Ghosh and Pal 2010

Crop Types	*Component Crops*	*Citrus Species*	*Effects*	*References*
Plantation crops	Tea	*Citrus spp.*	Reduced heavy metals like Hg, Ar, Cd, Pb, Ni and increase soil available N, P, K, Mn, Zn and Cu	Wen *et al.*, 2020
	Tea	*Citrus spp.*	Improve green tea quality with higher free amino acid, lower catechins and the ratio of phenol to amino acids	Wen *et al.*, 2020
	Tea	*Citrus spp.*	Increase the diversity and species richness of key natural enemies, Orb-weaver spiders (araneids) and suppress key tea pests, tea green leafhopper (*Empoasca vitis* Gothe)	Ye *et al.*, 2010
	Tea	*Citrus spp.*	Enhanced levels of amino acids in tea by 26.21 per cent.	Duan *et al.*, 2024
	Arecanut based HDMSCS	Assam Lemon (*Citrus limon*)	Highest benefit cost ratio of 3.22 - 3.60 at 2/3rd recommended fertilizer application	Sujatha *et al.*, 2016
Cover crops	Green manure	*Citrus spp.*	Improved biogeochemical cycles of C, nitrogen (N), and phosphorous (P) and soil organic matter	Castellano-Hinojosa and Strauss 2020
	Legume cover crop (*Arachis pintoi*)	*Citrus spp.*	Mitigate soil compaction and degradation, lower soil bulk density and increase soil porosity by 3.01 per cent	Dung *et al.*, 2022
	Urochloa ruziziensis mulch cover crop	Tahiti acid lime (*C. latifolia*)	Increase leaf potassium and soil exchangeable K, soil moisture content and lower soil resistance penetration and fruit yield by 56 per cent	Azevedo *et al.*, 2020
	Legume cover crops (mucuna, dolichos, cowpea)	Sweet orange (*Citrus sinensis*)	Improves fruit size and sugar composition	Mulinge *et al.*, 2018
	Green manure, sod, natural vegetation, grasses	*Citrus spp.*	Increased fruit yield, soluble solid content, sugar content, and vitamin C content and reduced fruit acidity	Fang *et al.*, 2021

Crop Types	*Component Crops*	*Citrus Species*	*Effects*	*References*
	Ageratum conyzoides	Sweet orange (*Citrus sinensis*)	Possess secondary metabolites, ageratochromene which inhibit the growth *Bidens pilosa, Digitaria sanguinalis* and *Cyperus difformis* and spores of *Phytophthora citrophthora, Pythium aphanidermatum* and *Fusarium solani*	Kong *et al.,* 2004
	Large green and *Mucuna pruriens*	*Citrus spp.*	Reduced broadleaf weed and grass weed density *like Gomphrena celosoides, Oldenladia corymbose* and *Tridax procumbens*	Olorunmaiye and Afolayan 2012
	Hemerocallisflava and white clover planted on contour hedgerow	*Citrus spp.*	Preventing surface runoff by reducing the degree of spattering and scouring of the soil by rainfall and kinetic energy of rainfall thereby reducing runoff, sediment, total N and total P losses	Li *et al.,* 2024
Vegetables	Potato	*Citrus spp.*	Increase the ground cover, reduce evaporation and soil erosion, increase soil organic matter, alkali hydrolysable N and available P	Guo *et al.,* 2020
	Groundnut (kharif) - onion (rabi)	Sweet orange (*Citrus sinensis*)	Maximum net profit (Rs. 68,000/ha)	Dalal and Gohil, 2009
	Watermelon	Kinnow (*C. reticulata*)	Increase number of fruits (920 vs 821), and yield 410.58 q/ha vs 352.58 q/ha) compared to sole Kinnow	Kumar *et al.,* 2022
	Potato	Sweet orange (*Citrus sinensis*)	Increase population of ladybeetle, *Coccinella undecimpunctata*	Mousa and Ueno, 2019
	Water melon	Kinnow (*C. reticulata*)	Maximum fruit weight (161.11 g) compared to sole kinnow (155 g)	Kumar *et al.,* 2022
	Okra and cowpea	Mosambi	High net return of Rs. 22,520.0/ha and Rs. 22,420.0/ha in okra and cowpea respectively	Ghosh and Pal, 2010
	Chilli pepper + amaranthus	Sweet orange (*Citrus sinensis*)	Highest yield efficiency/m^3 canopy (23.7) compared to mono-cropping (15.4)	Aiyelaagbe, 2001

Crop Types	*Component Crops*	*Citrus Species*	*Effects*	*References*
	Cucumber	Cleopatra mandarin (*Citrus reticulata reshni*)	Recorded 78 per cent higher profit margin at optimum cucumber population of 2500 plant/h	Olaniyan *et al.*, 2006
	French bean	Lemon (*Citrus limon*)	High BC ratio of 3.94	Hnamte *et al.*, 2013
	Okra-French bean crop rotation	Kinnow (*C. reticulata*)	High net return of Rs. 388629\ha with BCR of 4.02:1	Rohit and Singh, 2019
	Brinjal-Cauliflower crop rotation		Net return of Rs. 355190\ha with BCR of 4.12:1	
	Fodder maize-muskmelon—cabbage	Lemon (*Citrus limon*)	High net return/ha/year (Rs. 1,74,500) and B:C ratio (3.56:1	Dubey *et al.*, 2016
Oilseeds	Mustard (var. Giriraj)	Mandarin (*C. reticulata*)	High benefit cost ratio	Singh *et al.*, 2024
	Mustard	Mandarin (*C. reticulata*)	Maximum seed yield, biological yield and oil yield	Singh *et al.*, 2023
Spices	*Curcuma longa* L. var. GNT-1	Mandarin (*C. reticulata*)	Highest gross returns (Rs/ha), net returns (Rs/ha) and B:C ratio of sole mandarin	Ranawat *et al.*, 2018
	Curcuma longa L.var. GNT-1	Mandarin (*C. reticulata*)	High curcumin content (3.31 vs 2.54%) and high benefit cost ratio (3.15 vs 2.12) compared to sole cropping	Bahiriya *et al.*, 2019
Tuber	Cassava	Sweet orange (*Citrus sinensis*)	Tuber yield of Cassava planted at 2 and 3 m from trunk is higher than at 1 m away	Olaniyan *et al.*, 2018
	Sweet potato	*Citrus spp.*	Reduced broadleaf weed and grass weed density *like Gomphrena celosoides, Oldenladia corymbose* and *Tridax procumbens*	Olcrunmaiye and Afolayan, 2012

Crop Types	*Component Crops*	*Citrus Species*	*Effects*	*References*
Cereals	Wheat - groundnut rotation	Kinnow (*C. reticulata*)	High net returns (Rs. 82,000/ha)	Soni *et al.*, 2024
	RCPL-1-24 upland rice	Assam lemon (*Citrus limon*)	High total biomass production, maximum grain yield (41.66 q/ha) and high net monetary benefit (Rs. 13,787.60/ha)	Bhatt and Misra, 2003
	Perennial grass species, *Festuca rubra and Festuca ovina*	Citrumelo (*Poncirus trifoliata* x *C. paradisi*)	Preventing leaf chlorosis and improving their growth	Ammari and Rombola, 2010
Fruit trees	Guava	Sweet orange (*Citrus sinensis*)	Maximize plant height, maximum and number of fruits and increase mortality rates of psylla adults, reduced citrus greening incidence 90 per cent and citrus greening severity by 83 per cent	Islam *et al.*, 2020
	Guava	*Citrus spp.*	Possibly facilitate the growth of citrus by acting as a windbreak and effectively reduces greening during the first year	Ichinose *et al.*, 2012
	Guava	Sweet orange (*Citrus sinensis*)	Volatile terpenoid compounds, α-pinene exhibited repellent activity against the Asiatic citrus psyllid	Zaka *et al.*, 2015
Fodder	Berseem, ryegrass and lucerne	Kinnow (*C. reticulata*)	Higher green fodder yield	Ghasal *et al.*, 2024
	Barseem	*Citrus spp.*	Maximize diversity and community structure of mesostigmatid soil mites (Ameroseiidae, Ascidiae, Laelapidae, Pachylaelapidae, Parasitidae, Sejidae and Uropodidae)	Ghaffar *et al.*, 2019
	Alfalfa (*Medicago sativa*) and hairy vetch (*Vicia villosa*)	Mandarin (*C. reticulata*)	Increased soil total nitrogen, nitrate N, and available P, enhance soil fungi and bacteria including Mortierella and Streptomyces, and activity of soil sucrase and β-1,4-glucosidase	Xie *et al.*, 2024

Crop Types	*Component Crops*	*Citrus Species*	*Effects*	*References*
	Leguminous herb *Stylosanthes gracilis* inoculated with the AM fungus	Sweet orange (*Citrus sinensis*)	Mediates competition and optimized intercrop combination, increased the biomass, P content and proportion of legume roots entering the soil	Yao *et al.*, 2005
	Maize fodder	Kinnow (*C. reticulata*) and Mosambi (*Citrus sinensis*)	Conserving/augmenting predators and natural enemies, coccinellids and *Chrysoperla carnea* and reduce infestation of citrus miner (*Phyllocnistis citrella*)	Ahmed *et al.*, 2013
	Forage peanut or coast-cross grass	Natal sweet orange (*Citrus sinensis*)	Reduces the ascospores of *Guignardia citricarpa*, a causal organism of citrus black spot	Bellotte *et al.*, 2015
Flowering plants	Insectary plants, Sweet alyssum and buckwheat	Sweet orange (*Citrus sinensis*)	Strongly attracts syrphids hoverflies, *Allograpta obliqua*, the main predator of *D. citri*	Irwin *et al.*, 2021
Rhizo-mediators	Hyperaccumulators, *P. vittata*	*C. reticulata*	Lower net remediation cost (8.6 $/g) in remediation of heavy metal contaminated farmland compared to 13.3 $/g in P. vittata monoculture	Yan *et al.*, 2022

- ✰ Detailed planning
- ✰ Timely planting of each crop
- ✰ Adequate fertilization at optimal rate and times
- ✰ Effective weed, pest and disease control
- ✰ Efficient harvesting

Intercropping System in Citrus Orchards in NER

Khasi mandarin is a popular commercial citrus crop cultivated by farmers in NER. Farmers followed an indigenous way of orchard management, with few practicing an ecologically sustainable multi-cropping system and orchard intercropping. Soybean, turmeric, ginger, maize, arecanut, tea, beans and short-duration vegetables are commonly grown as commercial intercrops. Growing intercrops not only generates additional income and improves food security but also helps to check soil erosion through ground coverage and stabilize terrace walls. Short-duration crops are an ideal choice for intercropping mandarin orchards. Moreover, the region receives ample precipitation during the rainy season (June–August) which creates ample scope for growing short-duration crops. Among the intercrops, maize creates strong competition for nutrients and light and is the most detrimental to mandarin productivity. Hence, vegetables, leguminous crops and ginger are best suited as intercrops to generate improved economic returns (Patiram *et al.,* 1994).

The orange tree canopy modifies the microclimate and influences the physiological processes of story crops. As the tree canopy widens, the photosynthetic active radiation (PAR) and temperatures decrease, while humidity under the canopy increases. PAR interception under the canopy is crucial for both the main and component crops. The major intercrops used were ginger, ginger + maize, maize + finger millet, maize + cassava, maize + buckwheat, and other vegetables. These intercrops are grown in such a way that they take advantage of both horizontal and vertical spaces available between fruit trees. Study performed at Barapani, Meghalaya indicated that intercropping positively influenced growth parameters in different tree species like alder, albizia and cherry, though mandarin exhibited slower growth and lower biomass, consistent with its fruit-bearing nature. Notably, higher intercrop yields were observed with mandarin due to its compact canopy, enabling light penetration, which supports leguminous intercrops like soybean and groundnut during early orchard years, benefitting farmers with intermediate income and weed suppression (Bhatt *et al.,* 2001).

Khasi mandarin locally known as 'Suntala', is the primary commercial fruit cultivated in Sikkim and is integral to the local agroecosystem. Cultivation of mandarin occurs on both sides of the Tista and Rangit rivers and their tributaries, predominantly within sub-tropical climates at altitudes of 600 to 1,500 m above mean sea level. The unique topography, featuring gently sloping river valleys with deep, fertile, and well-drained soils rich in organic matter, provides optimal growing conditions for this crop. In Sikkim, mandarins are never grown in monoculture; instead, they are intercropped throughout the year with various crops, particularly legumes and ginger, which have been identified as the most economically beneficial intercrops (Lachungpa, 2004). The multicropping system, which includes ginger, finger millet, cassava, buckwheat, and vegetables, maximises spatial utilisation and contributes to soil conservation by providing ground cover and stabilising terrace walls (Lachungpa, 2004). This system not only generates additional income for farmers during the mandarin trees' lengthy gestation period but also enhances household food security. A less than 10 pe cent drop in mandarin yield was obtained with ladyfinger bananas and 10-14per cent with other intercrops. On this basis it can be concluded that vegetables, leguminous crops, and ginger can be used as an intercrop to generate improved economic returns (Patiram *et al.,* 1994).

Homestead forest/Home gardens in the region exhibit a rich diversity of plant species, predominantly woody perennials, which are arranged in a multistoried, stratified structure, closely resembling natural forests. These are managed ecosystems with multiple functions, offering a range of tangible and intangible benefits (Keerthika, 2024). Due to low productivity and a reduction in the duration of jhum fallows, some farmers have begun cultivating cash crops. Species such as citrus and *Cinnamomum obtusifolium* are frequently cultivated, often with an understory of ginger, banana or pineapple. While, broom grass (*Thysanolaena maxima*) and bamboo (*Dendrocalamus hamiltonii*) are harvested either from the wild or from cultivated areas (Majumder *et al.,* 2011). In the NER, dedicated horticultural land use can be established by planting mandarin oranges at 5-metre intervals, with semi-shade-tolerant pineapples intercropped within the same row and vegetable cultivation in the spaces between rows (Majumder *et al.,* 2011). Windbreaks, shelterbelts, or filler trees such as *Salix* spp., *Populus* spp., and *Alnus nepalensis* effectively protect orange trees from high winds without impacting fruit yield (Walia and Kaur, 2023).

In Lunglei district, Mizoram, the state government has introduced Sloping Agricultural Land Technology (SALT), which has been positively received by jhum farmers. A diverse range of crops, including citrus, banana, turmeric, maize, passion fruit, and pineapple is now cultivated using this approach (Grogan *et al.,* 2012). Grafted pear and valuable medicinal plants like *Clerodendron colebrookianum* have also been integrated. The combination of annual and perennial crops diversifies production, providing a buffer against adverse weather, disease outbreaks, and market fluctuations. This agricultural extension initiative has led to the adoption of SALT practices by hundreds of farmers across Southern Mizoram.

Citrus spp. are integral part of multiple traditional AFS in different stated of NER. For instance, the Agri-Horti-Silvi-Piscicultural system, an age-old agroforestry practice of the Kalita community in Assam, integrates multiple productive components on a single land unit. This system includes hybrid paddy, vegetables such as cabbage, carrot and chilli, timber and fruit trees *viz.,* Musa spp., Citrus spp., *Ananas comosus* and fish ponds encircled by trees to create a fully integrated biological system. It provides substantial economic benefits by generating income through crop sales, which reduces household spending on food, and ensures year-round employment for participating families (Deb *et al.,* 2009). Likewise, the Nyishi community in Arunachal Pradesh follows the Agri-Horti-Silvicultural and Horti-Silvi-Pastoral systems. The former involves crops like paddy, *Eleusine coracana*, and tubers (*Manihot esculenta*), with pineapple and banana as common crops. Trees including citrus spp. are cultivated along plot boundaries, supporting food, fodder, fuel, and timber needs, with medicinal plants and spices intercropped for additional utility. The Horti-Silvi-Pastoral system combines timber species (*Terminalia myriocarpa, Gmelina arborea*) and bamboo, with *Livistonia jenkinsiana* ('Toko') offering economic benefits due to its short gestation period and its use as roofing material. Livestock, including cows and mithun, are integral to nutrient recycling and enhance self-sufficiency and sustainability (Devendra and Thomas, 2002). Together, these systems diversify income sources, optimise land use, and improve farmer livelihoods through increased biomass production, especially during off-season periods when crops are less productive.

Intercropping of Khasi Mandarin in Tea Plantation in North-East India

Tea and arecanuts are popular cash crops grown in NER. The majority of tea gardens in Assam are raised in a monocropping system but are integrated with shade trees, as they are shade-loving plants. Tea can be interplanted with other economically significant crops such as fruit trees, vegetables, spices, timber or multifunctional trees, or cover crops to increase production in tea plantations and generate additional profit from the sale of intercrop (Baruah *et al.,* 2005). A beneficial effect of tea intercropping on tea planted as a sole crop has been reported. In Tinsukia, Assam, positive results for the tea + arecanut + Khasi mandarin cropping system were clearly seen with higher yield, gross return and B:C ratio (Tanti *et al.,* 2022). The intercropping system provides a source of additional income generation and also serves as insurance against price fluctuations in green tea leaves. Citrus–tea intercropping could improve the microclimate environment of the tea garden, efficiently integrate the use of water, soil, light, heat, and air in the tea garden, and improve land and light energy utilisation. Fruit tree canopies are conducive to resisting the damage caused by excessive direct sunlight on tea trees, reducing the evaporation and transpiration of tea plantations, increasing air humidity, and improving the environmental conditions of tea plantations. Simultaneously, tea-fruit intercropping enriches the biological community of tea plantations and provides a suitable habitat for natural enemies of tea tree pests. Moreover, it gives full play to the natural control of pests, significantly reducing the main pests of tea trees, thus contributing to improving tea quality. The interaction between fruit trees in intercropped tea plantations can increase the permeability of soil, alleviate the acidification intensity of soil, and improve the quality of tea leaves (Feng and Sunderland, 2023). In *Citrus reticulata* intercropping with tea, there was a significant increase in the content of organic matter, fast-acting P, fast-acting K, and alkali-hydrolysable N (Dong *et al.,* 2015). At the same time, soil pH varied according to the type of intercropped fruit trees. It also effectively increased the organic matter content of tea tree soil, which helped to increase the content of soil organic carbon. Intercropping maize with Khasi mandarin in hilly regions of northeast matter and available nutrients such as N, P and K (Shen *et al.,* 2002).

Figure 20.1: Multi Cropping System of Khasi Mandarin in Hilly Regions of North-East.

Figure 20.2: Multi Cropping System of Khasi Mandarin in Plain Regions of North-East.

Intercropping fruit crops in tea garden however, demands attention as tea has its own peculiarities in influencing soil properties. Long-term monoculture tea plantations can lead to high acidity and reduced local soil fertility, resulting in soil which can induces variable amount of changes upon the soil properties as well as on its nutrient concentration and distributions (Feng and Sunderland,

2023). Previous studies reflected the impact of tea plantation upon soil quality and showed depletion of soil nutrient content in tea soil (Kamau *et al.,* 2008). Soil condition and agronomy management practices are detrimental in tea cultivation. More than 70 per cent tea garden soil has P deficient (Zhang *et al.,* 1997). Besides, K deficient is also common in tea plantation soil due to imbalanced fertilization, which is characterized by increased N application with or without insufficient supply of other nutrients.

Figure 20.3: Intercropping of Maize with Khasi Mandarin in Hilly Regions of North-East.

Figure 20.4: Intercropping of Khasi Mandarin in Tea Plantation in Hilly Regions of North-East.

Tea being an acid loving plant requires acidic soil. Tea cultivation can have a significant impact on soil pH, causing soil acidification (Yan *et al.,* 2018). With decreasing soil pH levels, the organic matter content, cation exchange capacity and the available nitrogen, available phosphorus and available potassium contents in the rhizosphere soil of the tea trees decreased significantly which, in turn, fruit yield and quality (Manorama *et al.,* 2022). Soil acidity inhibits the functionality of arbuscular mycorrhizal fungi by reducing arbuscule formation and suppresses the colonisation of AMF in plant roots (Liu *et al.,* 2020). Han (2002) reported that 40 per cent of tea garden has a pH around 4. The soil under tea plantations is liable of having high BD over the years because of intensive leaching and trampling (Phukan and Baruah, 2015). Soil characteristics along with the proper nutrient management parameters play a good role in better tea production and maintenance of good soil health. Therefore, developing and implementing a sustainable management scheme is critical to keeping the stability of soil aggregates and maximally reducing the losses of soil exchangeable cations in the process of tea cultivation. Tea intercropping benefits tea plantations in terms of soil ecosystem service functions, such as water retention capacity, mineral contents, effects on energy transformation, and regulating environmental conditions. Intercropping tea plantations were more sustainable than regular tea plantations because of the different degrees of variability and benefits. However, tea - fruit trees intercropping often require careful planning and preliminary experimentation to determine the type of intercropping that will have positive impacts, especially in the long term.

Intercropping in Assam Lemon

Growth architecture of Assam lemon is most favourable for growth of component crops, as the canopy of the species allows sufficient light to reach to the ground, therefore, agricultural losses due to shade could be minimized. Moreover, the root system of lemon does not compete for moisture and nutrients with agricultural crops as it goes deep into the soil, hence absorbing the required moisture and nutrients from the lower soil horizon (Bhatt and Misra, 2003). Assam lemon-based AFS which integrates groundnut, chilies, soybean, turmeric and ginger recorded high monetary benefit in NER (Bhatt *et al.,* 2001). However, Bhatt and Misra (2003) reported higher net monetary benefit in Assam lemon intercropped with rice. In arecanut plantation, banana, ginger, chilli, colocasia, turmeric, flowers and vegetables can be successfully grown during the initial years. However, as the age advances, only few crops like pepper, cocoa, Assam lemon, lime, betel vine and partial shade loving vegetables and flowers can be grown profitably as mixed and intercrops (Bhatt *et al.,* 2001). In Assam, intercropping model consisting of arecanut, black pepper, banana and citrus were the best combination. Application of two-third recommended dose of fertilizer to arecanut and other components crops was sufficient to get maximum yield of different intercrops. Assam lemon and Gandharaj types can be successfully adopted (Acharya and Singh, 2014). Assam lemon or Kaji nemu trees are planted in the borders of tea plantation at 3 m × 3 m. This not only optimize land use and maximize productivity but also can act as a bio-fencing in the garden. Assam lemon adds value to farm's produce with its culinary uses, allowing farmers to explore opportunities for processing and value addition.

Study conducted in Mizoram to evaluate production potential and economic feasibility of intercropping lemon (*Citrus limon* Burm) with various vegetable crops indicated that lemon + French bean combination yielding the highest gross return, whereas sole lemon cropping yielded the lowest (Hnamte *et al.,* 2013). The highest benefit-cost ratio was recorded for the lemon + French bean system at 3.94, indicating superior economic viability. Findings suggest that the lemon + French bean intercrop is the most profitable, offering the best economic return in this region.

Bhatt and Misra *et al.* (2003) indicated that guava and Assam lemon based AFS are most suitable for the NEH region. Rice varieties-RCPL-1-24, RCPL- 1-25 and RCPL-1-29 can be intercropped success-fully with guava and Assam lemon. Both the AFS were found most profitable and sustainable for the NEH region. With the prevailing land tenure system in the region, short gestation crops, *e.g.*, fruit crops, can be recommended for farming. As tribal communities are interested in adopting improved/viable AFS even on jhum fallow lands and rice is the main food crop of the region, guava and Assam lemon based AFS are most suitable for rainfed conditions. Growth architecture of guava and Assam lemon is most favourable for intercropping of rice, as the canopy of both the species allows sufficient light to reach to the ground, therefore, agricultural losses due to shade could be minimized. More-over, the root system of both fruit tree species does not compete for moisture and nutrients with agricultural crops as it goes deep into the soil, hence absorbing the required moisture and nutrients from the lower soil horizon.

Benefits of Intercropping in Citrus

Citrus orchard has a long gestation period and takes at least 5-6 years' time for economic fruiting. Cultivation of fruits especially during initial stages of orchard establishment are highly capital-intensive. Intecropping system in citrus orchard can help a great deal in supplementing the initial expenditure needed for establishment of new orchard and provide as a supporting source of income in initial years. The beneficial effect of intercrop though not obvious during the vegetative growth phase, as the season advance the effect becomes apparent and quantifiable at the fruiting stage. In perennial fruit trees, the additive effects of environmental stress during the vegetative growth phase can be carried over to the fruiting stage, thus good management for high fruit yield should start from the juvenile stage. There has been an increase in grower interest in using intercropping since it could reduce management inputs that result in sustainable systems more efficiently using an even potentially replenishing natural resources used during crop production for long term management of farmland. The advantages of intercropping in citrus are briefly discussed below.

Promotes Soil Health

Monoculture farming often relies on intensive inputs, such as synthetic fertilizers, to meet the demands of large-scale production. Numerous studies have indicated that excessive use of fertilizers over a long period can lead to soil nutrient imbalances, soil acidification, and reduced soil microbial activity, ultimately degrading soil quality (Zhang *et al.,* 2022). Root exudates released under long-term monoculture increase pathogen accumulation and decrease plant beneficial microorganisms in the rhizosphere, playing a crucial role in soil health (An *et al.,* 2022). Moreover, intensive cultivation with the excessive use of chemical fertilizers has led to deleterious issues, such as a decline in fruit quality, altered fruit flavour, and a decrease in yields (Liu *et al.,* 2021). Incorporating green manure is a vital aspect of organic and sustainable agricultural practices. Green manure can enhance the ecosystem services in orchards, including the control of soil erosion, a reduction in nutrient leaching, improved sequestration of soil carbon (C) and nutrient availability, increased soil microbial diversity, and the suppression of pests and weeds (Kim *et al.,* 2020). Citrus being a perennial tree, intercropping with green manure has a positive impact on the cycles of C, nitrogen (N), and phosphorous (P), which is an effective strategy that can increase the abundance of the genes related to the soil organic matter (SOM), promote the degradation of plants, and ultimately increase the content of SOM (Castellano-Hinojosa and Strauss, 2020). Intercropping has long been recognized for its ability to improve the biodiversity and health of soil (Xie *et al.,* 2024).

Intercrop provides a good soil cover which relatively lowers soil temperature. This prevents burning of the organic matter in the soil and loss of nutrients. It also provides a microclimate that can be favourable for associated crops. Intercropping citrus with green gram-pea-green gram favours development of higher bacterial and fungi (actinomycetes) population. These legumes intercrop increase respiration of the rhizospheric soil, improve soil organic matter decomposition rates and eventually enhanced nutrient mineralization. The energy obtained from this decomposition can be utilized by microorganisms to decompose soil organic matter in order to release organic N when inorganic N is limiting (Fontaine *et al.,* 2011). Besides soil nutrients, legumes also improves activity of the enzymes in the soil and enriched the antagonistic microbiome in the soil, promoting soil health in the citrus orchard. Integration of alfalfa (*Medicago sativa*) and hairy vetch (*Vicia villosa*) between the rows of mandarin increased the total nitrogen, nitrate nitrogen, and available phosphorous in the soil. It also altered the diversity richness and composition of the soil microbial community and enriched the soil with beneficial fungi and bacteria, including Mortierella and Streptomyces. The activity of soil sucrase and β-1,4-glucosidase increases in alfalfa and hairy vetch intercropping respectively compared to the monoculture (Xie *et al.,* 2024). Intercrops like pulses, guar (*Cyamopsis tetragonoloba*), and senji (*Melilotus parviflora*) proved beneficial in citrus orchards due to their nitrogen-fixing properties (Singh *et al.,* 1999). Specifically, cowpeas were observed to increase the fruit yield of Valencia oranges by 26 per cent. Furthermore, intercrops were shown to improve soil quality by reducing bulk density, pH, and electrical conductivity (EC), while enhancing nitrogen (N) and potassium (K) availability, nitrification, and overall microbial activity.

Improve Soil Physical Properties

Soil bulk density is a vital indicator of soil degradation because it influences soil porosity, plant nutrient availability, and soil microorganism activity (Recha *et al.*, 2022). Soil conservation measures are known to decrease soil compaction, resulting in increased root development and length which in turn reduced root biomass (Shaheb *et al.*, 2021). Soil compaction decreased crop growth and yield due to poor nutrient availability and uptake, thus limiting/preventing root growth (Gursoy, 2021). Growing peanut in citrus orchard reduced bulk density and improved total porosity of soil at variable depth. Peanut increase the soil organic matter content in citrus orchards which is the main reason for the increase in total porosity (Dung *et al.*, 2022). In tea plantations, citrus as intercrop exerts a beneficial benefit in enhancing soil characteristics. Citrus intercropped in tea plantations increase the buildup of large-macroaggregates in the soil and enhance the soil's total available nutrients, available nitrogen and available phosphorus content, and exerts a significant impact on increased extracellular enzyme activity (Duan *et al.*, 2024).

Minimize Weed Growth

In young citrus orchards, high weed incidence is very common. Most growers usually followed manual weed control method which is only effective before weeds produce flowers. In addition, moisture through irrigation supplied to young citrus trees during the dry season are being exploited by these weeds for growth and development. Hence intercropping of young citrus trees with arable crops can exert a significant influence on weed growth. The growth of broadleaf weeds, grass weeds and sedges are usually encountered in citrus orchards. Higher frequency of occurrence of *Gomphrena celosoides, Oldenladia corymbose* and *Tridax procumbens* are reported in citrus orchards (Olorunmaiye and Afolayan, 2012) which is an indication of dominance over the other weed species. Intercropping citrus with sweet potato, cucumber, large green and *Mucuna pruriens* significantly reduced broadleaf weed and grass weed density in citrus. Fang *et al.* (2021) highlight the positive impact of intercropping citrus with vegetable crops, legume crops and clover as an eco-friendly management in checking weed growth weeds growth and minimized use of chemical.

The allelopathic potential of *Ageratum conyzoides* by releasing its volatile allelochemicals into the environment has been reported (Kong *et al.*, 2002). This plant has a wide range of secondary metabolites including flavonoids, chromenes, benzofurans and terpenoids. Among these secondary metabolites, three flavones, ageratochromene, and its two dimers has been isolated and identified from the *A. conyzoides* intercropped citrus orchard soil. These metabolites had different biological actions on major weeds and soil pathogenic fungi and could significantly inhibit the growth of weeds *Bidens pilosa, Digitaria sanguinalis* and *Cyperus difformis*, and spores germination of soil pathogenic fungi *Phytophthora citrophthora, Pythium aphanidermatum* and *Fusarium solani* (Kong *et al.*, 2004).

Impact of intercropping juvenile citrus trees with eight arable crops *viz.*, maize, cucumber, sweet potato, *Corchorus olitorius*, large green, grain amaranth, *Mucuna pruriens* var. utilis and groundnut on weed biomass and species diversity was studied in the Nigeria (Olorunmaiye and Afolayan, 2012). Intercropping with sweet potato, cucumber and large green significantly decreased broad leaf and grass weed density compared to citrus intercropped with groundnut. Additionally, citrus plots with sweet potato or *M. pruriens* var. utilis had notably lower grass weed dry weight than the control plot at six weeks after planting.

Improves Nutrient Recovery and Acquisition

In intercropping systems, when the soil N is limited, owing to low availability or increased N competition due to intercropping, legumes have been proven to be forced to fix the atmospheric N instead of competing with the crop for inorganic N the soil (Jensen, 1996). Plants have been shown to be able to receive environmental stimuli and modify the distribution of their roots depending on the availability of nutrient and competition with the root systems of other species (Cahill *et al.*, 2010). The increase in root length density and surface area under intercropping systems directly improves exploration of the soil, which thereby enhances the acquisition of P in the soil (Sun *et al.*, 2018). In addition, it has been generally accepted that intercropping, as a form of species diversity, can increase the activities of soil enzymes, which is a critical function in ecosystem productivity.

The competition for resources like light, heat, water, and nutrient are commonly encountered in intercropping system. The main interspecific relationships are facilitation and competition and they exist together in the intercropping system. Intercropping is advantageous when competition is lower than facilitation (Vandermer, 1989). In soybean monoculture, P recovery declined rapidly with increasing P application depth. It is possibly because the P diffusion coefficient was much smaller in soil (Dong, 1995), and P absorption by the plant depended mainly on root contact with P fertilizer. In soybean-citrus intercropping, P recovery increase significantly which was characterized by the distinct competition when P was applied in the soil. When P is applied on top soil, P recovery rate by soybean was higher. In contrast, P recovery rate was higher by citrus when P was applied in the deeper soil layer. This is because of the large amount of root biomass of soybean compared to citrus in the top 0–20 cm soil layer which favoured absorption of the P fertilizer. Secondly, plant nitrogen fixation relied more heavily on P fertilizer so competition for P was higher in soybean. The results suggested that the soybean-citrus intercropping system could promote shallow root crops to absorb and utilize P in deeper soil layers (Zhou *et al.*, 2009). It is well established that arbuscular mycorrhizal (AM) fungi benefits host plants by improving nutrient uptake from soil volume. In other words, AMF plays a contributing role in enhancing nutrients uptake by plants which reflect the mycorrhizal dependency of the plant species. The contribution of AM to plant P uptake of sweet orange and a leguminous herb, *Stylosanthes gracilis* was higher in monoculture as compared to sweet orange

– *S. gracilis* intercropping system which may be due to higher P application in monoculture. Also the AM contribution in *S. gracilis* is higher than in sweet orange. It is worth noting that Mycorrhizal inoculation do not decrease competition, but increased the biomass of the herb much more than that of sweet orange. Inoculation had little effect on P content of sweet orange, but significantly increased that of the legume roots (Yao *et al.,* 2005).

Preventing Surface Runoff

On slopy lands, *Hemerocallis flava* interplanted in contour hedgerows in citrus orchards effectively reduces soil erosion, especially in years with high rainfall. Under the same slope, soil type and land use conditions, the quantum of soil being eroded by water is influenced mainly by the kinetic energy of rainfall (Li *et al.,* 2024). Hedgerow planting of *H. flava* reduced the kinetic energy of rainfall by reducing the degree of spattering and scouring of the soil by rainfall by increasing ground cover and planting contour hedgerows (Li and Pan, 2018). The effectiveness of mulch and hedgerows in controlling soil erosion was much more visible under heavy rainfall (Shen *et al.,* 2010). The sediment particle size and weight are much larger than those of water molecules in runoff, so intercropping and mulching worked better at retaining sediment than at intercepting runoff. The hedgerows barriers formed in contour are also effective in controlling nutrient losses. The hedgerows barriers shortens the length of the eroded slope surface by cutting the slope surface into multiple sections, which, thereby reducing the local water catchment area and slowing the water flow speed, thus reducing the scouring capacity of the water and trapping sediment near the hedgerows (Do *et al.,* 2023). The improved no-tillage intercropping of potatoes in citrus orchards increase the ground coverage of citrus orchards, and reduce ground water evaporation and soil erosion, increase soil organic matter, alkali hydrolysable nitrogen and available phosphorus. This reflects that the amount of fertilizer applied to potato was sufficient to support its growth without replenishing the nutrients from the main crop (Guo *et al.,* 2020). Gordzhomeladze (1990) assessed the performance of 7-year-old 'Meyer' lemon and 'Kowano-Wase' satsuma mandarin trees planted on non-eroded, moderately eroded, and heavily eroded soils with varying inter-row soil treatments. The study found that the highest yield was achieved by sowing vetch, oat and soybean in the autumn and ploughing in the spring, particularly on non-eroded soil.

Economic Benefits

Intercropping in citrus offers economic benefits compared to citrus monoculture. Yield per se cannot be considered as the selection criterion for choosing the right intercropping models. The economic analysis is by far the most important index in screening the suitability and remunerative capacity of different intercrops. In juvenile citrus orchards, intercrop such as pepper, turmeric and basil responded positively in terms of growth and yield as compared to ginger (Oyedele *et al.,* 2015). In Kinnow Mandarin, pea intercropping generates maximum benefit cost ratio (2.41) and highest net returns (90,000 Rs/ha) due to suitable combination of satisfactory yield and market prevailing sale rate followed by brinjal and gram while lowest B:C ratio was reported in coriander (Singh and Kaur, 2017). Though in sweet orange, plant growth was highest in groundnut (*kharif*) - onion (*rabi*) intercropping sequence with maximum yield level, the B:C ratio was maximum (4.22) in green gram (*kharif*) - gram (*rabi*) intercropping sequence. However, groundnut (*kharif*) followed by onion (*rabi*) sequence generate the maximum net profit (Rs. 68,000/ha) compared to monoculture practices (Dalal and Gohil, 2009).

A high economic return is also obtained in citrus intercropping with cassava, corn, common bean, cowpea, yam, fava, pumpkin, okra, peanut, watermelon, papaya, and passion fruit (Martins *et al.*, 2021). Ghosh and Pal (2010) and Pal and Tarai (2015) obtained highest net return from Mosambi + groundnut combination (Rs. 35,000/ha) followed by Mosambi + okra and Mosambi + cowpea intercropping. In Kinnow, intercropping system had higher gross returns, net return and BCR as compared to sole Kinnow production. Wheat - groundnut rotation grown as intercrop in Kinnow earned higher net returns (Rs. 82,000/ha), followed by fenugreek-cluster bean intercropping system as compared to sole Kinnow.

Improve Plant Growth

Intercrops modify the microclimate of the field which may help in active intake of nutrients and their partitioning. The intercropping systems help in efficient utilization of inputs by improving the input use efficiency of the system as compared to sole plantation where the interspaces were left uncultivated and did not receive any additional input in terms of fertilizer, supplemental irrigation, additional biomass return through leaf litters *etc.* Kinnow raised under intercropping system with wheat-groundnut and fenugreek-cluster bean is more vigorous as compared to sole plantations (Soni *et al.*, 2024).

Improve Fruit Quality

Legume cover crops grown in citrus orchards increase the potassium content of Valencia orange leaves. Cover crops enhance the retention of organic soil carbon and nitrogen as well as soil aggregation and biological balance in orchard management practices. The legume cover crop plays a significant role in the provision or maintenance of soil phosphorus, potassium and calcium that are essential for fruit development, seed formation and fruit quality. On decomposition, they release nutrients and improved soil nutritional status, soil water holding capacities which support plant growth and enhanced fruit yield. Such an increase in nutritional supply from the soil especially potassium improves the fruit size and sugar composition of citrus fruits thereby improving the fruit quality and flavonoid composition of citrus (Mulinge *et al.*, 2018). Citrus-pea intercropping significantly increased fruit weight, size, juice content, TSS and carbohydrate and nitrogen percentage in leaves, followed by Pumpkin. Interplanting soybean with mandarin trees also increased fruit size, TSS and yields which may be attributed to higher atmospheric N fixing bacteria (rhizobia) in rhizosphere of mandarin roots by intercrop (Selim *et al.*, 2020).

Increase Yield

Growing of leguminous vegetables like pea as intercrop has beneficial effect on mandarin and orange which increased the leaf nitrogen content and thereby vigour and yield. Interplanting soybean with mandarin reduced soil pH and thereby increased availability of other elements in the soil. According to Lovatt *et al.* (1988), the higher contents of N in the buds and leaves increased the number of flowers per tree, which explained the increase in fruit yield per tree. Legumes enhance soil N by virtue of symbiotic atmospheric N fixation (Arshad *et al.*, 2016).

During the intermediate phase of Kinnow growth, the system productivity was at its mnimum especially in sole Kinnow as it was not in full bearing. Intercropping Kinnow with fenugreek-clusterbean and wheat-groundnut improved the system productivity. The annual system productivity of sole annual crops and intercropping system of Kinnow were comparable which shows that the area occupied by Kinnow plantations and sacrificed by the non-sowing of intercrops was compensated by the Kinnow yield (Soni *et al.*, 2024). Intercropping of legumes cover crop within the mandarin interspaces minimized fruit drop and increase fruit yield. Soybean interplanting also increase fruit yield in 3-year-old mandarin (Selim *et al.*, 2020).

A comparison of intercropped and monocropped Nagpur mandarin orchards in Central Indian condition showed that monocropped orchards yielded 68.5 kg per tree, while intercropped orchards yielded 51.4 kg per tree (Srivastava *et al.*, 2007). However, citrus orchards intercropped with legumes, such as soybean and chickpea, produced significantly higher yields (72.2 kg per tree) compared to those without intercrops (68.5 kg per tree). These legume-based orchards also had higher leaf nutrient levels, including nitrogen, phosphorus, potassium, iron, manganese, copper and zinc (Srivastava *et al.*, 2007).

Pest Suppression

Intercrops can safeguard the main crop and restrict pest colonization and growth by physically impeding their host foraging, visually camouflaging crops, emanating confusing chemical volatiles which mask the presence of the host crop for specialist insect pests and release repellent chemicals or chemicals that attract their natural enemies (Mousa and Ueno, 2019). When aerial pests search for a host plant in a diverse intercropped field, they are more likely to land on unsuitable plants and move on from the field. By decreasing the search efficiency of herbivores, intercrops can increase the likelihood that pests will be detected and consumed by predators before ever reaching a suitable host plant (Straub *et al.*, 2014).

Citrus orchards interplanted with guava had low level of psyllids population and incidence of citrus greening than non-intercropped orchards (Ichinose *et al.*, 2012). In monocultured orchards, insects can easily and accurately, locate and land on the host plants as compared to intercropped ones. The rate at which pests land on target host plants may depend on the presence of interplanted plants that interrupt the visual detection of hosts, thus reducing the success of the pest invasion (Floater and Zalucki, 2000). Guava emits volatile terpenoid compounds from leaves, which exhibited repellent activity against the Asiatic citrus psyllid. Terpenes are plant derived chemicals with potent insect antifeedant and toxic properties to herbivorous insects (Gonzalez Comoma *et al.*, 2005). Limonene alone or with other monoterpenes is deterrent (Ntiamoah and Borden, 1996) or repellent (Ibrahim *et al.*, 2001) toward insects. The α-pinene is a bicyclic terpene found in many essential oils and possesses various biological activities such as repellent and antifeedant activities (Fan *et al.*, 2011). The α-pinene alone exhibited a very weak repellent activity against the adults of citrus psyllids, but was effective when mixed with other compounds to significantly exert repellent activities against the adults of citrus psyllids. As such, α-pinene is dominant for the reduction of psyllid chemotaxis and may be synergistically enhanced by other volatile terpenes such as limonene. (Zaka *et al.*, 2015) Besides, the physiological interaction exerted by interplanted guavas may facilitate the growth of citrus, possibly by acting as a windbreak (Ichinose *et al.*, 2012).

Kong *et al.* (2005) investigated the impact of intercropping *Ageratum conyzoides* in citrus orchards on the populations of predatory mite *Amblyseius newsami* and citrus red mite *Panonychus citri* in South China. Results indicated that volatiles released by *A. conyzoides* attracted *A. newsami* while repelling *P. citri*, thereby promoting natural biological control. Field trials showed that *A. conyzoides* essential oil emulsion temporarily increased *A. newsami* populations in non-intercropped orchards; however, sustained release of volatiles in intercropped orchards maintained higher *A. newsami* densities. Thus, intercropping *A. conyzoides* fosters an ecosystem favourable to *A. newsami*, enhancing pest management through allelochemical activity.

Disease/Pathogen Suppression

Intercropping with green legume manures can reduce the damage caused by pests and pathogens by enriching the soil with microorganisms that accelerate nutrient cycling and with bacteria and fungi that have disease-resistant functions (Wang *et al.*, 2023). Intercropping can effectively reduce the incidence of disease in crop fields by modifying crop microclimates in ways that suppress pathogen growth and/or have allelopathic effects (Gomez-Rodriguez *et al.*, 2003). Intercrops also diversify soil microbe communities and cultivate plant-soil feedbacks that limit pathogen growth (Wang *et al.*, 2021). They can reduce exposed soil and the likelihood of splash-dispersed and soilborne disease, suppress virus vectors and ultimately dilute the pool of viable host plants for pathogens (Zhang *et al.*, 2019).

Citrus greening or Huanglongbing (HLB), the most destructive disease of citrus is transmitted by the Asian Citrus Pysllid (ACP), *Diaphorina citri*. ACP causes huge losses to citrus farmers through at least two commonly known mechanisms, namely sucking plant fluids, specifically on young leaves and stems in large quantities and becoming vectors of bacteria that cause HLB disease. Since citrus greening weakens citrus trees and induces them to be less tolerant of other diseases (Chang and Bay-

Petersen, 2003), the protection of trees from citrus greening, especially in the initial period of growth, will provide the citrus with a better chance to remain healthy and produce a greater fruit yield. In citrus intercropped orchards, the vegetation cover of intercrop contributes significantly in minimizing the incidence of ascospores of *Guignardia citricarpa,* a causal organism of citrus black spot, by inhibiting or reducing their reproduction. Forage peanut or coast-cross grass as intercrop between Natal sweet orange reduces the ascospores spreading and increases the percentage of fruits with a commercial standard (Bellotte *et al.,* 2015).

Citrus nursery trees interspersed with guava did show reduced HLB incidence and disease progression over time. However, vector and disease reduction resulting from guava intercropping in citrus nurseries was not adequate to recommend it as a management strategy (Gottwald *et al.,* 2014). Unfortunately, reducing disease-vectoring pests does not always eliminate disease since only a few individuals are needed to infect crops, and intercrops might even introduce some risk as pathogen reservoirs if they act as an alternative host for the disease (Ocimati *et al.,* 2018). Allelopathic compounds of oilseed/legume intercrops (*i.e.,* exuded secondary metabolites) suppress the growth of soil fungi, nematodes, pathogens, and weeds that would otherwise negatively impact legumes (Dowling *et al.,* 2021).

The study performed in the Brazil examined the impact of intercropping coast-cross grass and forage peanut between rows of Natal orange trees on the dispersion of *Guignardia citricarpa* ascospores and the subsequent control of citrus black spot (Bellotte *et al.,* 2013). Results indicated that coast-cross grass significantly reduced ascospore production, with lower mean values compared to conventional, non-fungicide treatments. Findings further revealed that neither coast-cross grass nor forage peanut affected the nutrient content of citrus foliage, and both crops proved compatible with citrus plants. Intercropping with these species curtailed ascospore spread and improved the proportion of commercially viable fruits, while fungicide application also effectively managed citrus black spot.

Conserving/Augmenting Predators and Natural Enemies

Intercrops can provide natural enemies with alternative food such as nectar, pollen, and prey. Intercropped flower strips in crop systems can increase the presence of natural enemies and reduce pest insect abundance (Hatt *et al.,* 2017). In potato - citrus intercropping system, there is a significant reduction in the infestation of whitefly, *Bemicia tabaci.* Natural enemies such as ladybeetle, *Coccinella undecimpunctata* densities were greater on intercropping than mono–cropping potato in both winter and summer seasons but *Orius insidiosus* was rather fewer in intercropping systems (Mousa and Ueno, 2019). According to Ye *et al.* (2010), intercropping tea with economic crops like citrus, waxberry, and snake gourd helped suppress key tea pests by increasing the diversity and species richness of key araneid natural enemies. The population of tea green leafhopper (*E. vitis* Gothe) was suppressed and diversity of key araneid natural enemies increased in intercropping tea plants with citrus, waxberry, and snake gourd helped. Furthermore, population density and spatial distribution patterns of the leafhopper and the Orb-weaver spiders (araneids) were regulated, thereby checking the population abundance of tea leaf hopper (*Empoasca vitis* Gothe).

Figure 20.5: Intercropping of Short Duration Vegetables with Khasi Mandarin in Plain Regions of North-East.

Maize raised in between rows of citrus orchard improve the predators' population that controls citrus leaf minor (Ahmed *et al.*, 2013). The diverse crops grown in between the interspaces increases the population of beneficial insects especially when flowering crops are included in the cropping system. Predatory hoverflies (*i.e.*, syrphids) are important natural enemies because of their high reproductive rates, high rates of prey search and consumption, and strong dispersal capabilities (Bugg, 1992). The enhancement of *D. citri* predators, especially syrphids, through habitat manipulation may increase biological control and subsequent population suppression of *D. citri*. *Allograpta obliqua* (Diptera: Syrphidae) larvae, the only hoverfly species attacking *D. citri* in the field, is predatory and likely feeds on a variety of citrus pests, like *D. citri*, as psyllids are a common prey for some species of syrphid (MacDonald *et al.*, 2016). In California citrus orchards, typically one or more hoverfly larvae in high density *D. citri* patches can consume more than 100 *D. citri* nymphs within 48 h (Kistner and Hoddle, 2015). Insectary plants like buckwheat, *Fagopyrum esculentum*, tansy phacelia, *Phacelia tanacetifolia* and sweet alyssum, *Lobularia maritima* (Burgio *et al.*, 2016) are especially attractive to adult syrphids which feeds on *D. citri*. In citrus orchards, planting of alyssum followed by buckwheat strongly attracts hoverflies which may be due to the higher floral area of this plant species attracting greater numbers of predators (Lundin *et al.*, 2018 and Irwin *et al.*, 2021).

Challenges of Intercropping System in Citrus in Northeastern Region

Although citrus-based intercropping offers ecological and economic opportunities, its successful adoption in Northeast India is constrained by a combination of agro-ecological, socio-economic, and institutional challenges. The region's diverse topography, fragile soils, and erratic rainfall patterns demand site-specific crop management, making optimization more complex than in the plains. Steep slopes restrict mechanization, while heavy monsoonal rainfall often leads to runoff, soil erosion, and nutrient depletion. Conversely, prolonged dry spells intensify soil moisture stress, affecting both citrus and intercrops. Shifting cultivation and shortened jhum cycles have further reduced soil fertility, creating a fragile production environment where intercrops must compete with citrus for already limited resources. Despite abundant potential benefits, intercropping can be challenging to implement because it requires unique management and optimization for each site. Adding intercrops to monocultures complicates planting, management, and harvest and adds to labour costs, which can sometimes lead to intercropping being an unprofitable solution. Such management complexity issues form barriers to adoption of intercropping among farmers.

Competition for light, moisture, and nutrients between crops remains a major technical constraint. Although several intercrop models have been proposed, their reliability is questionable, as they frequently yield inconsistent or contradictory outcomes on yield benefits of intercropping. The combinations of intercrops across other systems are less predictable and sometimes. A comprehensive, region-specific studies to develop robust intercropping strategies suited to local conditions is lacking. For instance, Ijaz et al. (2014) reported that a sole citrus orchard is more productive than citrus-based intercropped orchards. Likewise maize intercropped with mandarin in Sikkim has been shown to significantly reduce yields due to strong competition for nutrients and light (Lachungpa, 2004). Similar findings have been reported in sweet orange orchards, where cassava and maize reduced tree vigour by overshadowing young plants and intensifying water stress (Aiyelaagbe, 2001). Evidence from other regions supports these observations where intercrops such as wheat, maize, pearl millet, sugarcane, and sorghum have consistently been found to reduce citrus vigour and yield, even with additional fertilizer inputs (Singh et al., 1999). Some vegetables, including radish, cabbage, ginger, and potato, have also proven unsuitable due to their high nutrient demands (Srivastava et al., 2007).

Pest and disease interactions add further complexity. Certain intercrops may harbour aphids, fungal pathogens, or psyllids that can spread to citrus, thereby increasing the risk of infestations. Although floral intercrops have been shown to enhance pollination and natural enemy populations, they may also attract harmful adult lepidopteran pests, sometimes undermining pest management strategies (Melgarejo et al., 2024). Thus, the choice of intercrops requires careful consideration to ensure that ecological services are enhanced rather than compromised.

Beyond agro-ecological factors, socio-economic constraints significantly hinder adoption. Farmers in the region often operate on small and fragmented landholdings, which limit diversification options. Labour shortages due to seasonal migration further reduce the ability to maintain orchards and intercrops. Low awareness and limited training on intercropping practices result in suboptimal crop combinations that reduce citrus productivity and farmer income. Market access also remains weak, with inadequate linkages for intercrops such as legumes, vegetables, and spices, leading to reduced profitability even when yields are adequate. Institutional and policy challenges compound these difficulties. Extension support for intercropping systems remains inadequate, with few demonstration orchards available to guide farmers on suitable crop combinations. Credit and insurance products are largely designed for monoculture systems, leaving diversified farms without sufficient financial backing. This discourages adoption, as farmers perceive intercropping as both risky and labour-intensive.

Nevertheless, some evidence suggests that well-chosen intercrops can be neutral or even beneficial. For example, Bindra (1975) reported that fodder crops such as *Cyamopsis psoraloides* and *Melilotus parviflora* did not negatively affect the growth or yield of sweet orange, in contrast to maize–wheat systems which proved detrimental. Such findings highlight the importance of region-specific research to identify complementary and non-competitive intercrops.

Future Directions for Citrus-Based Intercropping in Northeast India (NEI)

The heterogeneity of Northeast India's landscapes ranging from high-rainfall valleys to steep, erosion-prone slopes, calls for the design of site-specific intercropping systems. Advancing citrus-based intercropping in Northeast India requires an integrated research agenda that addresses the region's complex agro-ecological variability, socio-economic challenges, and sustainability needs. While citrus occupies a central role in the horticultural economy of the region, productivity is often constrained by soil degradation, biotic stresses, and low adoption of improved orchard management practices (Sharma et al., 2024). In this context, systematic research and technology development are essential to optimize intercrop choices, improve resource use efficiency, and enhance long-term orchard sustainability.

1. Research and Technology Development

Improving intercropping systems in citrus requires a strong research foundation that accounts for the ecological, agronomic, and socio-economic diversity of the Northeast region.

Locally adapted intercrop models: Designing crop combinations tailored to specific micro-environments is a priority. The Khasi mandarin (*Citrus reticulata*) performs well when intercropped with soybean or groundnut in *kharif* and pea or gram in *rabi*, while Assam lemon (*Citrus limon*) shows compatibility with turmeric and ginger on sloping lands (Das et al., 2019). Such combinations optimize space, reduce pest pressure, and provide interim income. However, empirical evidence remains limited to isolated trials, emphasizing the need for multi-location, long-term studies that consider soil type, rainfall distribution, and irrigation access (Srivastava et al., 2007).

Shade- and slope-suitable intercrops: As citrus orchards mature, canopy closure reduces light penetration, restricting intercrop growth. Shade-tolerant crops such as legumes (cowpea, rice bean), short-duration vegetables, and spices (turmeric, ginger) are ideal candidates (Selim et al., 2020). Agroforestry studies in humid subtropics confirm that such crops improve soil cover, suppress weeds, and enhance biodiversity (Baumann et al., 2001). On hill slopes, crops like pineapple and ginger contribute not only to income but also to soil conservation (Majumder et al., 2011). Breeding programs focused on shade tolerance and slope adaptability could further expand the intercrop portfolio.

Long-term soil and biological monitoring: Intercropping influences soil fertility and microbial dynamics, yet most studies are short-term. Leguminous intercrops have been shown to enhance nitrogen fixation, increase soil organic carbon, and promote beneficial microbial communities (Aziz et al., 2008; Castellano-Hinojosa & Strauss, 2020). Establishing long-duration trials in NEI is crucial to quantify these impacts and develop soil health benchmarks for citrus orchards. Monitoring indicators such as microbial biomass, enzymatic activity, and rhizosphere diversity can provide insights into ecological sustainability.

Integration with digital tools and decision support systems: Future research should also emphasize the use of remote sensing, GIS, and decision support systems to map suitable zones for specific citrus-intercrop combinations. Mobile-based advisory platforms can facilitate real-time farmer engagement and provide feedback on intercrop performance under varying field conditions. Such integration of digital agriculture with intercropping research will accelerate technology dissemination and adoption, especially among smallholders with limited extension access (Ye et al., 2010).

2. Climate-Smart Practices

The fragile hill ecosystems of NEI face soil erosion, nutrient loss, and erratic rainfall. Integrating intercropping with climate-smart practices is therefore essential.

Slope-based conservation practices: Soil and water conservation techniques—such as contour hedgerows, terracing, mulching, and grass strips—help reduce erosion on slopes (Sharma & Prasad, 1995). When combined with intercrops like pineapple, turmeric, and legumes, these measures not only conserve resources but also provide additional outputs. Research in NE hill ecosystems demonstrates that integrated conservation practices reduce annual soil loss and improve nutrient retention (Das et al., 2019).

Legume-based rotations and cover crops: Incorporating legumes and green manures enhances soil fertility by fixing atmospheric nitrogen, adding organic matter, and improving soil structure (Srivastava et al., 2007). Reviews of fruit-based systems confirm that legumes improve yield stability and reduce fertilizer dependence (Morugan-Coronado et al., 2020). Cover crops such as *Arachis pintoi* and cowpea also suppress weeds and minimize soil compaction (Dung et al., 2022).

Water harvesting and contingency irrigation: Rainfed citrus orchards in Northeast India (NEI) are highly vulnerable to dry spells. Simple water harvesting structures, such as contour trenches and small ponds, can store monsoon runoff for supplemental irrigation during drought periods (Sharma et al., 2024). Integrating water harvesting with intercropping ensures consistent crop performance and reduces climate-induced risks.

3. Value Chains, Processing, and Market Development

The profitability of citrus-based intercropping depends on both production and market access. Smallholders in NEI face fluctuating prices, weak infrastructure, and limited value addition.

Value addition and local processing: Intercrops such as turmeric, ginger, pineapple, and lemon hold significant potential for value-added products like powders, pickles, dried slices, and juices (Barbhuiya et al., 2022). Small-scale processing not only stabilizes farm income but also generates employment opportunities, especially for women and youth. With the region's reputation for organic produce, branding and niche marketing can further enhance returns.

Collective marketing through FPOs: Fragmented landholdings and low volumes constrain market access. Farmer Producer Organizations (FPOs) and cooperatives enable aggregation, grading, and bulk marketing of intercrop produce, reducing transaction costs and improving bargaining power. Successful models in pulses and spices can be replicated for citrus-based systems in NEI.

4. Institutional Support and Capacity Building

Strengthening institutional mechanisms is vital for scaling intercropping practices.

Demonstration orchards: Establishing citrus-based intercropping demonstration plots across districts under ICAR institutes, KVKs, and state horticulture departments can provide visible evidence of benefits. Past demonstrations in the region have shown that hands-on exposure accelerates adoption (Das et al., 2019).

Training and extension: Farmers require skills in crop planning, integrated pest management (IPM), soil conservation, and post-harvest handling. Structured training programs, combined with field demonstrations, improve adoption rates and reduce risks associated with new practices (Shirgure, 2012). Training FPO leaders and extension agents further enhances outreach.

5. Policies, Incentives, and Financial Support

Policy interventions are necessary to encourage adoption at scale.

Incentivizing soil-friendly practices: Subsidies or input support for legumes, biofertilizers, and cover crops can lower initial costs and improve adoption. Aligning these supports with ongoing schemes such as MIDH and NMSA ensures efficient resource use.

Crop-specific finance products: Designing credit and insurance schemes tailored to diversified citrus-based systems can mitigate risks and encourage farmers to diversify. Coverage should account for both the main citrus crop and intercrops, protecting farmers from market and climatic uncertainties.

In summary, the challenges of citrus-based intercropping in Northeast India stem from a complex interplay of ecological fragility, socio-economic pressures, and weak institutional support. While the practice is likely to expand due to growing land-use pressures and farmer needs, its sustainability will depend on selecting suitable intercrops, addressing knowledge gaps through extension, and ensuring enabling policy and financial frameworks. Research and technology development for citrus-based intercropping in Northeast India must prioritize region-specific crop models, shade- and slope-adapted intercrops, long-term soil health monitoring, and the integration of modern digital tools. Together, these strategies will create a robust knowledge base for designing sustainable, climate-resilient, and economically viable citrus production systems in the region

Conclusion and Way Forward

Khasi mandarin and Assam lemon dominate commercial citrus cultivation in NER. However, despite its significant genetic diversity, citrus productivity in NER is constrained by several challenges, including erratic rainfall, soil erosion, and nutrient depletion, largely due to traditional shifting cultivation practices. This highlights the need for sustainable agricultural strategies including intercropping that can optimize land use and maintain the ecological balance. Intercropping within citrus orchards presents a promising solution to address these challenges. By incorporating compatible crops such as legumes into citrus orchards, farmers can improve soil fertility, enhance nitrogen fixation and suppress weeds. This approach not only diversifies farm income but also mitigates the risks associated with monocropping, which is vulnerable to pest and disease outbreaks. Furthermore, intercropping can provide interim income during the early years of citrus cultivation before the fruit trees reach full production, supporting farmers' livelihoods in the short term.

The adoption of integrated citrus-based systems, including AFS, also shows great potential for soil conservation and eco-restoration. Combining citrus with other tree species or crops can stabilise yields, reduce erosion, and restore degraded lands. Agroforestry strategies, particularly those that incorporate high-value crops with citrus, can contribute to the sustainable use of marginal lands, ensuring year-round employment and providing additional resources for rural communities. The success of intercropping in citrus orchards requires careful planning and consideration of the region's diverse agro-climatic conditions, socio-economic needs, and ecological stability. Awareness campaigns for farmers, focusing on the benefits of intercropping and agroforestry, as well as guidance on the selection of appropriate intercrops, are essential. In nutshell, intercropping in citrus orchards, combined with agroforestry, holds significant promise for enhancing the productivity and sustainability of agriculture in Northeast India. By diversifying income sources, improving land use efficiency, and contributing to ecological restoration, intercropping can help farmers adapt to the region's unique environmental challenges. With the right support and strategic implementation, this approach can play a crucial role in improving the livelihoods of farmers, ensuring long-term agricultural productivity, and safeguarding the region's natural resources.

REFERENCES

Acharya GC and Singh LS, 2014. Arecanut based cropping system: Alternative pathway to achieve sustainability in North Eastern India. *Indian J. Arecanut, Spices and Medicinal Plants*, **16**(1): 23-26.

Ahmed S, MA Khan and M Qasam, 2013. Effect of intercropping of maize in citrus orchards on Citrus Leaf Miner infestation and population of its natural enemies. *Pak. J. Agric Sci.*, **50**: 91-93.

Aiyelaagbe IO, 2001. Productivity of an Intercropped Sweet Orange Orchard in Southwestern Nigeria, *Biological Agriculture and Horticulture*, **18**(4): 317-325.

Ammari TG and Rombola AD, 2010. Overcoming iron deficiency chlorosis in citrus through intercropping with perennial grass species. *Acta Hortic.* **868**: 327-332.

An SH, Wei YL, Li HM, Zhao ZJ, Hu JD, Philp J, Ryder M, Toh R, Li JS, Zhou Y *et al.*, 2022. Long-term monocultures of american ginseng change the rhizosphere microbiome by reducing phenolic acids in soil. *Agriculture* **12**: 640.

Andrews DJ&Kassam AH, 1976. The Importance of Multiple Cropping in Increasing World Food Supplies. In: Multiple Cropping. (Ed. Papendick, R.I., Sanchez, P.A. and Triplett, G.B.), ASA Special Publication 27, *American Science of Agronomy*, Madison, WI, USA.

Arshad M, Hussain N, Schmeiskyc H and Rasheed M, 2016. Enhancing Soil Fertility through Intercropping, Inoculation and Fertilizer. *Pakistan Journal of Scientific and Industrial Research*, **59**: 1-5.

Azevedo FA, Almeida RF, Martinelli R, Próspero AG, Licerre R, Conceição PM, Arantes ACC, Dovis VL, Boaretto RM, Mattos D Jr, 2020. No-tillage and high-density planting for Tahiti acid lime grafted onto Flying Dragon trifoliate orange. *Front. Sustain. Food Syst.* **4**: 1–14.

Aziz RAA, SE Salem and L Al-Bitar, 2008. Effect of intercropping cover crops on citrus orchards growth and fruiting under Toshka conditions. *J. Agric. Vet. Sci.* **1**: 101-110.

Baruah S and Ahmed N and Saikia S, 2005. Intercropping in young tea plantation, *Natural Product Radiance* 4(1): 35-39

Baumann DT, Bastiaans L and Kropff MJ, 2001. Competition and Crop Performance in a Leek-Celery Intercropping System. *Crop Science*, **41**: 764-774.

Bellotte JAM, Kupper KC, Rinaldo D, de Souza A and de Goes A, 2013. The effects of inter-crop cultivation between rows of citrus crop on spreading of *Guignardia citricarpa* ascospores and in the citrus black spot occurrence. *Revista Brasileira de Fruticultura* **35**: 102-111.

Bhatt BP and Misra LK, 2003. Production potential and cost- benefit analysis of agrihorticulture agroforestry systems in Northeast *India. J. Sustain. Agric.* **22**(2): 99-108.

Bhatt BP, Singh R, Mishra LK, Tomar JMS, Singh M, Chauhan DS, Dhyani SK, Singh KA, Dhiman KR and Datta M, 2001. Agroforestry practices and research: An overview. In. Steps towards modernization of agriculture in NEH region. (eds.) N.D. Verma and B.P. Bhatt. ICAR Research Complex for NEH region, Umiam, Meghalaya. pp. 365-392.

Bugg RL, 1992. Habitat manipulation to enhance the effectiveness of aphidophagous hover flies (Diptera: Syrphidae). *Sustain. Agric. Tech. Rev.* **5**(2): 202-208.

Burgio G, Marchesini E, Reggiani N, Montepaone G, Schiatti P, Sommaggio D, 2016. Habitat management of organic vineyard in Northern Italy: the role of cover crop plants management on arthropod functional biodiversity. *Bull. Entomol. Res.* **106**: 759–768.

Cahill JF, McNickle GG, Haag JJ, Lamb EG, Nyanumba SM, Clair CCS, 2010. Plants integrate information about nutrients and neighbors. *Science,* **328**: 1657.

Castellano-Hinojosa A and Strauss SL, 2020. Impact of cover crops on the soil microbiome of tree crops. *Microorganisms* **8**: 328.

Chang WN and Bay-Petersen J, 2003. Citrus Production: a Manual for Asian Farmers. FFTC, Taipei, Taiwan, p. 85.

Dalal NR and Gohil SN, 2009. Study of intercropping sequence at pre-bearing stage of sweet orange under Western Maharshtra conditions. *The Asian Journal Horticulture*, **4**(1): 219-220

Dhyani SK, Tripathi RS, 1998. Tree growth and crop yield under agrisilvicultural practices in north-east India. *Agroforestry Systems* **44**: 1–12.

Do VH, La N, Bergkvist G, Dahlin AS, Mulia R, Nguyen VT and Oborn I, 2023. Agroforestry with contour planting of grass contributes to terrace formation and conservation of soil and nutrients on sloping land. *Agric., Ecosyst. Environ.* **345**: 108323.

Dong M, Gu J, Liu T, Yang D, Zhang G, Lu H and Qian H, 2015. Differences in soil mineral nutrients and their correlation in Dongting Biluochun tea-fruit intercropping garden. *J. Zhejiang Agric. Sci.*, **56**: 812–816.

Dong YY, 1995. Diffusion coefficient and transportation of phosphorus and potassium in red earth. J. Huazhong Agr. Univ. (in Chinese). **14**(2): 247–251.

Dowling A, VO Sadras P, Roberts A Doolette Y Zhou and Denton MD, 2021. Legume-oilseed intercropping in mechanised broadacre agriculture—a review. *Field Crops Res.* **260**: 107980.

Duan Y, Wang G, Liang L, Wang M, Jiang J, Ma Y, Zhu X, Wu J and Fang W, 2024. Intercropping fruit trees in tea plantation improves soil properties and the formation of tea quality components. *Plant Physiology and Biochemistry*, **210**: 108574

Dubey SK, Sharma N, Sharma JP, Sharma A and Kishore N, 2016. Assessing citrus (lemon) based intercropping in the irrigated areas of northern plains of Haryana. *Indian J. Hort.* **73**(3): 441-444.

Dung TV, Ngoc NP, Dang LV and Hung NN, 2022. Impact of cover crop and mulching on soil physical properties and soil nutrients in a citrus orchard. *PeerJ* **10**: e14170 DOI 10.7717/peerj.14170

El-Mehy-Amira A and El-Badawy HEM, 2017. Evaluation of intercropping corn, soybean and cowpea with Washington navel orange orchard under different N fertilizer levels. *Middle East Journal Agriculture Research*, **6**(2): 513-533.

Fan LS, Muhamad R, Omar D and Rahmani M, 2011. Insecticidal properties of Piper nigrum fruit extracts and essential oils against Spodoptera litura. *Int. J. Agric. Biol.*, **13**: 517–522.

Fang L, Shi X, Zhang Y, Yang Y, Zhang X, Wang X, Zhang Y, 2021. The effects of ground cover management on fruit yield and quality: A meta-analysis. *Arch. Agron. Soil Sci.* **68**: 1890–1902.

Feng Y and Sunderland T, 2023. Feasibility of Tea/Tree Intercropping Plantations on Soil Ecological Service Function in China. *Agronomy*, **13**: 1548.

Floater GJ, Zalucki MP, 2000. Habitat structure and egg distributions in the processionary caterpillar Ochrogaster linifer: lessons for conservation and pest management. *J. Appl. Entomol.* **37**: 87e99.

Fontaine S, Henault C, Aamor A, Bdioui N, Bloor JMG, Maire V, Mary B, Revaillot S and Maron PA, 2011. Fungi mediate long term sequestration of carbon and nitrogen in soil through their priming effect. *Soil Biology Biochemistry* **43**: 86–96.

Ghosh SN and Pal PP, 2010. Effect of intercropping on plant and soil of Mosambi sweet orange orchard under rainfed conditions. *Indian J.Hort.*, **67**(2): 185-190.

Gill MS, Khehra S and Gupta N, 2018. Impact of intercropping on yield, fruit quality and economics of young Kinnow mandarin plants. *Journal Applied and Natural Science*, **10**(3): 954-957.

Gomez-Rodrýguez OE, Zavaleta-Mejýa VA, Gonzalez-Hernandez M, Livera-Munoz and Cárdenas-Soriano E, 2003. Allelopathy and microclimatic modification of intercropping with marigold on tomato early blight disease development. *Field Crops Res.* **83**: 27–34.

Gonzalez-Comoma A, Guadano A, Tonn CE and Sosa ME, 2005. Antifeedant/ insecticidal terpenes from asteraceae and labiatae species native to Argentinean semi-arid lands. *Z. Naturforsch.*, **60**: 855-861.

Gottwald TR, Hall DG, Kriss AB, Salinas EJ, Parker PE, Beattie GAC *et al.*, 2014. Orchard and nursery dynamics of the effect of interplanting citrus with guava for huanglongbing, vector, and disease management. *Crop Prot.* **64**: 93–103.,

Guo Y, Hu T, Ji Q, Guo L, Jiang H, Yang F and Cao X, 2020. Production Mode and Economic Benefit of Improved No-tillage Intercropping with Potato in New Mode Citrus Orchard/*Asian Agricultural Research,* **12**(9): 51-57.

Gursoy S, 2021. Soil compaction due to increased machinery intensity in agricultural production: its main causes, effects and management. In: Ahmad F, Sultan M, eds. Technology in Agriculture. London: IntechOpen.

Han WY, 2002. "The major nutrient limiting factors in tea soils and development of series tea speciality fertilizers." *J. Tea Sci.,* **22**: 70-74. Available: https: //cir. nii.ac.jp/crid/ 1570572699366310656

Hatt S, Lopes Boeraeve TF, Chen J, and Francis F, 2017. Pest regulation and support of natural enemies in agriculture: experimental evidence of within field wildflower strips. *Ecol. Engin.* **98**: 240–245.

Hnamte V, Gopichand B and Lalnunmawia F, 2013. Study on economic feasibility of intercropping of lemon (Citrus limon Burm) with vegetables in the hilly terrain of Mizoram. *Science Vision.* **13**: 40-44.

Ibrahim MA, K Ainulainen P and Aflatuni A, 2001. Insecticidal, repellent, antimicrobial activity and phytotoxicity of essential oils: With special reference to limonene and its suitability for control of insect pests. Agric. *Fd. Sci. Finl.*, **10**: 243–259..

Ichinose K *et al.,* 2012. Limited efficacy of guava interplanting on citrus greening disease: Effectiveness of protection against disease invasion breaks down after one year. *Crop Prot.,* **34**: 119-126.

Ijaz A, Adil SA, Hassan S, Bakhsh K, Bashir MK and Abbas Q, 2014. Agro-economic dimensions of intercropping in citrus farms: The case of district Toba Tek Singh, Pakistan, *Pak. J. Agri. Sci.,* **51**(3): 763-767.

Irvin NA, Pierce C and Hoddle MS, 2021. Evaluating the potential of flowering plants for enhancing predatory hoverflies (Syrphidae) for biological control of *Diaphorina citri* (Liviidae) in California. *Biological Control* **157**: 104574.

Islam MR, Islam MN, Alam MZ, Hossain MM, Hussen MAM and Tipu MMH, 2020. Chemical inducers, nutrient management, guava intercropping and insecticides can reduce Huanglongbing incidence and severity in Sweet orange. *Archives Agriculture and Environmental Science,* **5**(4): 436-446.

Jensen ES, 1996. Grain yield, symbiotic N2 fixation and interspecific competition for inorganic N in pea-barley intercrops. *Plant Soil,* **182**: 25–38.

Kamau DM, Spiertz JHJ and Oenema O, 2008. "Carbon and nutrient stocks of tea plantations differing in age, geno- type and plant population density." *Plant Soil,* **307**: 29-39.

Kim N, Zabaloy MC, Guan KY, Villamil MB, 2020. Do cover crops benefit soil microbiome? a meta-analysis of current research. *Soil Biol. Biochem.* **142**: 107701.

Kistner EJ and Hoddle M.S, 2015. Life of the ACP: Field experiments to determine natural enemy impact on ACP in southern California. *Citrograph,* **6**: 52–57.

Kong CH, Hu F and Xu XH, 2002 Allelopathic potential and chemical constituents of volatile from *Ageratum conyzoides* under stress. *J. Chem. Ecol.* **28**: 1173–1182.

Kong CH, Liang WJ, Hu F, Xu XH, Wang P and Jiang Y, 2004. Allelochemicals and their transformations in *A. conyzoides* intercropped citrus orchard soil. *Plant Soil.* **264**: 149-157.

Kumar D, Singh S, Arora I, Ram JS and Pankaj R. 2022. Effect of Mulched and without Mulched Intercrops on Growth Yield and Fruit Quality of Kinnow Mandarin Plantation under Semi-Arid Irrigated Conditions. *International Journal Environment and Climate Change,* **12**(11): 1394-1400.

Kumar SR, Kumar H and Kumar SA, 2010. Brassica Based Intercropping Systems-A Review. *Agricultural Reviews,* **31**(4): 253-466.

Li H, Zhu N, Qiao J and Tang J, 2024. Evaluating the long-term effects of best management practices on pollution reduction and soil quality improvement in sloping farmland of the Three Gorges Reservoir area. *Agricultural Water Management* **297**: 108844.

Li C and Pan C, 2018. The relative importance of different grass components in controlling runoff and erosion on a hillslope under simulated rainfall. *J. Hydrol.* **558**: 90–103.

Linares J, J Scholberg J, Boote K, Chase CA, Ferguson JJ and McSorley R, 2008. Use of the cover crop weed index to evaluate weed suppression by cover crops in organic citrus orchards. *Hort Sci.* **43**: 27-34.

Liu X, Feng Z, Zhao Z, Zhu H and Yao Q 2020. Acidic soil inhibits the functionality of arbuscular mycorrhizal fungi by reducing arbuscule formation in tomato roots, *Soil Science and Plant Nutrition*, **66**(2): 275-284.

Liu LY, Zheng XQ, Wei XC, Kai Z, Xu Y, 2021. Excessive application of chemical fertilizer and organophosphorus pesticides induced total phosphorus loss from planting causing surface water eutrophication. *Sci. Rep.*, **11**: 23015.

Lovatt CJ, Zheng Y and Hake CD, 1988. Demonstration of a Change in Nitrogen Metabolism Influencing Flower Initiation in Citrus. *Israel Journal of Botany*, **37**: 181-188.

Lundin O, Ward KL, Williams NM, 2018. Identifying native plants for coordinated habitat management of arthropod pollinators, herbivores and natural enemies. *J. Appl. Ecol.* p. 1-12.

MacDonald FH, Connolly PG, Larsen NJ, Walker GP, 2016. The voracity of five insect predators on *Bactericera cockerelli* (Sulc) (Hemiptera: Triozidae) (tomato potato psyllid; TTP). *New Zealand Entomol.* **39**(1): 15–22.

Manorama TM, Gurung S and He B, 2022. The Effects of Tea Plantation Upon the Soil Properties Based Upon the Comparative Study of India and China: A Meta – *Analysis Journal of Agriculture and Crops*, **8**(4): 309-322.

Martins CR, Rodrigues GS and Inacio de Barros, 2021. Assessment of economic and environmental performance in citrus-based intercropping systems. *Rev. Bras. Frutic., Jaboticabal*, **43**(1): (e-463)

Melgarejo SP, González-Cabrera J, Miranda-Salcedo MA, Alvarez-Hernandez JC, Morales- Alonso SI and Morales-Hernández F, 2024. Does Citrus Intercropping Reduce the Population of *Diaphorina citri* 1 in the Citrus Area of the Apatzingán Valley, in Michoacan, México? *Southwestern Entomologist*, **49**(1): 402-415.

Morugán-Coronado A, Linares Gómez-López CMD, Faz and Zornoza R, 2020. The impact of intercropping, tillage and fertilizer type on soil and crop yield in fruit orchards under Mediterranean conditions: a meta-analysis of field studies. *Agric. Syst.* **178**: 102736.

Mousa K and Ueno T, 2019. Intercropping Potato with Citrus Trees as Ecologically–Based Insect Pest Management. *J. Fac. Agr., Kyushu Univ.*, **64**(1): 71–78.

Mousavi SR and Eskandari H, 2011. A General Overview on Intercropping and Its Advantages in Sustainable Agriculture. *Journal of Applied Environmental and Biological Sciences*, **1**: 482-486.

Mulinge JM, Saha HM, Mounde LG and Wasilwa LA, 2018. Effects of Legume Cover Crops on Orange (*Citrus sinensis*) Fruit Weight and Brix. *International Journal of Plant and Soil Science*, **21**(4): 1-9.

Navpreet Singh and Amarjeet Kaur, 2017. Study on economic feasibility of intercropping of Kinnow (*Citrus reticulata* Blanco) with legume crops in the sub tropical region of Punjab. *International Journal of Development Research*, **7**(12): 17556-17559.

Ntiamoah YA, borden JH and Pierce HDJR, 1996. Identity and bioactivity of oviposition deterrents in pine oil for the onion maggot, Delia ntique. *Ent. Exp. Appl.*, **79**: 219–226.

Oad FC, Siddiqui MH and Buriro UA, 2007. Agronomic and Economic Interference Between Cotton Gossypium hirsutum L. and Pigeon Pea Cajanus cajan L. *Journal of Agronomy*, **6**: 199-203.

Ocimati W, Were E, Groot JC, Tittonell P, Nakato GV and Blomme G, 2018. Risks posed by intercrops and weeds as alternative hosts to *Xanthomonas campestris* pv. musacearum in banana fields. *Front. Plant Sci.*, **9**: 1471.

Olaniyan AA, Fagbayide JA, Oladapo MO and Amih CA, 2006. Productivity of Cleopatra Mandarin Rootstock Seedlings Intercropped with Cucumber. *Asian Journal of Plant Sciences*, **5**(3): 534-536.

Olaniyan AA, Fagbayide JA, Ayodele VI and Alamu OO, 2018. Productivity of sweet orange (*Citrus sinensis* (L.) Osbeck 'Agege 1') intercropped with cassava. *Acta Hortic.* **1225**: 167-172.

Olorunmaiye PM and Afolayan SO, 2012. Weed Biomass and Weed Species Diversity of Juvenile Citrus Trees Intercrop with some Arable Crops. *Notulae Scientia Biologicae*, **4**(1): 131-136.

Ouma G and Jeruto P, 2010. Sustainable horticultural crop production through intercropping: The case of fruits and vegetable crops: A review. *Agric. Biol. J. N. Am.* **1**: 1098-1105.

Oyedele OO, Olorunmaiye PM, Odeleye OMO, Alamu OO, Adebayo OS, Bamimore KM, Onyegbule U and Olaniyan AA, 2015. Productivity Responses of Spice and Vegetable Crops in Citrus Juvenile Orchard. *Journal of Biology, Agriculture and Healthcare*, 5(10): 46 - 50

Pal PP and Tarai RK, 2015. Viable vegetable based intercroping system in Sweet orange cv. Mosambi. *International Journal Advanced Research in Biological Sci*ences, **2**(12): 126-129.

Patiram Upadhyaya RC and Subba JR, 1994. Annual Report 1994. Department of Horticulture, Government of Sikkim, Krishi Bhawan, Tadong, Sikkim.

Phukan IK and Baruah A, 2015. "Studies on soil physical, chemical and microbiological properties under compacted and non-compacted tea soils of south bank." *International Journal of Science, Environment and Technology*, **4**: 253-263.

Recha JW, Olale KO, Sila AM, Ambaw G, Radeny M, Solomon D, 2022. Measuring soil quality indicators under different climate-smart land uses across east African climate-smart villages. *Agronomy* **12**(2): 530.

Rohit A and Singh SK, 2019. Economic Feasibility of Intercropping of Vegetable rops Under Kinnow Orchard. *Think India Journal*, **22**(30): 541-552.

Sarathambal C, Singh VP, Barman KK, Raghuvanshi MS and Dubey RP, 2015. Intercropping and weed management effect on soil microbial activities in newly planted mango and citrus orchards. *Indian Journal of Weed Science* **47**(2): 178–182.

Selim MAF, Hefny YAA, Abdel-Wahab EI and Mohamed MKA, 2020. Interplanting Some Soybean Cultivars with Mandarin Trees in Sandy Soil. *Agricultural Sciences*, **11**: 88-110.

Shaheb MR, Venkatesh R, Shearer SA, 2021. A review on the effect of soil compaction and its management for sustainable crop production. *Journal of Biosystems Engineering*, **46**(4): 417–439.

Shen, J, Yang J, Wang J, Wei X, Zheng W, Zhu, Z and Zhou F, 2002. Ginkgo and tea tree intercropping technology. *China Tea*, **5**: 30–31.

Shen ZY, Gong YW, Li YH and Liu RM, 2010. Analysis and modeling of soil conservation measures in the Three Gorges Reservoir Area in China. *Catena* **81**: 104–112.

Shirgure PS, 2012. Effect of different intercropping systems on soil moisture conservation, fruit yield and quality of Nagpur mandarin (*Citrus reticulata*) in central India. *Scientific Journal of Agricultural*, **1**(7): 168-176.

Singh B, Pandey SBS and Vaishnav Ramniwas, 2024. Economics of cultivation of mustard varieties (*Brassica juncea* L.) under Nagpur mandarin (*Citrus reticulata* Blanco.) based agroforestry system in South-Eastern Rajasthan. *International Journal of Research in Agronomy*, **7**(11): 5-9.

Soni ML, Birbal, Nangia V, Saxena A, Yadava ND and Subbulakshmi V, 2024. Performance of Kinnow (*Citrus reticulata* Blanco) based Agri-horticultural Systems during Intermediate Stage of Fruiting in Light Textured Soil of Arid Rajasthan. *Annals of Arid Zone*, **63**(3): 51-59.

Srivastava AK, Huchche AD, Ram L and Singh S, 2007. Yield prediction in intercropped versus mono-cropped citrus orchards. *Scientia Hort.* **114**: 67-70.

Straub CS, Simasek NP, Dohm R, Gapinski MR, Aikens EO and Nagy C, 2014. Plant diversity increases herbivore movement and vulnerability to predation. *Basic Appl. Ecol.* **15**: 50–58.

Sujatha S, Bhat R and Chowdappa P, 2016. Cropping systems approach for improving resource use in arecanut (*Areca catechu*) plantation. *Indian Journal Agricultural Sciences*, **86**(9): 1113–1120.

Sun, YB, Bi HX, Xu HS, Duan HQ, Peng RD and Wang JJ, 2018. Variation of Fine Roots Distribution in Apple (Malus pumila M.)-Crop Intercropping Systems on the Loess Plateau of China. *Agronomy*, **8**: 280.

Vandermer J, 1989. The Ecology of Intercropping. Cambridge University Press, Cambridge.

Wang CX, Liang Q, Liu JN, Zhou R, Lang XY, Xu SY, Li XC, Gong AD, Mu YT and Fang HC *et al.,* 2023. Impact of intercropping grass on the soil rhizosphere microbial community and soil ecosystem function in a walnut orchard. *Front. Microbiol.,* **14**: 1137590.

Wang G, Bei S, Li J, Bao X, Zhang J, Schultz PA, Li H, Li L, Bever JD and Zhang J, 2021. Soil microbial legacy drives crop diversity advantage: linking ecological plant–soil feedback with agricultural intercropping. *J. Appl. Ecol.* **58**: 496–506.

Wen B, Zhang X. Ren S. Duan Y *et al.,* 2020. Characteristics of soil nutrients, heavy metals and tea quality in different intercropping patterns *Agroforest Syst.,* **94**: 963–974

Xie Y, Jing Y, Wang Y, Zheng R, Xu Q, Sun Z and Duan T. 2024. Leguminous Green Manure Intercropping Promotes Soil Health in a Citrus (*Citrus reticulata*) Orchard. *Agriculture,* **14**: 1897.

Yan P, Shen C, Fan L, Li X, Zhang L, Zhang L and Han W, 2018. Tea planting affects soil acidification and nitrogen and phosphorus distribution in soil. *Agriculture, Ecosystems and Environment,* **254**: 20-25.

Yao Q, Zhu HH, Chen JZ and Christie P, 2005. Influence of an Arbuscular Mycorrhizal Fungus on Competition for Phosphorus Between Sweet Orange and a Leguminous Herb. *Journal of Plant Nutrition,* **28**(12): 2179-2192.

Ye H, Cui L, He X and Han B, 2010. Effect of Intercropping Tea with Citrus, Waxberry, or Snake Gourd on Population Density and Spatial Distribution of the Tea Green Leafhopper and Araneids. *Acta Ecol. Sin.* **30**: 6019–6026.

Yunxian Yan, Jun Yang, Yue Guo, Junxing Yang, Xiaoming Wan, Chen Zhao, Junmei Guo and Tongbin Chen, 2022. Potential evaluation of different intercropping remediation modes based on remediation efficiency and economic benefits – a case study of arsenic-contaminated soil. *International Journal of Phytoremediation,* **24**(1): 25-33.

Zaka SM, Zeng XN, Wang HT, 2015. Chemotaxis of adults of the Asiatic citrus psyllid, Diaphorina citri Kuwayama, to volatile terpenes detected from guava leaves. *Pak. J. Zool.,* **47**: 153–159.

Zhang C, Dong Y, Tang L, Zheng Y, Makowski D, Yu Y. and van der Werf W, 2019. Intercropping cereals with faba bean reduces plant disease incidence regardless of fertilizer input; a meta-analysis. *Eur. J. Plant Path.* **154**: 931–942.

Zhang YL, Luo SH, Zeng YH and Pengo PY, 1997. "Study nutrient scale of sufficiency or deficiency in tea soils in Hunan province and fertilizing recommendation." *J. of Tea Soil,* **17**: 161-170.

Zhang YJ, Ye C, Su YW, Peng WC, Lu R, Liu YX, Huang HC, He XH, Yang M and Zhu SSm 2022. Soil acidification caused by excessive application of nitrogen fertilizer aggravates soil-borne diseases: Evidence from literature review and field trials. *Agric. Ecosyst. Environ.*, **340**: 108176.

Zhou WJ, Zhang YZ, Wang KR, Li HS, Hao YJ and Liu X, 2009. Plant phosphorus uptake in a soybean-citrus intercropping system in the red soil hilly region of South China. *Pedosphere,* **19**(2): 244–250.

Index

A

B

U

V

W

X

Y